# GIRLS COMING TO TECH!

**Engineering Studies series**

EDITED BY GARY DOWNEY

Matthew Wisnioski, *Engineers for Change: Competing Visions of Technology in 1960s America*

Amy Sue Bix, *Girls Coming to Tech! A History of American Engineering Education for Women*

# GIRLS COMING TO TECH!

## A HISTORY OF AMERICAN ENGINEERING EDUCATION FOR WOMEN

AMY SUE BIX

THE MIT PRESS

CAMBRIDGE, MASSACHUSETTS

LONDON, ENGLAND

© 2013 Massachusetts Institute of Technology

All rights reserved. No part of this book may be reproduced in any form by any electronic or mechanical means (including photocopying, recording, or information storage and retrieval) without permission in writing from the publisher.

This book was set in Bembo by the MIT Press.

Library of Congress Cataloging-in-Publication Data

Bix, Amy Sue.
Girls coming to tech! : a history of American engineering education for women / Amy Sue Bix.
    pages   cm. — (Engineering studies)
Includes bibliographical references and index.
ISBN 978-0-262-01954-5 (hardcover : alk. paper), 978-0-262-54651-5 (paperback)
1. Women in engineering—United States—History. 2. Women in higher education—United States—History. 3. Engineering—Study and teaching—United States—History. I. Title.
TA157.5.B59   2013
620.0071′073—dc23
2013004385

In memory of Michael Bix, the most wonderful elder brother anyone could have.

# CONTENTS

We live in highly engineered worlds. Engineers play crucial roles in the normative direction of localized knowledge and social orders. The Engineering Studies Series highlights a growing need to understand the situated commitments and practices of engineers and engineering. It asks, What is engineering for, and what are engineers for?

Drawing from a diverse arena of research, teaching, and outreach, the work that is done in engineering studies raises awareness of how engineers imagine themselves in service to humanity and how their service ideals help define and solve problems with multiple ends and variable consequences. It examines relationships among technical and nontechnical dimensions and the ways that these relationships change over time and from place to place. Its researchers often are critical participants in the practices that they study.

The Engineering Studies Series publishes research in historical, social, cultural, political, philosophical, rhetorical, and organizational studies of engineers and engineering, paying particular attention to normative directionality in engineering epistemologies, practices, identities, and outcomes. Areas of concern include engineering formation, engineering work, engineering design, equity in engineering (gender, racial, ethnic, class, geopolitical), and engineering service to society.

The Engineering Studies Series pursues three related missions: (1) to advance understanding of engineers, engineering, and outcomes of engineering work; (2) to help build and serve communities of researchers and learners in engineering studies; and (3) to link scholarly work in engineering studies to broader discussions and debates about engineering education, research, practice, policy, and representation.

Amy Sue Bix's *Girls Coming to Tech! A History of American Engineering Education for Women* offers a valuable contribution to this series and engineering studies scholarship by mapping how increasing numbers of women gained access to engineering education in the United States during and after World War II. Recent policy initiatives to increase enrollments by women beyond the perplexing and frustrating ceiling of 19 percent largely fail to ask how women achieved that level in the first place. Bix carefully examines the stretching of gender boundaries, challenges to sex discrimination, activism for coeducation, appeasement of men students, resistance by men students, and the widely varying ways women students have framed their identities as prospective engineers. *Girls Coming to Tech!* thus makes visible the range of normative considerations and strategies through which women engineering students have successfully challenged the dominant masculine predisposition to retain ownership of engineering life and, hence, work. Engineering has been, and is, for women.

Welcome to the Engineering Studies Series.

Gary Downey, Editor

# ACKNOWLEDGMENTS

In working on this book, I have been fortunate to discover that the study of women's engineering education is a topic that can elicit interest from many individuals and groups—engineering students and educators, scholars of science and technology studies, and historians, especially those who specialize in women's history, the history of education, and the history of technology. Recent years have brought an exciting revitalization of the history of engineering, including new books, articles, and conferences, especially those hosted by the Society for the History of Technology (SHOT) and the International Network for Engineering Studies (INES). I owe a great deal to many members of the broad engineering studies community, especially Gary Downey, who offered me vital advice and insights as editor of the Massachusetts Institute of Technology's Engineering Studies Series. I am particularly grateful to Atsushi Akera, Ann Johnson, Amy Slaton, and Matthew Wisnioski for useful intellectual feedback. Within the historical and academic community more broadly, I have enjoyed productive intellectual dialogues with many colleagues and encouragement from friends, especially Alan Marcus, Hamilton Cravens, David Wilson, Rosalind Williams, Tim Wolters, Steve Usselman, Bruce Seely, Terry Reynolds, Bill Leslie, Jeff Schramm, Volker Janssen, Donna Andreolle, and Alice Pawley. At the MIT Press, Marguerite Avery, Katie Persons, and Deborah Cantor-Adams deserve special thanks for their much appreciated patience and practical assistance. I also want to thank three anonymous manuscript reviewers at the MIT Press for their very helpful suggestions. This research and writing was partly supported by the National Science Foundation, the National Academy of Education, and Iowa State University, including its College of Liberal Arts and Sciences, its Center for Excellence in the Arts and Humanities, and its Women's Studies Program. I want to thank my colleagues in the history department at Iowa State University, especially chairs

Pamela Riney-Kehrberg, Charles Dobbs, Andrejs Plakans, and George McJimsey, as well as Jennifer Rivera. A special thanks to Cynthia Bennet for her assistance in helping to organize the illustrations for this book. Of course, responsibility for any mistakes remains mine alone.

As is common in historical research, this book would not have been possible without the time, effort, and generous assistance of librarians, archivists, and other scholarly staff. My research started at Rensselaer Polytechnic Institute and moved to the National Archives and the archives at MIT, Georgia Institute of Technology, California Institute of Technology, Purdue University, Cornell University, Pennsylvania State University, University of California at Los Angeles and at Berkeley, Wayne State University, University of Illinois, University of Minnesota, University of Texas, University of Michigan, Missouri University of Science and Technology, and Iowa State University. I am indebted to too many people there to mention them all by name, but I want to offer special appreciation to Tanya Zanish-Belcher and Becky Jordan at the Iowa State University Special Collections Department, who have been of assistance to me on multiple projects.

For living with my absorption in this subject of women's engineering education, my husband, Taner Edis, deserves perpetual thanks. My family has been an invaluable source of encouragement and love, particularly my mother, Mary Bix, and my brother, Robert Bix. Thanks to my cousin, Beryl Arbit, for helping me perfect the acknowledgment in memory of my beloved older brother, Michael Bix. As ever-present members of the household, first Wimsey and Poirot and now Quark, Muon, and Symm(etry) have contributed by messing up my papers, taking over my desktop, walking across my computer keyboard, and providing all the other help that only cats can give.

Engineering education in the United States has a gendered history that until relatively recently prevented women from finding a comfortable place in the predominantly male technical world. Throughout the nineteenth century and most of the twentieth, American observers treated the professional study of technology as men's territory. For decades, women who studied or worked in engineering were popularly perceived as oddities at best and outcasts at worst because they defied traditional gender norms. By the 1950s, women still made up less than 1 percent of students in U.S. college and university engineering programs. Sixty years later, in academic year 2010–11, women earned 18.4 percent of engineering bachelor's degrees, 22.6 percent of master's degrees, and 21.8 percent of doctorates, and female faculty members held 13.8 percent of tenured or tenure-track positions in engineering departments.[1] Although those levels of female representation fell far short of demographic parity, by the twenty-first century, women's participation in American engineering programs had become accepted, even officially encouraged. Such a welcome should not be taken for granted. In historical terms, it represented a dramatic shift in the essence of engineering education, which was long presumed to be inherently, purely masculine.

This book examines the issues and tensions that were generated as female students sought to enter engineering programs in the United States, starting from the appearance of a few "engineeresses" at public land-grant schools and small private institutions in the late 1800s. The first women who pursued engineering were labeled as *others*, a small, threatening group of outsiders "invading" a man's world. In many important organizations, formal barriers maintained engineering as a male-only preserve. Up to World War II and beyond, some of the nation's foremost technical institutions refused to enroll female undergraduates. Many male students, faculty, and alumni at elite schools openly criticized or ridiculed the idea of women engineers. Leading national engineering scholastic honor societies refused to grant full membership to women who matched or exceeded men's academic performance. Unwritten rules also discouraged women from attempting to begin an engineering education. The few young women who were admitted to the Massachusetts Institute of Technology before World War II struggled against a hostile intellectual and social environment. Those who resisted coeducation feared that girls could not fit "naturally" into technical programs, with their weight of masculine traditions.

At the most basic level, changing the gendered dimensions of technical studies required that all-male programs begin to grant access to female students and that coeducational colleges agree to expand beyond a grudging token enrollment of women in engineering. During World War II and throughout the 1960s, key institutions gave in, one at a time, under various legal, political, and social pressures. Advocates of coeducation insisted that given a fair chance, a sizable number of women could hold their own in engineering classes. They had to convince skeptics that admitting women would not undermine the quality of technical training and that enough women would be interested in engineering to make college recruitment worthwhile. The emergence of second-wave feminism and the broader social evolution of American gender roles during the second half of the twentieth century helped to break down legal and cultural barriers that had long limited women's employment and education.

Even after female students gained new opportunities to enroll in engineering, however, many were deterred, disheartened, or driven away by a cold or even antagonistic atmosphere. The debate over women's place in engineering served as a barometer of gender biases, displaying society's limits on what it accepted as masculine and feminine cognitive categories. Powerful cultural systems legitimated and facilitated "proper" choices for college majors for men versus women. The authority and tradition that had long made engineering a virtual male monopoly often made women feel like uninvited intruders in classrooms, laboratories, and residence halls. Marginalization, isolation, and various degrees of harassment made everyday campus life difficult for many (though by no means all) female engineering students. Sexual and intellectual tensions overshadowed and complicated their relationships and interactions with male classmates, other women, male professors and administrators, and even some family members and casual acquaintances who reacted with disbelief, laughter, or scorn to the phrase *woman engineer*.

Individual resources, luck, and personal support helped a growing number of women complete engineering degrees, often with pleasure and distinction, during the decades following World War II. More than that, engaged female students, women professors and professionals, and male allies started vigorously promoting the campus and cultural dialogues that were necessary to battle discrimination against women in engineering. In the process, they raised thorny but profound issues about the gendered nature of childrearing and early education, the legal and cultural manifestations of second-wave feminism, the acceptability of women's aspirations, and the nature of engineering itself.

This book details the intellectual, institutional, and social revolution in gender dimensions of American engineering education from the late 1800s through most of the twentieth century. For focus, it centralizes gender over other vital demographic variables such as race and class. When female white engineers were scarce, female black engineers and those from other minority backgrounds were even rarer. Amy Slaton's pathbreaking analysis of the history of engineering education for African Americans in the United States after World War II examines detailed case studies of three key institutions.[2] Although her work shines a spotlight on race, gender remains mysterious, and the net effect regrettably pushes African American women to the sidelines. Other scholars will also need to track important questions of how women's postgraduation employment and professional identity in engineering developed from the nineteenth century through today. This analysis covers only the United States; in Europe and elsewhere, statistics for women's presence in engineering studies do not automatically parallel U.S. figures. Many countries, including Norway, France, Portugal, and Bulgaria, have witnessed a sharp rise in women's presence in engineering over recent decades. Although gender discrimination has by no means been entirely eliminated, the experiences of these women in non-American engineering studies have followed a distinctly different history.[3]

This work does not pretend to encapsulate the stories of all female engineering students in the country over more than ten decades. Instead, its chapters seek to capture the voices of numerous female engineering students and those who interacted with them within a certain set of historical contexts. Of these women, some felt thrilled and rewarded by the process of coping with a traditionally male field, while others became frustrated or infuriated by the struggle. Readers who are engineers themselves or witnessed the history of engineering coeducation may find some stories that resonate; others will inevitably have a different perspective. Each woman entering engineering studies brought a unique psychological and personal background to her experience; on campus and off, each interacted with different individuals in different circumstances that together colored elements of her overall life as a female engineering student as good, bad, or just different.

This book is for the generations of students, faculty, administrators, and professionals who lived through the shifting gendered climate in engineering education and who
observed or helped shape developments at MIT, Georgia Tech, Caltech, Iowa State,
RPI, and other schools around the country. In talking with women who currently are
majoring in engineering at different institutions, I have met many whose satisfying experiences make them almost unable to comprehend the depths of past discrimination,
as well as others who remain upset by the ongoing manifestations of sexism that they
encounter in college settings. This historical perspective may benefit both groups, helping the first set appreciate just how much has changed over the decades for women in
engineering studies, while giving the second group hope that conditions can and will
keep improving.

## ILLUMINATING THE HISTORY OF GENDER IN ENGINEERING EDUCATION

Margaret Rossiter has written that before 1940, "women scientists were . . . caught between two almost mutually exclusive stereotypes: as scientists they were atypical wom-
en; as women they were unusual scientists."[4] Such a statement applies many times over
to the condition of female engineering students in the United States over a far longer
period of time. Indeed, some scholars would argue that even today, female engineers are
still popularly perceived as more atypical than female scientists. Most ordinary citizens
have heard of Marie Curie, at least, but might be hard-pressed to name a single woman
engineer. Women's engineering ambitions were of a more deeply transgressive nature
because technical knowledge—with its ties to industry, heavy manual labor, and the
military—was a far more masculine domain than science was. As would-be engineers,
women faced wider resistance, both formal exclusion and casual discrimination, especially at the most elite levels of education. Throughout most of the twentieth century,
they faced distinct challenges on the grounds that as women, they were not "proper"
engineers and as engineers, they were not "proper" women.

Historically, women in engineering programs stood out due to their rarity, even
more than in science or medicine. In seventeenth-, eighteenth-, and early nineteenth-
century Europe, upper- and upper-middle-class families often allowed and even encouraged girls and adult women to pursue certain scientific interests, providing they did
not take their scholarship so seriously as to endanger their prospects for marriage and
motherhood. Progressive educators such as Jan Comenius, Anna Maria von Schurmann,
Bathusua Makin, and Mary Astell argued that female minds were not inherently inferior to those of males and campaigned for reforming girls' education to include more

challenging and significant subjects. Advocates praised scientific inquiry as a valuable method for cultivating virtues such as patience, perceptive observation, and a reverence for God's creation. Society considered botany to be an acceptable extension for a feminine love of nature; the study of flowers offered artistic women scope for illustration and painting. *The Ladies' Diary* magazine offered readers the chance to solve algebra problems and speculate about scientific puzzles. In 1686, Bernard de Fontenelle wrote an astronomy book (translated into English by famous female playwright Aphra Behn) that was structured as a fictional flirtation between a gallant young scientist and his noble female pupil. Their conversational exchange was designed to interest and instruct female readers, in particular, about the Copernican universe, mechanistic physics, and even speculations about extraterrestrial life. Also adopting dialogue form, in 1805, Jane Marcet published the book *Conversations on Chemistry, Intended More Especially for the Female Sex.* In the early 1800s, Margaret Bryan opened an elite London girls' school that taught science and mathematics.[5]

During the colonial era of American life, "dame schools" offered young girls elementary literacy, while boarding schools taught wealthier girls to raise their matrimonial prospects by becoming proficient in attractive arts, including embroidery, drawing and painting, music, dancing, and French. But by the Revolutionary and early national periods, influential figures such as Judith Sargent Murray, Abigail Adams, and Benjamin Rush argued for extending young women's education beyond such "ornamental" skills as a political and social asset to the country. Representing what historian Linda Kerber has described as the philosophy of "Republican motherhood," Rush linked future national liberty to the contributions of patriotic, cultivated women who would instill in sons the moral and practical lessons to turn them into contributing citizens and strong leaders and would train daughters to become the next generation of virtuous wives and mothers. In his 1787 speech to the Young Ladies' Academy of Philadelphia, Rush argued that young women should master good grammar, handwriting, bookkeeping, and essential mathematics so that they could maintain household budgets and help a husband with business. To help a girl mature into "an agreeable companion for a sensible man," Rush also regarded some training in religion, geography, and history as valuable, alongside "a general acquaintance with the first principles of astronomy and natural philosophy, particularly . . . to prevent superstition." Murray and other advocates emphasized that young women deserved the chance to gain self-confidence and the skills that would help them to sustain a well-run home and, if left a widow, to support themselves and their children.[6]

Critics warned that educating young ladies beyond the mere basics would render them unmarriageable, destroy their natural femininity, tempt them to neglect domestic duties, and undermine man's God-given position as a superior creature. Still, by the 1790s, female academies were flourishing in New England, as scholars such as Nancy Cott and Mary Beth Norton have documented. In cities and smaller towns, these academies spread and regularized young women's training in composition, arithmetic, history, and geography. Educators generally thought it inappropriate for women to study Greek and Latin. Although the vast majority of families could not afford to grant their sons the luxury of pursuing higher education, in the eighteenth century, Harvard College and the College of William and Mary trained ministers, lawyers, and men of leisure, occupations that were firmly barred to females.[7]

By the 1820s and 1830s, national expansion opened up one new rationale for "female improvement." Increasing numbers of white children attended school, at least part-time, not just in urban areas but in Midwestern frontier territory, and local officials realized that it made sense to replace male instructors with young women, who eagerly worked at a lower pay. Given assumptions that women had a natural skill and interest in childrearing, it did not radically threaten established order to have them teach boys and girls, especially if those teachers resigned after a few years to get married. Arguing that better-prepared female schoolteachers would handle classrooms more adeptly, entrepreneurial advocates founded ambitious new institutions for women's education—Emma Willard's Troy Female Academy in Troy, New York (1821), Catharine Beecher's Hartford Female Academy in Hartford, Connecticut (1823), and Mary Lyon's Mount Holyoke Seminary in South Hadley, Massachusetts (1837). All drew increasing numbers of middle-class daughters. According to Kathryn Kish Sklar, Beecher in particular promoted teaching as a desirable uplift for young ladies that gave them a worthy purpose in life and a chance to contribute to community life without contesting man's public role. Such principles reflected the rhetoric of advice books, popular women's magazines, and the writings of public figures. That prescriptive literature equated feminine nature with an idealized selflessness, emotional nurturing, and a pious purity that molded children's characters and influenced the world for good. Teaching offered women a morally irreproachable opportunity to support themselves without the hard physical stress of farm work or factory jobs such as those in the Lowell, Massachusetts, textile mills.[8]

Although relatively well-off white women gained opportunities to attend all-female institutions during the early nineteenth century, coeducation at the higher levels of study appeared far more slowly. Oberlin College, founded in 1833, soon admitted female students and African Americans alongside white men, living up to the institution's progressive outlook on learning as a vehicle for social justice in causes such as Christian missionary outreach and the abolition of slavery. State universities such as those in Wisconsin and Michigan moved toward coeducation over the next few decades, a trend that gained momentum in 1862, when Congress passed the Morrill Act creating the mechanism to support public land-grant education. The legislation did not require land-grant institutions to allow in female students; those in southern and some eastern states often denied or delayed women's access, while other schools, such as Cornell and Nebraska, became coed relatively early.

Trustees at Iowa State College, which admitted women from its opening in 1869, declared, "If young men are to be educated to fit them for successful, intelligent & practical farmers & mechanics, is it not as essential that young women should be educated in a manner that will qualify them to properly understand and discharge their duties as wives of farmers & mechanics? We must teach the girls through our Agricultural College to acquire by practice a thorough knowledge of the art of conducting a well-regulated household, practiced in our Farm House, Boarding Hall, garden, dairy & kitchen." First president Adonijah Welch spoke about giving female students "increased facilities for scientific instruction," with a "prominent" place for "the study of domestic economy." His wife, Mary Welch, helped create Iowa State's first official class in scientific homemaking in the 1870s, called "Chemistry as Applied to Domestic Economy." But that study of household matters represented only one course in the "ladies' course of study" in 1871, which otherwise required female students to take a year's worth of inorganic and organic chemistry or qualitative analysis, botany and physics, psychology, comparative anatomy and physiology, geology, meteorology, history, and the "study of Shakespeare," with options in Latin, French, literature, music, and drawing.[9]

As Andrea Radke-Moss has documented, coeducation did not automatically imply integration. Nebraska physically separated women's and men's seating in classrooms and other college functions in the 1870s, and authorities imposed rules to minimize unnecessary conversations and interactions between the sexes. Nevertheless, some nineteenth-century advocates of coeducation hoped that the female presence could exert a civilizing influence, calming the unruliness, drinking, and even violence among male students. As is discussed below, promoters used similar arguments to press for admitting women to Caltech's engineering and science environment in the 1960s.[10]

Across the United States, by 1870, women comprised about 21 percent of total undergraduate enrollment, rising to 32 percent in 1880, just under 40 percent in 1910, and 47 percent by 1920. Their growing presence sparked a backlash, as critics warned that female dominance would destroy the strength of American higher education.[11] Amid these gender tensions, men on some campuses found many ways to marginalize or even harass their female counterparts. As Rosalind Rosenberg has discussed, male students at

Wesleyan University once punished classmates who were seen conversing with female students and rendered women invisible by excluding them from the yearbook. Some professors remained skeptical about women's presence and sometimes pretended that they did not exist, addressing them as "Mister" or using last names.[12]

In 1873, Edward Clarke, a Boston physician who taught at Harvard Medical School, wrote *Sex in Education, or, A Fair Chance for the Girls*. Although Clarke did not flatly reject the idea of women pursuing intellectual work, he warned that for physical reasons, they should not follow an educational course identical to and as demanding as men's. Medical theory of the day often conceived of the body in terms of specific amounts of resources, and Clarke claimed that if women strained their biological systems by studying too long, especially during each month's menstrual period, they would divert too much energy to their brain and deprive their reproductive organs of essential support. He reported that he knew a number of cases of female students who pushed themselves too hard and collapsed from exhaustion and physical distress before graduating, or who finished college and found themselves unable to bear children and be proper wives to their husbands, or who died after destroying their brains. Clarke's book went through numerous reprintings, and many American readers took his concerns seriously, especially when the eugenics movement sounded an alarm that white women were not bearing enough children to counterbalance the immigration and reproduction of "unfit" minorities and ethnic groups, who supposedly were prone to physical weakness, mental instability, criminality, and socially expensive vices. To fight criticism that higher education ruined female health, women's colleges required students to study anatomy and take regular gym classes. Advocates encouraged young women to pursue ladylike extracurricular sports such as golf, skating, calisthenics, swimming, and basketball (under special rules that limited speed and contact) that were meant to foster exercise without developing the muscles or aggressive competitiveness that properly belonged only to men. Women's colleges and the Association of Collegiate Alumnae (which later became the American Association of University Women, still in existence) collected data to document that female students did not suffer from serious illness and proved perfectly capable of bearing and raising healthy children.[13]

Despite the ongoing levels of discomfort with higher education for American women, female students could, with varying levels of difficulty, find entry points to get a foothold in science, especially after the mid-1800s. Following Mary Lyon's path in establishing Mount Holyoke, the rest of the Seven Sisters (and some other women's colleges) kept raising academic standards for female students, encouraging them to pursue and succeed in science disciplines as well as in the humanities, arts, and social sciences.

Vassar invested resources in building its own medical lecture room and geology collections. In 1865, Vassar hired internationally famous researcher Maria Mitchell to head its astronomical observatory. Mitchell became the first woman member elected to the American Academy of Arts and Sciences and at Vassar trained a new generation of respected female astronomers. Vassar required all students to take at least one semester of chemistry, plus botany, zoology, geology, and physiology. Charismatic chemistry professor Charles Farrar influenced numerous students, including Ellen Swallow, who went on to become the first woman admitted to MIT and, as a chemistry instructor there, helped open its doors to other female students. At all-female Wellesley College, Sarah Frances Whiting taught astronomy, physics, and meteorology, introducing students to spectroscopy, X-ray photography, and other novel scientific advances. In 1878, Wellesley encouraged Whiting to follow MIT's model and establish what became (after MIT) the country's second undergraduate experimental lab dedicated to physics. Barnard College allowed women who were based in New York City to study chemistry and biology, while Bryn Mawr advanced the principle of opening graduate-level education to women. Meanwhile, by the early twentieth-century at coeducational state land-grant colleges, home economics programs required women to study physics, physiological and nutritional chemistry, food analysis, research statistics, and scientific writing.[14]

Margaret Rossiter has described an "infiltration" process before 1940 that allowed individual women, who could find sympathetic male faculty as sponsors, to work their way into doctoral science programs. Those who earned degrees immediately confronted an entrenched discrimination against female scientists that presented numerous obstacles in the educational system, in the scientific profession itself, and in the general culture. As Rossiter has thoroughly detailed, patterns show that many graduates were channeled into "women's work"—research-assistant jobs with relatively low pay, low recognition, and little opportunity for advancement. Yet as illustrators of scientific books, museum catalogers or curatorial associates, lab technicians, and number-crunching "computers" in astronomical observatories, women at least maintained a relatively permanent presence in science. Especially in biology and specialized subfields of chemistry, early twentieth-century women made incremental inroads into the scientific establishment, although many still had to fight to win grudging or belated credit for their talent.[15]

Similarly, although many American medical schools and teaching hospitals limited female admission well into the twentieth century, other routes almost always offered some women access to practice medicine. Since ancient times, women throughout the Western world often served as informal healers and midwives within the family or community, drawing on shared knowledge of homemade herbal remedies and folk cures.

Women continued to fill those roles in early America and later frontier life at a time when medical licensing laws were lax and many male doctors opened practices without having completed much formal training. Popular "alternative" treatment philosophies, such as homeopathy and hydropathy, cultivated a clientele of female patients and often welcomed female practitioners especially to treat other women.[16]

As new mainstream medical schools arose and spread in the United States during the nineteenth century, a few persistent women managed to claim places, starting with Elizabeth Blackwell, who earned a degree from Geneva Medical College in 1849. But because most American medical schools excluded women and hospitals routinely denied female physicians access to internships, Blackwell and her few female medical colleagues created their own female-run medical colleges and female-oriented hospitals in Philadelphia, New York, Boston, Chicago, and other cities. Those institutions were devoted to training the next generation of women physicians and supporting their instruction in clinical practice.[17]

Gatekeepers sought to discourage women, warning that medical studies would be too taxing for their weak nature and that topics such as reproductive medicine and venereal disease were too unseemly for ladies' ears. But even as critics denounced it as inappropriate for female physicians to handle bloody injuries and the bodies of strange men, more women began serving as nurses. Newspaper and magazine writers romanticized the "angels" who combined feminine ideals of sentimental compassion and gentle caretaking. Their portrayals depicted nursing as a higher calling of selfless service. Especially during and after the Civil War, female nurses became an indispensable mainstay of battlefield medical tents, hospitals, and other care facilities. Observers considered nursing a "proper" station for women, which relegated them to a professionally and socially subordinate station behind male doctors. Nevertheless, during the late 1800s, women who wanted to become physicians gradually gained new access to medical training at public institutions in states such as Michigan. In the 1880s, Blackwell and her allies successfully pressured the Johns Hopkins Medical School to become coeducational. Hopkins subsequently trained a significant number of women, including Florence Sabin, whom Hopkins hired as its first female faculty member. According to some historians' estimates, women comprised roughly 18 percent of the doctors who practiced in Boston in around 1900.[18]

Virtually none of that record for women in science and medicine held true for women in engineering. Women had long been inventors and innovators, even if they did not apply for legal patent protection. Across cultures and through generations of farming, food preparation, and home making, women for centuries had acquired and refined a wide range of technical skills and experience. Women produced textiles and embroidery, both in the household and in early industrial employment, and women in seventeenth-century France found employment as laborers who built canals.[19] But the modern Western concept of engineering grew out of the Renaissance, when men such as Leonardo da Vinci and Francesco di Giorgio Martini allied art and architecture to mechanics, specializing in the design of siege engines, cannon-resistant fortifications, waterwheels, and geared machinery. France's Ecole Polytechnique systematized the foundations of formal instruction in military and civil engineering. In the United States, West Point was founded in 1802 and hired key professors from France, endorsing training that was rooted in Newtonian mechanics, descriptive geometry, and other higher mathematics. Graduates worked for the U.S. Army Corps of Engineers or with private companies and applied that knowledge to lay out new roads, plan canals and bridges, and improve harbors and other infrastructure.[20]

Opportunities for men to pursue formal studies in technical subjects slowly expanded. In upstate New York, the school later known as Rensselaer Polytechnic Institute began offering civil engineering degrees in 1835. Harvard set up the Lawrence Scientific School in 1847, at about the same time that the Sheffield Scientific School opened in connection with Yale. Instruction in mathematics and theoretical science shaded into work in practical chemistry and applied science. As Terry Reynolds has shown, numerous other antebellum institutions, especially in the South and frontier West, incorporated a partial technical instruction into more broadly defined curricula. Congress passed the 1862 Morrill Act to provide federal assistance for the teaching of topics that were "related to agriculture and the mechanic arts." Advocates promoted the ideal of a democratically accessible education that offered practical value. It aimed to set young people on rewarding paths in life while advancing knowledge of farming and manufacture for state and national economic benefit. By 1872, the United States had about seventy departments or schools that taught engineering. Although some degree programs increasingly stressed research and theoretical principles, other programs maintained extensive shop-work requirements to ensure that graduates would fit into industry and not appear to be abstract bookworms who could not tell one tool from another.[21]

Even as American colleges and universities began producing more engineering graduates, many more men still entered the field without degrees. For them, the drive to build the Erie Canal, survey railroad lines, design water and sewer systems for expanding cities, and advance steam-powered business all provided immediate technical experience. Such hands-on training cultivated experts in practical problem solving who mastered the basics of mapping, math, and mechanization on the job. A number of nineteenth-century mechanical engineers, for example, began as apprentices in small machine shops or railroad yards and then worked their way up to machinists. Through repairing locomotives and servicing equipment, participants acquired sophisticated experience with engines. As historian Monte Calvert has suggested, leaders of this "shop culture" indoctrinated young men into the traditions of work and an entrepreneurial orientation.[22]

As the twentieth-century began, an increasingly standardized "school culture" came to dominate engineering, fostering the spread of programs that offered advanced degrees. Graduates often secured jobs with major corporations such as General Electric, DuPont, and AT&T rather than with small businesses. The engineering "organization man," to use a later nickname, gained the opportunity to enter management, reflecting the development of large-scale technological systems in the big-business world of mechanical, electrical, and chemical industries. Training became specialized into a rising number of subdisciplines, each with separate professional organizations, publications, and requirements for membership.[23]

For women in the United States in the 1800s and early 1900s, the U.S. Army Corps of Engineers, the Erie Canal, railroad shops, mines, and corporate engineering all remained remote. There was no pure engineering equivalent to domestic science or homeopathy, an alternative philosophy or direction of practice outside the professional mainstream that proved comparatively open to women. In contrast to the world of popular science, publishers did not turn out engineering books or magazines that were devoted to translating the latest discoveries for a female audience. Educational reformers did not call for engineering classes for women, and the popularity of science as a lady-like pursuit in upper-class society did not extend to applied technology. The all-female medical schools and the science curricula offered by women's colleges had no parallel in engineering; there were no separate engineering programs set up by and for female engineers that hired other women as instructors. Even secondary paths to technical employment were thoroughly blocked. In medicine, nursing offered an accessible alternative for some women who could not afford or arrange for a doctor's training. Women with science degrees could hold associate researcher positions or find employment at women's colleges in teaching female students. As discouraging as those low-appreciation paths in medicine and science might be, engineering lacked even those fallback options for giving women a place on the sidelines.

Instead, male engineers deliberately cultivated a macho image for their field, as scholars Carroll Pursell, Sally Hacker, Cynthia Cockburn, and others have detailed.[24] Marketers encouraged boys to play with Erector sets and model trains and marketed toy kitchens and dolls for girls. Few women were welcome in factories or machine shops. Engineering schools portrayed their field's all-male nature as inherently obvious and necessary. In her work *Making Technology Masculine*, historian Ruth Oldenziel paints a clear picture of a "male romance with technology" and discusses the factors that made engineering historically masculine. Through self-promoting autobiographies, professional rhetoric, and fraternal language, middle-class white men of the late 1800s and early 1900s claimed for themselves an engineering identity. Popular writers celebrated the vigorous heroism of engineers who tamed nature through ingenuity, courage, and sheer effort.[25]

All this made it unusually difficult for women to break into engineering, and the story of how they did so tells us as much about engineering itself as about the women. As Karen Tonso has written, "Engineering education, as one facet of engineering culture, is not simply training in a prescribed set of appropriate, academic courses, but is enculturation into a well-established system of practices, meanings, and beliefs . . . 'passed down' from mature practitioners to novices . . . in activities through which engineering traditions are propagated. Thus, as engineering students progress through their undergraduate engineering education, among the things they learn is what it means to be an engineer."[26] Although historians, sociologists, and other observers have produced many books about female scientists and healers and about gender issues in the practice of science and medicine, astonishingly little has been documented about the history of women and engineering. In 2001, Betty Reynolds and Jill Tietjen published *Setting the Record Straight: The History and Evolution of Women's Professional Achievement in Engineering*, a short survey of the story from ancient times (with Hypatia and Miriam the alchemist) to 2000, with an emphasis on key female figures. Other writers have offered biographical and autobiographical perspectives on individual female engineers. The lion's share of attention has gone to Lillian Gilbreth, well-known even among nonhistorians for the popular movies and stage plays that were loosely based on her children's accounts of their unusual family life—*Cheaper by the Dozen* and *Belles on Their Toes*. Jane Lancaster has written a valuable biography of Gilbreth titled *Making Time: Lillian Moller Gilbreth—A Life beyond "Cheaper by the Dozen,"* and Laurel Graham has written the more focused work *Managing on Her Own: Dr. Lillian Gilbreth and Women's Work in the Interwar Era*. But although Gilbreth, as "the first lady of engineering," was unquestionably influential in shaping the future for other female engineers, her story is unique in so many ways that it does not begin to address the broader history of women's entrance to the field.[27] The historical field has reached a point that calls for a substantive narrative shift, extrapolating beyond individual accounts and anecdotes to analyze institutions, aggregate trends, and systemic concepts such as the evolution of engineering-school gendered climate.

Beyond the biographical approach and isolated studies such as the work of Oldenziel, Hacker, and Reynolds and Tietjen, literature on the history of gender and engineering in the United States remains sparse. The field widens, though only somewhat, by expanding the geographic scope. Most notably, *Crossing Boundaries, Building Bridges* compares the history of women, men, and engineering across Europe and elsewhere. Editors Annie Canel, Ruth Oldenziel, and Karin Zachmann discuss how "the timetables by which women entered engineering deviated so dramatically" depending on social context and national and international political events. For example, feminist and reform sentiment in Russia inspired the creation of a female school for engineering and applied science school in 1905, the St. Petersburg Polytechnic Institute, and France established the Ecole Polytechnique Feminine in 1925.[28]

Other than such special monographs, standard works on the history of engineering often take for granted (and thereby reinforce) the sense of technical work as a disciplinary presence defined by masculinity. Studies of the history of specific American colleges and universities, such as Georgia Tech and MIT, concentrate on the development of engineering at a particular institution. However, most of these books either ignore gender issues in the history of engineering programs or acknowledge in passing the presence of a few female students and staff.[29]

Recent years have brought an extensive and varied literature on the conditions that face women in American engineering today, including works sponsored by organizations such as the National Academy of Science, the National Academy of Engineering, and the Society of Women Engineers. Books, articles, working papers, and other publications in this genre typically explore various psychological, social, political, economic, and intellectual barriers that contribute to female underrepresentation in engineering, studying strategies to overcome such factors and support greater educational and workplace inclusivity.[30] These valuable materials approach current issues in engineering education and practice from the perspectives of sociology, gender studies, and education studies, but they generally do not analyze the historical background that set the stage for the ongoing tensions that face women in engineering. A parallel set of works assesses the broader situation for women in United States STEM fields (science, technology, engineering, and mathematics), including topics such as curriculum reform and career-pattern complications, as seen in the writings of Sue Rosser, Mary Frank Fox, and many others.[31] Feminist historians and philosophers, such as Evelyn Fox Keller, Sandra Harding, Londa Schiebinger, and others, have inquired whether it is possible to create a gender-free science and, if so, what would shape its characteristics.[32] Although discussions focused on science can offer substantial insights into modern questions of gender equity in engineering, the contrast between women's relative levels of representation in engineering and biology is substantial. Such differences are no accident. In many ways, they reflect separate historical paths in the development of scientific versus engineering education for women and also in the different professional climates and gendered culture.

## GENDER STUDIES AND HISTORICAL CONTEXT FOR WOMEN PURSUING "MEN'S WORK"

This account of women's entry into engineering studies reflects a broader historical context and the interplay of gendered dynamics within the changing patterns of American life. Just as in other social institutions (such as the workplace, religious groups, media, the political system, and the family), individual women's experiences in higher education were mediated by the stereotypes, expectations, and realities of gender identity. The men who were the central actors at engineering schools—the students, teachers, and administrators—carried normative power to define, assert, and police standards. Their dominant values, dictating what men and women should want and could do, matched and justified the institutional and social barriers that separated women from technical expertise. Gendered consciousness and control shaped discussions of engineering coeducation. According to conventional wisdom, male and female identities provided an elementary filter that defined the appropriateness of whether a particular student should or could enter engineering. Engineering culture contained embedded stereotypes about gender, infusing assumptions about technical knowledge and masculinity into the cognitive machinery of successive generations. Even when male and female classmates sat in the same lecture halls, completed identical homework, and took exams together, gender inescapably penetrated that shared surface experience sooner or later, with blatant or subtle ramifications. Individual choices, actions, and interactions never came in a vacuum; they often, though not inevitably, reflected underlying gendered practices. Questioning masculine and feminine norms, either explicitly or implicitly, required effort and bravery.

Language matters. For decades, the commonly used term *coeds* differentiated women from the default *college student*, especially since the awkward antonym *eds* for male students never really caught on. The nicknames Joe College and Betty Coed encoded men's position at the center of campus life and women's role as an asymmetric accompaniment. Female engineering students always remained a subset of the full population of campus women. They shared many college experiences with women in other majors, including sorority life, music groups, the dress code, and other rules governing women's behavior. Yet for women studying engineering, many personal aspects of college life, from dating to decisions about how to present themselves in everyday appearance, acquired an extra layer of complexity. Their identity as female engineering majors uniquely colored perceptions about them, both as individuals and as a group. Like their dorm-mates pursuing literature, librarianship, education, arts, home-economics, psychology, and social work, female engineering students were perceived in terms of broader cultural impressions of "college women" and in turn responded to such expectations. During World War II, patriotic messages pushed young women to contribute to national defense. After the war ended, a supposed return to normalcy fed a trend toward early marriage and conventional division of labor between the male breadwinner and female homemaker. In practice, class, race, and other factors prevented large numbers of women from conforming to this conventional "Leave It to Beaver" model even at the height of its cultural dominance, while other postwar women appreciated the sense of accomplishment and financial independence that they gained from working outside the home. Yet the vast majority of college-trained women continued to enter conventionally gendered (and often relatively low-paid) fields, including teaching, bookkeeping, nursing, retail, libraries, and social agencies. Even during the decades when total female participation in the labor force rose, women in engineering remained starkly separate from their far more numerous counterparts who chose the paths of less resistance and so occupied a different space within the "college woman" context.[33]

A chain of gender stereotypes left female engineering students of the postwar decades boxed into one of two disagreeable roles. The standard premise was that a normal woman would prefer to study almost any field other than engineering. Starting from this premise, many classmates, teachers, and outside observers concluded that the few college women who defied such norms could not be serious about their work and had entered engineering only to catch a husband—in which case, faculty and male classmates did not need to respect the women's intellectual ambitions. The second corollary suggested that those women who were *too* serious about their work must therefore be inherently unfeminine—in which case, college men felt free to slander them as unattractive. Whenever a female engineering student dropped out of college or switched majors (especially when frustrated by an unfriendly reception), observers might interpret her individual "failure" as confirming the wider gender truth that engineering belonged to men. The assumption that women naturally tied their emotions and values more to family than to occupation (and the reverse for men) fed suspicions that top-quality engineering training would be wasted on women. When female engineering students married before or soon after graduation, they seemed to validate the notion that women were more dedicated to intimate relationships and babies than to "real" work. Those female engineering students who did not marry at the culturally standard age seemed to verify doubts about whether they were "real" women. Such gendered assumptions spilled over into decades of workplace discrimination when female engineering students encountered some potential employers who refused to take them seriously or questioned whether women could commit themselves to a job as intensively as a man supposedly did for any length of time.

In practice, over the decades, many individual women who entered engineering studies did manage to escape this rhetorical trap. Given the obstacles that they faced, an impressive number succeeded in securing decent employment while getting married (often to men they met as engineering classmates or coworkers), raising families, or otherwise enjoying fulfilling personal lives. But it was no accident that female engineering students repeatedly asked their mentors about the "secrets" to combining employment with child bearing and child rearing and about the best strategies for reentering the labor force after temporarily stepping out of it. Male engineering students rarely, if ever, explicitly confronted or agonized over questions about family and career balance. Prevailing cultural standards dictated that women who held employment outside the home should continue to assume the bulk of responsibility for domestic chores and child care—the "second shift" of shopping, laundry, bedtime routines, and more.[34] Although society might tolerate and sometimes celebrate eccentrics, women (and men) generally earned higher approval and rewards by following gendered models, and women who opted to follow "men's work" in fields such as engineering defied expectations. To offset such nontraditional choices, some women might deflect implied or explicit criticism by paying special attention to performing femininity in their personal lives. In other cases, journalists and other observers might impose visions of idealized gender on nontraditional women, assessing and emphasizing the cooking or sewing skills, family devotion, appearance, or fashion choices of female engineers. Consciously or unconsciously, individual women in engineering studies had to decide how to position themselves—as "one of the guys" who sidelined her femininity during work hours, as an engineer who took pride in combining a displayed femininity with technical competence, or some alternative tactic. But inevitably, some responses to female engineering students lay outside their control. Parents, friends, classmates, teachers, potential employers, journalists, and the culture in general all had certain visions of "what women did," and engineering was not one of them.

In the face of such challenges, female engineers of the postwar era formed organizations such as the Society of Women Engineers to give each other mutual support and to provide career counseling and mentoring, financial assistance, and psychological and social reinforcement for the next generation. The second-wave feminist movement of the 1960s and 1970s offered a broader basis for institutional change and women's individual empowerment. Activists, especially with the National Organization for Women, pushed to ensure that the Equal Employment Opportunity Commission actually enforced provisions of the 1964 Civil Rights Act that made it illegal to discriminate by sex when hiring or firing or in actual conditions of employment. Further legislation, policies, affirmative-action rulings, and numerous court battles followed, all seeking to clarify procedures for resolving complaints of unequal pay, hostile work environments, and sexual harassment. Beginning in the 1970s, discussions about workplace discrimination often grew heated, yet the balance of change clearly swung toward helping women close the gap in professional, economic, and social opportunities. Similarly, political and cultural pressure motivated many Americans to work toward resolving problems with women's access to equal educational conditions. Title IX, part of the Education Amendments Act of 1972, mandated that no federally supported educational activities could engage in discrimination on the basis of sex. Subsequent legislation, such as the Women's Educational Equity Act of 1974, expanded discussion of female students' well-being from primary schools through graduate training. Organizations such as the American Association for University Women lobbied to maintain this focus on equity. Its 1992 assessment, *How Schools Shortchange Girls,* warned about the cumulative effects of even seemingly small disparities, such as teachers who directed more attention to boys.[35]

Feminist consciousness raising, popular discussion, and academic studies deepened awareness of the complexities that lay behind gendered choices of university majors and professions. Increasing numbers of women and men alike explored the implications of a culture in which, long before women arrived at college, many social and psychological influences had already delivered the message that engineering was most appropriate for males. Interactions with parents, teachers, friends, and classmates conditioned norms that often linked masculinity to mathematics, hard sciences, and hands-on technical experience.[36] By deliberately or unconsciously treating young girls as less interested and less able than boys in science, math, and mechanical and building work, adults reinforced such presumptions both in individual children and as a broad generalization. Social learning separated carpentry sets, radio kits, model trains, and other boys' playthings from girls' dolls, sewing kits, and toy kitchens. Psychologists and sociologists documented how adolescent young men and women tended to become absorbed in adopting gender frameworks and defining personalized expression, a pattern that coincided with crucial years of gatekeeping in secondary schools. Teenage girls who were subtly discouraged or openly counseled against enrolling in college-track mathematics were left at a disadvantage for considering engineering as a future option. Just as the feminist movement of the 1970s sought to break the barriers that deterred many young women from joining sports teams or male-only clubs, advocates mobilized to promote engineering as a visible, viable option for young women to consider before they reached higher education. Through outreach efforts, targeted workshops, elementary schools, Girl Scout programs, and more, female engineers presented themselves as models and as mentors who were eager to foster girls' curiosity about technology and awareness of career options.

## CASE STUDIES WITH MEANING FOR ENGINEERING COEDUCATION

To raise the curtain, chapter 1 of this book details the handful of individual women who, as "rare invaders," breached the boundaries of masculine engineering education and work in the late 1800s and early 1900s. World War II proved a crucial transition, as chapters 2 and 3 document. National emergency led the federal government, schools, and industry to undertake coordinated efforts urging women to serve their country by entering engineering. The unusual pressures of wartime justified stretching gender boundaries, at least temporarily. Case studies detail the intense interest in women's technical potential and the complex path toward engineering coeducation in the United States during this crucial transition period. Manpower shortages led major American companies, including General Electric and Grumman Aircraft, to recruit and train women as engineering aides, channeling them directly into defense work. Although training programs for women were designed as temporary war expedients, they helped break down both formal and informal barriers to women's participation in the campus engineering culture. Most of those who participated in special wartime programs did not ultimately make full-time postwar careers in technical fields, but their numbers reflected positively on the growing number of women who were studying engineering for regular degrees during World War II. A critical mass made life easier, and high numbers helped validate the notion that women could handle technical subjects. World War II brought a number of "firsts" for women in engineering, including an increase in the number of women who were initiated into student honor societies (at least those that allowed female membership). At individual institutions, the difference was apparent; wartime permanently pushed open the doors of a few engineering programs to women, most notably, Rensselaer Polytechnic Institute. In 1949, across the entire country, there were 763 female students enrolled in engineering; by 1957, that total had more than doubled to peak at 1,783. True, given that the number of male engineering students also soared, female students remained less than 1 percent of total enrollment. But their numbers proved sufficient to draw attention, especially in the broader context of gender-related changes in the workplace, higher education, and national society.

Chapters 4 through 6 here offer detailed case studies of the post–World War II debate over women's place in engineering at three significant institutions—the Georgia Institute of Technology, the California Institute of Technology, and the Massachusetts Institute of Technology. Covering the history of coeducation at every U.S. engineering program is beyond the scope of this book, but this suggestive sampling serves to highlight trends of change.

The in-depth case study of Georgia Tech offers an excellent perspective on the postwar period's heated debate over coeducation, since the institution's culture, customs, and community so closely connected engineering to masculinity. The college humor magazine ran entire issues that poked fun at the notion of female engineers. Nevertheless, the issue of coeducation appeared, with rumors in 1948 that women's organizations in Atlanta were preparing a court test case, a well-qualified young woman whose sex excluded her from any access to public engineering education in her native state. Georgia Tech's president advocated for gendered fairness, and facing that pressure, the Georgia Board of Regents in 1952 passed a measure (over much internal resistance) admitting women to Georgia Tech under limited circumstances. The first female students to appear on campus caused a sensation, and newspapers ran photographs showing women trying on Tech's traditional freshman "rat caps." These early generations of female students walked a tightrope politically, tactfully denying that women's presence would force any real change in the male-centered culture of their engineering school.

Caltech offers a valuable contrasting case study. As a private school, Caltech did not face the same threat of lawsuits over coeducation as had emerged in Georgia. Instead, in the mid-1960s, Caltech experienced internal pressure to begin admitting undergraduate women. Male undergrads had become restless, complaining that "Millikan's monastery" was an unhealthy social and intellectual anachronism that graduated brilliant scientists and engineers who were social idiots, unable even to talk to real women. In a 1967 survey, 79 percent of undergrads said they wanted Caltech to admit women. They promised administrators that a civilizing female presence would improve men's academic performance and even suggested that if Caltech relieved campus boredom by admitting women, male students would be less likely to indulge in marijuana. Some professors remained dubious about making Caltech coeducational, but administrators faced disturbing evidence that good high school men were rejecting Caltech in favor of coed Stanford or Berkeley. In 1968, Caltech trustees voted to admit female undergraduates. Although school leaders spoke about fairness in giving women access to an excellent school, the change was not inspired by a passionate dedication to women's rights. Those leading Caltech did not believe that large numbers of young women would be interested in or qualified to study high-powered engineering. At least initially, Caltech's primary motivation for accepting women was to appease and retain its male students by providing them with girlfriends and social "normality." Picking up on that mentality, many of Caltech's first female students had to fight to be taken seriously as students, fend off unwanted sexual approaches, and struggle to develop their own identity as women engineers.

The final focus, MIT, also offers rich grounds for comparison. Unlike Caltech or Georgia Tech, MIT had been coeducational since the 1870s but treated women's presence as a minor afterthought. Before the 1940s, MIT never had more than sixty-five female students at any time out of a total student population of just under five thousand. Some faculty and administrators resented giving any seats at MIT to women, assuming that they would either fail to complete the program or squander their degrees by marrying and leaving the workforce. Many at MIT still regarded women's presence as provisional; the post-WWII baby boom and national cult of white middle-class homemaking added ammunition for those who wished to revert to conventional gender norms. But alternate voices advocated for the value of coeducation. For all the talk that engineering was contrary to women's nature, MIT's admissions office had in fact artificially kept down the number of women accepted, evaluating them more selectively than men. In a typical year, MIT rejected at least four qualified women solely because it had no place to house them. In 1960, alumna Katherine McCormick pledged $1.5 million to build MIT's first on-campus women's dorm. The resulting publicity splash made many people aware for the first time that MIT was coed, and in 1964, the number of women applying jumped by 50 percent. Throughout the late 1960s and early 1970s, the few women professors at MIT—Emily Wick, Mildred Dresselhaus, Vera Kistiakowsky, and Sheila Widnall—promoted the cause of women's engineering education. Female students campaigned to change the MIT environment, broaching the subject of sexual harassment and the many negative comments that women encountered in classrooms and laboratories. Activists kept raising the issue of numbers, worried that female engineering students would never win respect if they remained below a critical mass. Between 1963 and 1973, the number of women enrolled at MIT tripled.

Each of these three schools serves as a unique lens that has been shaped by its institutional history and character, by geographical context and era, and by the political and personal inclinations of key figures in college leadership, including students, alumni, professors, administrators, and the general public. Georgia Tech's debate over coeducation reflected, among other factors, the clash between Southern traditions and a pressure for modernization to accommodate women and racial minorities in the nascent civil rights movement. Caltech's consideration of female undergraduate admission occurred a decade and a half later when southern California was modeling a hedonistic lifestyle and the student movement of the 1960s promoted youthful sexual liberalization and challenges to authority. At MIT, internal arguments about whether to end or extend coeducation peaked during America's cold war race to keep ahead of the Soviet Union in science and technology, adding international political weight to the question of whether trying to train more women in engineering was a logical advance or an inappropriate waste of time.

At the same time, across these campuses in different years, discussions of coeducation at each school raised overlapping questions, comments, and concerns about the issue of women as engineers. Georgia Tech, Caltech, and MIT all possessed ingrained norms that equated engineering study and technical knowledge with masculinity. Subscribing to theories of gender dichotomy stretching back to the ancient Greeks, popular campus assumptions linked men with rationality, action, and strength and women with emotion, passivity, and flightiness. In a century of technological advances from the Wright brothers to jet aircraft and intercontinental missiles, male engineering students symbolized intellectual and, by extension, national progress. Defined by their absence from that world, women represented, at best, future wives whose appealing femininity and steadying home influence would help engineer husbands succeed and, at worst, temptresses whose frivolity risked distracting good, sound men. The traditional engineering campus actively resisted female influence and indoctrinated men through an atmosphere charged with fraternity, empowering them by marginalizing women. The annual St. Patrick's ball and other social functions relegated women to the function of dates and prizes to be paraded and admired. In the engineering school, women otherwise were most visible in their capacities as secretaries, low-level research assistants, administrators in female niches such as the library, and professors' helpmates.

To skeptics, coeducation threatened to upset a smoothly functioning status quo. For them, a feminine presence would destroy the cohesion that was created through male bonding and destabilize effective teaching. All three schools raised scholastic pressure to a mandatory virtue, nurturing a conviction that professors could mold students into the best engineers and scientists only by driving them to the limits of endurance. Graduates took a virile, professional pride in having survived that trial by fire. Some male students, alumni, faculty, and administrators feared that adding female students would dilute that classroom intensity and rite of passage, compromising their institution's reputation. The ongoing use of language painting women as invaders underlined the masculine desire to retain ownership of engineering life.

As Georgia Tech and Caltech became coeducational and as MIT moved toward admitting more women, such assumptions intensified gender tensions surrounding the engineering campus. Throughout the 1950s and into the 1960s and beyond, men (and often other women) continued to regard female engineering students as curiosities. Ironically, even as many of the first women at these three schools were swamped by unabashed stares or unwanted attention from their male counterparts and inquiries from the media, they appeared to be a virtual footnote within the official institution. The default student in admissions bulletins and campus rhetoric remained male. It took uncomfortably long for administrators to figure out how to provide women with adviser support or even decent athletic facilities.

Feeling like second-class citizens on campus, the first generations of entering women felt perpetually on trial, under scrutiny to prove not only their individual ability but also the worth of their entire sex. In a commonality across all three schools and elsewhere, female students found that many Americans often refused to believe that women could be serious about pursuing technical work. Male classmates and casual acquaintances suggested that women enrolled in engineering simply to "catch a husband." Especially during the 1950s, such assumptions were no coincidence, given that the nationwide dropout rate for female students overall was relatively high, as was the rate of women's marriage either during or immediately after attending college. In fact, a number of women in that era admitted that one of their major goals in their higher education was to obtain a "Mrs. degree." But female engineering students who had demonstrated their brainpower and determination found such repeated questioning infuriatingly patronizing.

Although many faculty members proved to be supportive of female students, others displayed either hostility or an embarrassing favoritism. Even a professor's casual comments in class could reveal an awkward sexism. Certain male classmates simply were not ready to respect women in engineering as intellectual equals, future professional colleagues, or even normal fellow students. Instead, a good part of campus culture still reduced women to sexual objects who were evaluated on hair and eye color, body shape, and girlfriend potential. Official college engineering magazines and advertising by the country's largest advanced-technology corporations portrayed women as pinups in tight sweaters or swimsuits and ran raunchy cartoons that glamorized men's pursuit of sexual conquests. Especially in years when gender ratios in enrollment were particularly skewed, female students reported being bombarded with romantic approaches. In a youth-dominated environment that often seemed to facilitate relaxation of sexual mores, some women felt that men perceived them almost as prizes to be distributed.

Ironically, even as some male students made female classmates the targets of sometimes oppressive social and sexual pressures, popular lore at MIT, Caltech, and Georgia Tech also commonly disparaged the entire category of female engineers as ugly and undesirable. Some women ignored, challenged, or laughed off such slurs, but the often vicious portrayals could reinforce the worst fears of vulnerable young women. Those negative images both reflected and perpetuated the impression that normal women did not want to pursue technical subjects, a message that often was blamed for discouraging girls. Stereotypes thus denigrated female engineering students on both intellectual and personal levels, calling into question their worth both as females and as engineers.

Despite such obstacles, some of the first women at Georgia Tech, Caltech, and MIT and in engineering studies elsewhere survived and even thrived. Those students were fortunate or resourceful enough to find or establish a comfortable atmosphere that helped build their self-confidence, often with encouragement from male peers and professors, female friends, and outside backers such as the Society of Women Engineers (SWE). For the women who did graduate, often at or near the top of their engineering classes, academic success and personal fulfillment combined to make education in a male-oriented institution not only bearable but often richly pleasurable. Some insisted (at least publicly) that they themselves had never experienced any discrimination. Others took pride that their persistence ultimately overcame the doubters and won them acceptance among faculty and classmates. In negotiating survival strategies, some female engineering students embraced a self-identity as "one of the guys," studying alongside men, often amid friendly teasing. Other women (or sometimes the same women at other times) deliberately displayed their femininity, proud of being interested in fashion, skilled in the kitchen, and enjoying hobbies such as sewing. Virtually no woman in engineering studies could completely ignore her gender. Even those who fared well had to consider, reconsider, and reposition their attitudes toward femininity, other women, male classmates, teachers, and potential employers.

For other female students, an initial choice of engineering enrollment ultimately entailed more misery than reward. Without comprehensive statistics, it is impossible to tabulate definitively the number of women who transferred out of engineering, switched to other colleges, or dropped out of school. For some, the move may have been a positive step in their ongoing maturing as college students as they discovered other paths that were more personally appealing to them than engineering. But for others, the loneliness and uncertainty that faced a woman who was trying to break into a man's field multiplied the stresses caused by difficult classes and the coldness (sometimes outright antagonism and harassment) from some male classmates and teachers. Issues surrounding gender and engineering complicated the already disorienting process of leaving home to start college, with its mixture of anxieties and possibilities. Their decisions were not made in a vacuum. Personal campus experiences were embedded within the wider realities that faced women in engineering, which inevitably influenced individual academic outcomes.

As the final chapter here details, the American campus environment gradually but significantly evolved, especially in the late 1960s and 1970s, as a result of developments inside engineering programs and in society as a whole (including the rise of second-wave feminism, civil rights legislation, and decisions by national institutions to promote the success of women in science). Female students, faculty members, and other activists, in alliance with male advocates of liberalization, mobilized to pressure educators for change. For them, numbers made a difference. Their campaigns stressed the need for engineering colleges to work on attracting more women to attain a critical mass. After the female presence rose to a tipping point, those women gained power to support each other, both informally and through organizations such as SWE chapters.

For political and personal reasons, a number of women in engineering remained reluctant to identify themselves as feminists. Yet the feminist movement provided essential ideological, philosophical, and community support for a vital core of campus women who gained the language, voice, and courage to protest against sexist comments and demand an end to sexual harassment in classrooms, labs, and dorms. Those leaders insisted on holding schools accountable for breaking down institutional barriers and making visible changes for women in engineering programs.

College students themselves, along with established professionals, created other subversive initiatives aimed at encouraging young women to pursue technical interests. In the process, they expanded the discussion of how teachers, peers, the media, and the general culture affected the socialization of girls, leading them to internalize both positive and negative ideas about their life options. SWE and smaller groups of female engineers tried to reshape the messages projecting what academic, career, and personal directions were standard or unusual for women.

The Society of Women Engineers provided avenues for women to mobilize and provide each other with professional, social, psychological, and financial assistance. Women studying engineering at individual schools constructed their own interest groups and support strategies, adding up to some tangible improvement in the gendered conditions of engineering study. Starting in the 1960s, activists organized dozens of conferences, open houses, and other public events across the country to celebrate and advance women's achievements in engineering. Through networking, panel discussions, and publications, established professionals sought to leverage individual successes to build solid mechanisms that they hoped would ease the path for women coming along later. In consulting mentors and role models, younger women sought advice, including answers to persistent concerns about juggling career and family. Female engineering students during this era remained optimistic that excellence would win out and that talented women generally would not face any crippling discrimination in securing good jobs.

For all their value in providing professional advice, psychological reinforcement, scholarships, and other tangible assistance to young women in engineering, SWE and campus-based organizations could not totally transform the gender environment of the field overnight. Even the most supportive faculty and administrators could not wave a magic wand and bring female enrollment in technical majors up to 50 percent, nor could they guarantee a campus environment that was perfectly free of all gender stresses. The continued effort to reshape the national picture of women in engineering is the subject of the final chapter here. In retrospect, the postwar decades brought an often-impressively rapid increase in female engineering students, especially in certain majors and at specific institutions. But more recently, some observers have worried that women's representation in the discipline has hit a plateau, with enrollment leveling off or even declining. They interpret such trends as disturbing evidence that efforts to encourage women to consider nontraditional fields have stalled and that fundamental problems of gender inequity have lingered.

As feminist activists of the 1960s and 1970s pointed out, advocates can take many steps to improve women's college-level experiences in engineering and thus improve their retention rates and subsequent success, but intervention at the level of higher education often comes too late. Over recent decades, many elementary schools, middle schools, high schools, and communities have made heroic efforts, often working with the Girl Scouts and many other groups to nurture girls' interests in engineering and science. Through gender-neutral or girl-positive language and images, reformers hoped to meet the broad challenge of reshaping stubbornly ingrained notions of gender appropriateness and the cultural meaning of engineering.

By the start of the twenty-first century, almost all schools that offer serious programs in engineering had formally incorporated women into those disciplines and indeed initiated special efforts to recruit and support female students. As at Georgia Tech, Caltech,

MIT, and other schools in previous decades, female students' individual experiences after enrollment can vary dramatically. Some current women students take for granted that their years studying engineering have not included any major intellectual and social hassles, while others in different (or even the same) programs relate ongoing stressors, including demeaning comments or disrespectful treatment. Although introductory-level engineering classes may include a relatively substantial number of women, female students still stand out by their rarity in more specialized advanced courses. The account of how women moved into once-male territory, permanently changing the social, intellectual, and institutional nature of American engineering, is a complex tale. But in the end, that rich history of ongoing evolution adds a unique perspective to discussions about ongoing issues of diversity, balance, and fairness that today face women in engineering and science.

Decades before major American universities began welcoming significant numbers of women into their technical programs in the late twentieth century, a handful of women appeared, registering a unique presence in the field. Like men of the late 1800s and early 1900s, a rare group of women during that same era simply worked their way into engineering through observation, persistence, and the happenstance of being in the right place at the right time. Most notably, Emily Warren Roebling oversaw roughly ten years of construction work on the Brooklyn Bridge starting in about 1872 or 1873 after her husband, chief engineer Washington Roebling, was injured on the job. When Washington was inspecting underwater caissons, he traveled up and down too rapidly and suffered severe pain due to decompression stress (now commonly known as "the bends"). While he remained bedridden and partially incapacitated, his wife helped manage the technical correspondence regarding construction instructions. He coached Emily in bridge design so that she could inspect and help supervise work onsite. At the bridge's dedication in 1883, officials publicly honored Emily Roebling for her role in ensuring the success of her husband's signature accomplishment.[1]

As another example, Ethel Bailey turned a high school interest in radios and motorboats into a job during World War I as an assistant government inspector for Liberty airplane engines at an Indianapolis plant test field. Bailey took classes at Detroit's Michigan State Automobile School and special training through George Washington University. In 1920, she became the first female full member of the Society of Automotive Engineers and joined its staff as a research engineer.[2]

In the late nineteenth century, a number of American institutions of higher education, including Harvard, Yale, and Princeton, remained male-only. Schools created to shape sons of wealthy families into cultivated gentlemen could not accommodate women in that model. Similarly, coeducation seemingly had little place at colleges that prepared young men to become economic and political leaders, businessmen, clergymen, lawyers, and other professionals. Engineering education largely also fit that pattern; institutions or programs that specialized in training technical experts for industry, agriculture, government, or the military catered exclusively to men. When Rensselaer Polytechnic Institute received its first official admissions application from a woman in 1873, the school's director pressured her into dropping the request. He wrote, "The institute makes no discrimination in regard to sex, but . . . Miss Buswell's position as the only lady student would not be pleasant. If three or four other ladies were willing to join her, it is probable they would be welcomed." That promise remained untested.[3]

Although a large number of public land-grant colleges either were established purposefully as coeducational or generally soon admitted women, the majority of female students entered the humanities, home economics, and teacher-preparation tracks. Local and national educators designed "mechanic arts" and agricultural programs to give a state's future male citizens modern knowledge that would enable them to attain new heights of economic productivity. The few female students who might express interest in engineering were deterred, both officially and informally. From its start in the 1860s, Iowa State College (later University) followed a philosophy of this coeducational but gender-separated training. Trustees suggested, "If young men are to be educated [as] intelligent . . . farmers and mechanics, is it not as essential that young women should be educated . . . to properly understand . . . their duties as wives of farmers & mechanics . . . conducting a well-regulated household." In the 1870s, Iowa State taught "Chemistry as Applied to Domestic Economy" and created the first college-based experimental kitchen. The school soon developed related laboratory instruction in "scientific" cookery, sewing, laundry, and household management. The 1880s catalog explained that the domestic-economy curriculum was intended "to prepare young women for the highest demands of home life," following "the assumption that a pleasant home is the surest safeguard of morality and virtue." Iowa State's full "course for ladies" also included studies in English literature, languages, zoology, botany, and chemistry or geology. By 1912, home economics had grown into its own Iowa State division, which expanded rapidly in connection with the national home-economics movement.[4]

Still, it was a few of these state schools that provided the United States with its first female engineering graduates. In 1876, Elizabeth Bragg Cumming earned a civil engineering bachelor's degree from the University of California at Berkeley, writing a thesis that addressed a technical issue in surveying. She apparently did not practice engineering, however, and instead married soon after graduation and devoted her time to the family, community, and the civic activities that were popular among middle- and upper-income women in the late 1800s and early 1900s.[5]

For centuries, drawing had been a prized traditional accomplishment for well-bred girls in Europe and America, and possibly due to this artistic connection, skill in drafting became one of the earliest routes to justify women's entering engineering in the United States. At Iowa State College, Iowa native Elmina Wilson completed a bachelor's degree in civil engineering in 1892, alongside nine male civil engineering graduates. She stayed in Ames to earn her civil-engineering master's degree in 1894. Starting in 1892, Wilson joined Iowa State's staff, first as an assistant assigned to supervising the school's drafting room for a salary of $300. The school later promoted her to instructor and, in 1902, to assistant professor of civil engineering. She helped officials draft plans for a desperately needed campus water system, and when finished in 1897, the elevated steel-tank water tower became the first that was constructed west of the Mississippi River. Wilson published several technical articles, including a 1902 article in the *Iowa Engineer* about the testing of hydraulic cement formulas. Perhaps underlining the expectation that it was natural for women to gravitate toward domestic applications of technical knowledge, Wilson wrote a booklet titled *Modern Conveniences for the Farm Home*. Distributed by the U.S. Department of Agriculture in 1906 as part of the country's engineering experiment station extension work, her work discussed technical details of household heating systems and water supply (including plumbing traps and vents, waste-water pipes, and sewage disposal), earth closets, and garbage disposal. In 1904, Elmina Wilson left teaching to move into private employment with various companies, including Chicago's Purdy & Henderson structural engineering firm.[6]

Elmina Wilson's sister Alda was three years younger. She completed her civil engineering bachelor's degree at Iowa State in 1894, took some graduate architecture classes at MIT, and secured work with Chicago- and New York–based architectural firms. After World War I, Alda Wilson joined the Iowa Highway Commission, the state's very active road-planning agency, where she headed the Women's Drafting Department. Commenting on the sisters' careers, the journal of their women's fraternity Pi Beta Phi editorialized, "Probably no other women have done so much work in engineering lines as these two. . . . No better examples can be found of those refined, intellectual American girls who today are capable of engaging with the highest credit in the most scientific professions."[7]

Toward the end of the 1800s, a few other women from various institutions joined the slender ranks of female engineering graduates. Bertha Lamme completed an 1893 degree in mechanical engineering from Ohio State University with a specialization in electricity studies. She then designed motors and machines at Westinghouse for about twelve years. After Lamme married her supervisor, Russell Feicht, the Westinghouse director of engineering, in 1905, corporate anti-nepotism rules required her to retire. According to some accounts, she continued to serve as assistant to both Feicht and her brother Benjamin, who served as chief engineer at Westinghouse from 1903 to 1924. At the University of Michigan, Marian Sarah Parker earned an engineering degree in 1895 and then worked as a structural designer with Purdy & Henderson, the company that later hired Elmina Wilson. Parker transferred from its Chicago office to its New York office, where she planned foundations and steel framing for urban office buildings until she married and stopped working in 1905.[8]

Similar to the experience of many male engineers in the late nineteenth century, a handful of women entered practice after completing some classes but without a degree. For instance, Edith Julia Griswold took courses in civil, mechanical, and electrical engineering and around 1886 started her own New York City–based drafting practice, with a focus on drawing machinery for patent applications.[9]

Like male students in these decades, some women who enrolled in college engineering dropped their studies before completing degrees for a variety of reasons involving choice, necessity, or both. Especially in this era, however, leaving academia did not disqualify men and even a few women from engineering-related employment. The most notable example was Catherine (Kate) Gleason, who had grown up with tomboyish inclinations and became familiar with the Rochester, New York, machine shop owned by her father. After his oldest son died, Kate, then about age twelve, volunteered to take his place as an assistant in the plant. Both her parents endorsed the mid-nineteenth-century women's rights movement that had spread from nearby Seneca Falls; Kate's mother had befriended Susan B. Anthony, which might at least partially account for the family's acceptance of the unusual situation. Throughout high school, Gleason helped in the machine shop and took charge of bookkeeping. Some accounts credit her with playing a role in her father's creation of specialized gear-cutting tools that became widely used in automobile making and other valuable manufactures. At age eighteen, she became the first woman to matriculate in Cornell's mechanical arts program. Although apparently registered as a "special student," Gleason took shop classes with male classmates, substituting overalls for skirts. But within months, she returned home because her father still needed her business services. Gleason reentered Cornell in 1888, only to cut short her first year again due to health issues. Although she later took some part-time classes through the local Sibley College of Engraving and the Mechanics Institute (now the Rochester Institute of Technology), Gleason focused her energy on the family firm. In the 1890s, she officially became the secretary and treasurer for the Gleason Tool Company and also acted as its primary sales agent, securing important deals for sizable orders across the United States and Europe. In 1914, the American Society of Mechanical Engineers elected Gleason to full membership, the first woman to earn that honor. Later in life, she led construction projects to develop affordable housing in New York and California. She also discovered innovative standardized methods and patented new machinery for pouring concrete, becoming the first woman to hold membership in the American Concrete Institute.[10]

But in looking over late-nineteenth-century engineering education, the general absence of women is readily apparent. As just one example, early engineering classes at the University of Illinois appear to have been universally male, with one striking exception. The class of 1885 alumni register lists Josephine M. Zeller as that year's sole graduate in electrical engineering, which would be unusually early for that combination of gender and specialty. Examination of Zeller's transcript reveals that she primarily took classes in French, German, English, history, and various arts (painting, drawing, and clay modeling), with zoology and physiology as her only science classes. Given the apparent absence of any studies in physics or engineering, the logical conclusion seems to be that at some point in the school's record keeping, Zeller's actual category of "elective" studies may have been abbreviated and then later misread as "electrical engineering." The first true female graduates in engineering do not appear until four decades later. Louise Pellens earned an agricultural engineering degree in 1909, Beryl Bristow combined physics and engineering for a 1919 degree, and Grace Spencer received a chemical engineering degree in 1922. Before World War II, at least four other women graduated from Illinois in engineering.[11]

In the early twentieth century, the field of engineering still did not mandate rigid academic qualifications. Entrants came from a variety of backgrounds, a flexibility that created an avenue for one of the most visible female engineers in American history, Lillian Moller Gilbreth, who earned an undergraduate degree in English literature from the University of California Berkeley in 1900. Although skilled enough in high school science and math, she had no overt interest in anything related to engineering until her 1904 marriage to construction company owner Frank Gilbreth. Expecting his new wife to become a competent professional assistant, Frank coached Lillian in building-trade techniques and during her first pregnancy took her up ladders on construction-site visits. As the ambitious inventor of a time-saving scaffold for bricklaying and other ingenious construction devices, Frank soon became the disciple of Frederick Winslow Taylor, the famous and influential "father of scientific management." Lillian edited and coauthored (often anonymously) Frank's numerous publications and speeches on the subject of time-and-motion study and the quest for "the one best way"—the most efficient method of performing almost any job. As Frank developed growing international business as an industrial consultant, Lillian became an expert in the rapidly evolving field of industrial engineering. Lillian served as Frank's intellectual partner, helping him conduct experiments and make "micromotion" films of workers that could be analyzed second by second to eliminate all wasted movement. Frank never completed an undergraduate degree, so he appreciated Lillian's superior writing skills and pushed her to pursue a doctorate. He believed that by acquiring a Ph.D. Lillian could add prestige to the Gilbreth reputation and also claim credibility for herself as a woman among male professionals, few of whom held such advanced credentials but who "belonged" in engineering by virtue of their sex and hands-on mechanical experience. Lillian earned her doctorate in 1915 from Brown University in the area of psychology, uniting a special interest in industrial and educational psychology with engineering considerations to yield the then-novel field of applied management. Working together, the Gilbreths increasingly moved away from Taylor's pure scientific management system to emphasize the human element, creating a philosophy that defined workers' mental welfare and relief from physical fatigue as prerequisites for maximal productive efficiency.

After Frank's early death from heart disease in 1924 abruptly cut off the Gilbreths' links to consulting clients, Lillian struggled to support her eleven children and win business commissions in her own right. Frank's name had given her an entrée to the engineering world, but in the end, engineering colleagues and key employers still perceived her as secondary to Frank. Given those realities, Lillian Gilbreth was in a unique position to comment on the prospects for female engineers. In connection with the 1923 convention of the Women's Engineering Society in Great Britain, she wrote an article declaring that "Opportunities for women in industrial engineering in America are practically limitless, so far as training and cooperation from those already in the field is concerned." But when citing particulars, Gilbreth only underlined the scarcity of women's actual presence; she praised the one female "pioneer" who was then studying industrial engineering at Penn State and the lone female graduate of the Babson Industrial Institute, who happened to be the founder's daughter. Gilbreth repeatedly expressed optimism about engineering as a future field for women but tempered that enthusiasm with a realistic description of obstacles. In 1940, she wrote, "Standards of selection and placement are often far more rigid for women than for men, which results in a need for greater qualifications. . . . The chief need is to do away with sex discrimination."[12]

Gilbreth publicly thanked male engineers for extending "only the most kindly and cooperative treatment" to their few female colleagues. In fact, however, Gilbreth had to fight to win formal acceptance from her peers even after she had established a solid career on her own after Frank's death. In 1925, Gilbreth "slowly and carefully" sounded out leading members of the American Society of Mechanical Engineers about whether she had any chance to win full membership status in that group, as her husband, Frank, had. Kate Gleason represented the sole female precedent, and according to one former council member, influential leaders at the ASME raised "considerable objection" to the prospect of welcoming any others, feeling that the "admission of lady members has not been an entire success." Frederick Waldron confirmed that his ASME membership committee felt "well grounded" reservations about the worthiness of other recent applicants from "the fair sex." Hosea Webster, an engineer at Babcock and Wilcox, advised Gilbreth to postpone her request for consideration, writing, "Based solely upon experience with and observations of the personality and qualifications or lack of qualifications of women applicants who have been presented, there is a very decided feeling that it is not wise to further consider admission of women for membership in any way in this Society." But other contacts promised to provide glowing references of Gilbreth's ability and sponsor her petition as a symbol of encouragement to future female engineers. John Freeman wanted the ASME to grant women "equal privileges" inside the profession, telling Gilbreth, "I am glad to see you and Kate Gleason lead the way toward new openings for young women who have a mechanical turn of mind. Charles Newcomb of the Holyoke Steam Pump works during the war took a dozen high school women into his office and trained them for draftsmen, and after the war was over found them so useful that he continued to have a large proportion of young women in his drafting room. Two of our best draftsmen . . . of the Factory Mutual Insurance Company in Boston are young women. . . . I am most heartily in favor of opening every door for women with ambitions in this line." R. A. Wentworth added, "Undoubtedly the American Society of Mechanical Engineers would be honored by your membership. I will be glad to serve as one of your references in your application. . . . Certainly, in the future, there will be many women members." In 1926, the ASME elected Gilbreth to society membership.[13]

Even as Lillian Gilbreth slowly worked her way into an industrial engineering career of her own, the broader world of engineering was changing. Formal credentialing would increasingly come to play a greater role, a shift that ultimately paved the way for women's entry. True, early twentieth-century college engineering programs universally either excluded or marginalized female students, and for many powerful reasons, few women chose to pursue technical studies. But for women, engineering degree programs were at least relatively more accessible than the challenge of finding opportunities to work their way into the field through the unwelcoming route of industrial hands-on experience. Factories and construction sites proved inherently gender-hostile to women, except for a few who capitalized on family connections (such as Emily Roebling, Kate Gleason, and Lillian Gilbreth). By contrast, academia increasingly distanced engineers from machismo-laden field sites and placed new emphasis on theoretical desk work, enabling women to compete based on brains rather than brawn. Although discrimination continued for female engineering students, grades, examination results, diplomas, and civil service employment qualifying examinations at least in principle offered objective evidence of talent.

In the early twentieth century, a number of institutions added to the small roster of female engineering graduates, one or two at a time. When she entered the University of Kentucky, Margaret Ingels wanted to study architecture, but because the school had no such degree then available, the dean reportedly convinced her to try her hand at mechanical engineering. After graduating in 1916, Ingels then established a thirty-year career as the nation's first female air conditioning engineer, working mainly for the Carrier-Lyle Engineering Corporation. Ingels also wrote "Petticoats and Slide Rules," one of the earliest articles that documented the first female engineers in the United States, published in the *Midwest Engineer* in 1952.[14]

Prior to World War II, several female students enrolled at the Newark College of Engineering in New Jersey (now known as the New Jersey Institute of Technology). In 1930, for example, Newark College had two women studying chemical engineering, one a senior whose high grades made her a scholarship recipient. Newark's most distinguished alumna from that era was Beatrice Hicks, the daughter of an engineering executive who earned her bachelor's degree in chemical engineering in 1939. Hicks then completed a physics master's degree at Stevens Institute of Technology as well as graduate electrical engineering courses at Columbia. During World War II, Hicks became the first female engineer hired by Western Electric, where she worked on the research for and production of quartz crystal oscillators. After the war, she became chief engineer at New Jersey's Newark Controls Company, a business that had been founded by her family and that manufactured environmental sensing devices for the military and electronics industries as well as fire-control mechanisms for antiaircraft guns. Hicks received a patent on her invention of a special gas-density monitor, a technology that proved to be valuable in the rapidly growing American space program. In 1955, Hicks became the firm's president and director of engineering. Publicity material from the National Society of Professional Engineers described Hicks as "a very attractive woman, less than forty years old, with dark blue eyes, light brown hair and a svelte figure only five foot three." She married a fellow professional, Rodney Chipp, who served as director of engineering for ITT Communication Systems: "Altogether the kind of woman-executive

Hollywood likes to portray, . . . [Hicks] pleasantly surprises the most poised officers when she negotiates contracts with the Armed Forces." In the early 1950s, Hicks served as the president of the newly established Society of Women Engineers and traveled extensively to deliver lectures about engineering and women's role in the discipline.[15]

## EARLY EFFORTS TO ESTABLISH A PROFESSIONAL SISTERHOOD

Before World War II, simply being a woman who was studying engineering was still unique enough to rate a photograph that might appear on the front page of campus papers at Cornell, Iowa State, Minnesota, and elsewhere. The media coverage treated each woman engineer individually, as if each case were unusual—which it was. Under the headline "Beauty Meets Resistance," the *Penn State Engineer* noted in 1934 that Olga Smith had become the first female who enrolled in electrical engineering. At the State College of Washington, Clara Seaman received a degree in metallurgical engineering in 1928, reportedly with high honors; one account credited her with earning a 94 percent four-year scholastic average, "even though she worked in an office to pay her college expenses." Treating a female mining engineer as a pure curiosity, the *San Jose News* in California ran a short piece headlined "She's a Miner!" alongside amusing squibs about a woman who kept her marriage secret for four years and about the skeleton of a newly discovered elephant species. The article opened, "Clara Seaman . . . is not a gold-digger, but a miner. She's been taking the 'hard rock' course at State College." Most female engineering students talked about their experiences in the singular; they didn't know enough other female students to refer to themselves as a group.[16]

A rare and therefore significant exception to that rule came in 1914, when the University of Michigan had thirteen female students in engineering and architecture, enough of a cohort that the women sensed both an opportunity and the desirability of forming a club. The *Michigan Technic* explained that although the wives of engineering professors had displayed "a great deal of interest in the women and . . . entertained them a great deal at their homes," the unwieldy size of such gatherings led the female undergraduates to create their own campus group, the T-Square Society: "When women first enrolled in the Engineering Department they naturally felt somewhat out of place among so many men . . . , [so] one main idea of the society was to make the freshmen women feel at home." Monthly meetings also aimed "to promote sociability and discuss things of general interest." Member Bertha Yerex Whitman explained that one aim was to "make up for the lack of social relationships with the other women on campus, prevented by the nature of their work." T-Square automatically considered all women students in the department of engineering and architecture as its members. The organization designed its own membership pin, with an insignia of crossed T-square and triangle, which claimed a share of identity of male professions by embracing the iconic representations of their tools. The editorial staff of the *Michigan Technic* poked some gentle fun at the women's group, remarking that after a recent T-Square session, the engineering meeting room was littered with toothpicks and chewed paraffin. The first T-Square president was Hazel Quick, and its vice president was Alice Goff. Both received Michigan civil engineering degrees in 1915 and then entered the workforce. Whitman became the first female graduate of Michigan's College of Architecture in 1920; she had belonged to an earlier class but paused her education during wartime like her male classmates, so they could later graduate together. During World War I, the Dodge Brothers Company hired Whitman as its first female to work in engineering drafting.[17]

Other University of Michigan engineering graduates in that period included Dorothy Hall, Dorothy Hanchett, and Helen Smith. Their T-Square Society lasted, documented in photographs in the university's 1920s yearbooks alongside photos of male-dominated engineering student groups. Their club often included faculty wives and engineering-school secretaries as honorary members. Women students in architecture and engineering shared space on campus, which reinforced common experiences of marginalization. Architecture major Delight Sweeney recalled that in 1915, on her "first morning in the engineering school . . . when I had to walk from one end of the building to the other . . . there were *billions* of men . . . all staring. I can still remember how my knees banged together." Several T-Square alumnae established long careers in engineering and architecture. After finishing her Michigan degree, Yerex Whitman thrived in architectural work for three decades, although at first some prominent firms "flatly refused" to consider her for a job. Her Michigan contemporary, Juliet Peddle, became the second woman to finish the Michigan architecture program; she had learned drafting while young from her father, who taught machine design at Rose Polytechnic. In 1928, when Peddle and Whitman were both working in Chicago, those T-Square alumnae helped establish the Women's Architectural Club of Chicago. Sweeney, who also earned a 1920 architecture degree, became the first female member of the New Jersey Society of Architects and worked with the Federal Housing Administration in the 1930s.[18]

A similar nascent sense of sisterhood and common professional interests led several University of Colorado women to create a national group of female engineers and architects in December 1918. Lou Alta Melton and Hilda Counts announced formation of the American Society of Women Engineers and Architects, serving as its first president and vice president, respectively. Melton and Counts surveyed American engineering schools by mail to collect data on female students, past and present. Their inquiries revealed that the Universities of Michigan, Kansas, Ohio State, Illinois, and Texas seemed to lead the way in offering engineering training to women. At each of those five schools, the total of female students and graduates in engineering had reached double digits by academic year 1919–20. Putting those figures together, Melton and Counts estimated that about two hundred women might be eligible to join their new group, with membership open to any woman who had either degrees or work experience in either engineering or architecture. Announcements of this new organization ran in several national technical publications. *Engineering News-Record* explained that according to Hazel Quick, "the organizers felt the need of a society and were not received into the existing societies organized by and for men." By summer 1920, the group claimed to have eleven members already, from the branches of civil, mechanical, chemical, electrical, and architectural engineering. After Counts received her Colorado electrical engineering degree in 1919, she worked for Westinghouse and later for the Rural Electrification Administration. Melton, class of 1920, found employment as a bridge designer with the U.S. Bureau of Public Roads. A contemporary, Elsie Eaves, also helped initiate organizing efforts. As a civil engineering major at Colorado, Eaves was elected in 1918 to become the first female president of the school's student engineering society, the Combined Engineers (a fact reported in engineering publications nationwide). After completing her degree with honors in 1920, Eaves established a thirty-year career analyzing construction-cost economics and industry standards. Although the American Society of Women Engineers and Architects never grew into a major national organization, Eaves and Counts Edgecomb continued to encourage fellow female engineers and after World War II helped promote the creation of today's Society of Women Engineers.[19]

## INVASION RHETORIC: DEFINING THE BOUNDARIES OF MEN'S SACRED DOMAINS

Land-grant schools and public colleges such as Michigan remained the best places for women to establish a small but real presence as engineering students in the early twentieth century, although a few studied at other schools, such as MIT. One at a time, a handful of women earned degrees, and some moved into paid engineering employment. The account here is intended to be suggestive rather than a comprehensive documentation of every woman who completed engineering education before World War II. To note just a few examples, the University of Colorado awarded a civil engineering degree in 1903 to Minnette Ethelma Frankenberger, and the school's engineering journal listed her in its alumni directory for several years as working as a draftsman in

Boulder. At Cornell, Nora Stanton Blatch earned a bachelor's degree with honors in civil engineering in 1905. Extending the women's rights activism of her famous grandmother Elizabeth Cady Stanton, Blatch became an advocate for suffrage and the Equal Rights Amendment. She explained at one point that she decided to become an engineer precisely because so few women had entered the field. Blatch worked as a draftsman and assistant engineer with New York City's water-supply board and Public Service Commission and became chief draftsman for the Radley Steel Construction Company. In 1906, Blatch joined the American Society of Civil Engineers (ASCE) as a junior member. After reaching that category's age limit in 1915, she applied for an associate membership, as men normally did. The ASCE refused her request and dropped her from its rolls, alleging that despite her employment record, she was too inexperienced to qualify for an upgrade. In the ensuing lawsuit, the New York State Supreme Court decided that as a private body, the ASCE was not compelled to grant her any particular privileges. It was not until 1927 that the ASCE granted its first associate membership to a woman, Elsie Eaves.[20]

After Blatch, Cornell hosted a few other early twentieth-century female engineering students, including Olive Dennis. Before arriving in Ithaca, New York, Dennis completed a bachelor's degree in science and mathematics at Baltimore's Goucher College for women, followed by a master's degree as a fellowship student in math and astronomy from Columbia University. While teaching high school math, Dennis studied civil engineering in extension classes, adding surveying studies and other summer courses through the University of Wisconsin. Dennis then spent one year at Cornell to finish her training, receiving a civil engineering degree in 1920. She worked for over thirty years at the Baltimore and Ohio Railroad, first as a draftsman in the bridge engineering department, then as an engineer of service and a research engineer who designed some of the railroad's terminals. More in line with expectations about feminine nature, Dennis also designed the china used in the train line's restaurants and the interior decoration for passenger cars. To help make riders more comfortable, Dennis improved the quality of seating and lighting. She invented and patented a window ventilator that allowed riders to control their individual seating area's air flow. Dennis traveled as much as forty thousand miles on the B&O line annually as part of her mission to cultivate passenger loyalty, especially among females. As the first (and for many years, the only) female member of the American Railway Engineering Association, Dennis served on its committee supervising the economic study of railroad location and operation.[21]

At land-grant schools like Cornell, women such as Blatch and Dennis drew a certain attention because they were an intriguing rarity, a curiosity. Remarking on that female

# Three Co-eds Invade Engineering Courses
# And Compete With Men at Cornell University

Jeannette Knowles, a junior in mechanical engineering, operates a compression testing machine in the "mech lab."

## STAND WELL IN THEIR STUDIES

### All Plan to Enter Business on Graduation

ITHACA.—If it weren't for Jeannette Knowles, Jane Morse and Barbara Hart, there would be an even 850 students enrolled in the famous college of engineering at Cornell university.

These three young women, the greatest number of women students ever enrolled in engineering here at one time, seek to find a place for themselves in the fields of structural and mechanical engineering. All three are considered good students, fully capable of mastering the heavy studes which are the lot of the engineering student.

JUNIOR IN ENGINEERING

Miss Knowles, a native of Richmond, Ind., is a junior in mechanical engineering. Tho she made it

FIGURE 1.1

At Cornell University, as at so many other American schools in the twentieth century, the mere presence of female students in engineering counted as headline-worthy news. This article reported that Jeannette Knowles, Jane Morse, and Barbara Hart were all "considered good students, fully capable of mastering the heavy studes [sic] which are the lot of the engineering student." *The Syracuse Post-Standard*, November 14, 1937, p. 15, Cornell University archives, #16/3/860.

presence, a 1930s New York newspaper ran the headline "Three Co-eds Invade Engineering Courses and Compete with Men at Cornell University: Stand Well in Their Studies." The article featured a photograph of mechanical engineering junior Jeannette Knowles working on a compression-testing machine and noted that the three students represented "the greatest number of women students ever enrolled here at one time," attending classes alongside the 850 men who were enrolled in Cornell engineering.

In this era, the word *invade* appears repeatedly in popular references to the presence of a few female engineering students at a number of schools. Next to photographs of the two women enrolled at the University of Minnesota in 1925, that campus's engineering magazine ran the headline "Co-ed Engineers: Man's Domains Are Again Invaded." The piece noted that when Esther Knudsen and Ursulla Quinn first arrived in 1921, male students in the engineering classrooms heard "the click-click of women's heels upon the tiles of man's last retreat at the University" and helpfully rushed to redirect the presumably lost female students to their proper building. Doubters expected that within one quarter or, at most, one year, Quinn and Knudsen would "change their minds and leave the Engineering School to men alone." But the two "courageous" and "determined" women had advanced to graduation, "whereas most of the sceptics have dropped by the wayside." An accompanying cartoon undercut the tone of respect, showing two women doing outdoors surveying work in oddly short skirts that were riding up to show their legs. But the article praised Knudsen and Quinn for their excellence in practical surveying and design laboratory work as well as theoretical courses: "It is safe to predict that through their efforts and their succeeding, classes of engineering will have larger and larger quotas of girl students in the ranks. And Man's sacred domains will be sacred no longer." As it turned out, Knudsen and Quinn remained "determined," and both received their civil engineering degrees from Minnesota in 1925. Knudsen then became the first woman civil engineer hired by Wisconsin's highway commission, as the *Milwaukee Journal* explained under the headline, "Wisconsin Girl Engineers Like Their Jobs: Miss Esther Knudson [sic], of the State Highway Commission, and Miss Emilie Hahn, Graduate Mining Engineer, Attain Success and Retain Their Feminine Charm." Photographs showed Knudsen wearing what she called "sensible" masculine clothing (pants) while out in the field surveying, but also wearing a skirt and feminine shoes while posing on the steps of her drafting office. The article pointedly stated that "Knudsen abhors masculinity in women" and devoted several paragraphs to stressing the "feminine tastes" of Knudsen and Hahn. Both had studied music. Knudsen also enjoyed batik work and painting and had been a model. In noting that Hahn had just become the first woman to graduate with a bachelor's degree in engineering at the University of Wisconsin, that same *Milwaukee Journal* piece remarked, "The invasion of the university by women which began in 1860 with the entrance of one co-ed is now complete."[22]

Similarly, after Jeanne Chandler graduated from Purdue in 1941, the campus newspaper reported on her job as the only female engineer with Newark's electric and gas company under the headline "Coed Invades 'Man's World'; Finds Success in Engineering." And in 1940, Wayne State in Detroit described four women's presence in the engineering college with the phrase "Female Engineers Invade Chem Labs and Airplanes." The common use of invasion rhetoric following World War I and during World War

II is not entirely surprising. But the dictionary definition of "invasion" accentuates the negative connotations of that word, implying a hostile incursion by a threatening horde, with destructive consequences. As an exaggerated overreaction to the arrival of just one or two women, this choice of "invasion" language dramatically underlines the extent to which women in engineering appeared as the *other*, entering a field that everyone assumed was and must be male territory.[23]

Adopting another metaphor of violent attack, one 1930 newspaper interviewed fifteen female engineers and then declared that "In knickers and in printed chiffon, wearing long hair and hair closely cropped, modern women have stormed their way into the field of engineering, while men could only gape in surprise." The woman reporter opened with the question of fashion, noting that most female engineers preferred clothing that was neither overly tailored nor frilly, mirroring Gilbreth's philosophy that "a woman engineer should dress in a way that is neither too masculine nor too feminine. She should try to avoid appearing in any thing conspicuous or inappropriate, in order to help those who will follow her." The article did not cover up professional obstacles, quoting Laura Munson as saying that to overcome doubters, a female engineer "must be quicker, more accurate and more intelligent than the average man" who is doing a similar job. The women interviewed judged the typical female engineer as "superior to her male rivals" in persistence, tact, delicate touches, and "attention to the finer details which a man finds so boring and dull he lets them slide by." Aeronautics specialist Elsa Gardner, trained at St. Lawrence University and working for a U.S. torpedo firm, asserted that engineering "requires a breadth of vision which I believe women possess in greater measure than men."[24]

Throughout the early twentieth century, the few female engineering students scattered across various schools (while completely excluded from others) seemed like such an anomaly that no one seriously anticipated that a large number would choose to copy their venture. At Cornell, a mere hint that greater numbers of women might someday enter engineering spurred an undercurrent of uneasy jesting among their male counterparts. Before speaking to the Cornell Society of Engineers about financial problems in 1938, the vice chair of the board of trustees noted with amusement that when he arrived at Cornell in 1886, he heard rumors that during the two previous years, "there had been a woman who had had the temerity to register as an undergraduate in Sibley College," Cornell's engineering school. Justin DuPrat White went on to remind listeners that Cornell's later few female engineering students had earned the nicknames of "Sibley Sue" and "Slide Rule Sadie," a reference that elicited knowing laughter from his audience of engineering alumni. White proceeded to "congratulate" Cornell men "for their great tolerance [for] . . . the entrance of women in the Engineering College. . . . Lord knows that the lawyers have almost been swamped with the influx of women in their profession. . . . My prediction, as I gather it from the medical men, is that [women] are going to handle [engineering] with a good deal of challenge to you men if you are not careful."[25]

Leaders of engineering schools generally did nothing to promote or ease women's enrollment during this era and indeed, often apparently sought to discourage female students. When first-year student Gladys Tapman initially talked with Cornell's engineering dean sometime around 1931, "he pointed out the difficulties in the way of a woman engineer. In reply, she quoted the motto on the university shield which promises instruction in any subject regardless of sex." The *New York Times* commented, "To vindicate . . . her ambitions, she completed the course in three and a half years." After receiving her civil engineering degree in 1934, Tapman found employment helping design Mississippi River dams and locks, factories and oil refineries on the East Coast, and sanitation works and traffic lighting with the Civil Service in Manhattan.[26]

At Cornell alone, by 1938, Blatch, Tapman, Dennis, and more than eighteen other women had received engineering degrees, one by one over the years. Isolation made their experience hard. One "Slide Rule Sadie" (as they were nicknamed) said:

A girl has to want . . . pretty badly to go through with the course in spite of the unconscious brutality of . . . [male] classmates. . . . She must be ready to be misunderstood, as . . . many . . . will conclude that she took engineering . . . to catch a husband. She must do alone lab reports and other work men do in groups—because men who are willing to face the scorn of their peers and . . . work with her are more interested in flirting than in computations. She must be prepared for a lonely academic career; she cannot approach her classmates to exchange notes without appearing bold.[27]

But one hint of change came at Purdue, where Frank Gilbreth had long been connected. After his death, his wife, Lillian, stepped into his place as a visiting lecturer. In 1935, the School of Mechanical Engineering hired her to teach management, making her Purdue's first female engineering professor. Not coincidentally, that same year, Purdue appointed famed aviator Amelia Earhart to serve as a career consultant for female students. Purdue had recently opened its first residence hall for women and wanted to fill it. With Earhart's high-profile appointment, female enrollment jumped 50 percent, and the new dorm overflowed. For both Earhart and Gilbreth, their ties to Purdue reflected the direct influence of progressive-minded president Edward Elliott, who supported bold thinking about opportunities for women in American society. After hearing

Earhart deliver a speech encouraging women to become pilots and calling on the aviation business to welcome more women, Elliott courted her to come to his campus. Purdue had an ambitious aeronautical engineering program and was the only university in the country to own a fully equipped airfield, where Earhart worked on her latest planes. Engineering dean A. A. Potter grumbled about Earhart's appointment to Purdue's aeronautics staff; he reportedly objected to her lack of academic credentials and complained that she confused engineering with the job of a mechanic. But Earhart interacted well with both engineering students and female students. She scandalized conservative faculty wives but intrigued young women by wearing pants in public. Earhart warned Purdue women against sacrificing their own interests for a husband and encouraged them to think about combining marriage and career. Students recalled that she encouraged women to enter the fields of engineering and science, regretting the gendered separation of occupations that limited students' choice of classes. Earhart wrote, "Today it is almost as if the subjects themselves had sex, so firm is the line drawn between what girls and boys should study." One of Purdue's few female engineering majors of the time, Marguerite Call, praised Earhart for having "explained to us very clearly what some of the obstacles were in the way of women who went into what has always been known as a man's field. She was encouraging, too. She didn't see why, if a woman has special talents . . . she couldn't go out and show 'em!" One Purdue staffer described Earhart's role as "motivating the girls to do something more than take home economics," and the same could be said for Gilbreth's appointment. Elliott explained that he had brought Gilbreth and Earhart to Purdue "with the intention of introducing new forces for the study of the most important modern unsolved problem of higher education—the effective education of young women." In between other personal commitments and work-related travel, both Gilbreth and Earhart lived in Purdue's women's housing for several weeks each month and ate in campus dining halls, where they spent substantial time talking both formally and informally with female students. Gilbreth continued to teach at Purdue until retiring in 1948 at age seventy, the first woman to reach full-professor rank in engineering in any American institution.[28]

Although Gilbreth and Earhart hoped to encourage and inspire female students of any major, their presence did not immediately convert Purdue engineering into a more gender-balanced field. As at other coeducational schools before World War II, very few women at Purdue enrolled in engineering, and among that handful, attrition rates proved high. At least in public, some put on a brave face, stressing the positive aspects of their academic experience and the common experiences that they shared with male engineering classmates. In summer 1940, Ellen Zeigler and Kathleen Lux joined seventy

Purdue men at the school's regular civil engineering camp, where the "camper-ettes" joined men in playing baseball, swimming, and performing field observations and computations. In fall 1941, Lux joined male counterparts on a weeklong field trip to a student American Society of Civil Engineering conference in Chicago, where they toured filtration and sewer plants, a Gary cement company, the Hammond waterworks, and Carnegie Steel's Bessemer converters and open-hearth furnaces.[29]

Those industrial connections and the hands-on applied aspect of engineering still raised some of the deepest doubts about women's place in the discipline. In 1931, one University of Minnesota observer reported that female students in the college of engineering and architecture and in chemical engineering recorded lower grade-point averages than male counterparts. As a possible cause of women's "inferiority," the article suggested that "men are much more interested, and therefore scholastically better, in the practical courses. They have had closer contact with the practical side of engineering before their entrance into college and have some background for their theoretical studies. . . . Convention has told the women to leave this type of work to the masculine world, with the result that even vitally interested women were unable to gather that experience through work or observation."[30]

Perhaps to counter such lines of criticism, Purdue's Charlotte Bennett maintained in 1935 that she had completed shop class perfectly well. Although she initially had a hard time figuring out how to design a foundry, "I doubt if the fact that I was a girl made the work more difficult . . . because several of the students in the class seemed as ignorant of foundries as I was—if not more so." In a distinctly defensive tone, Bennett reported that she had not created any disasters in class. Although instructors initially refused to let her pour metal, "finally after much persuasion I was allowed to . . . pour several . . . molds. They said I did it as well as anybody." Her welding practice had gone without "any accident," and although she had accidentally made the top fly off her indicator when opening steam cocks in mechanical lab, "I got the steam shut off and someone caught the part that came loose and it wasn't damaged. It was one of those things that happen to eds and coeds alike and make one more careful." Bennett completed her degree in 1936, Purdue's first female graduate in chemical engineering, and then became a secondary-school math teacher.[31]

Purdue's Ellen Zeigler admitted that women's lower physical strength tended to keep them out of many industrial jobs but declared that for positions in drafting, consulting, designing, and statistical analysis, women with talent could compete through "extra enthusiasm and ability" plus hard work. Success in engineering came not through masculine identity, she insisted, but through nongender-linked qualities such as open-mindedness, insight, adeptness in social interactions, and manual skills for quick problem solving. Zeigler worried that too many female engineering students suffered from a lack of confidence and for that reason, she praised university co-op programs for giving women the experience to know "that she can and will succeed in her line of work." She warned fellow female engineering students who were seeking employment to be better-prepared than male classmates, since "if the woman desiring the job has the same record as the man, the man will undoubtedly get the job . . . [but] a woman of unusual ability has as much chance of success as any man."[32]

There was no point in denying that they were unusual, but these scarce female engineering students sought to stress that they approached their work in just as straightforward a manner as their male classmates and honestly enjoyed technical subjects. At the same time, media portrayals depicted women engineers as relishing their femininity. Purdue's campus newspaper told readers that electrical engineering graduate Jeanne Chandler enjoyed knitting sweaters for herself, assisted her mother in house cleaning, and envisioned herself eventually getting married. Penn State praised industrial engineering major Anne Very as a social butterfly who was fond of both mathematics and dancing. Significantly, Very was following in the academic footsteps of her father, Dexter, who had earned his Penn State degree in highway engineering in 1913, while also becoming football team captain and a national star. His daughter originally enrolled in liberal arts and then transferred, becoming Penn's first female graduate in industrial engineering in 1939.[33]

EXPERIMENTAL CURIOSITIES: WOMEN AT MIT, 1871–1941

The record of female students who entered MIT extends back to 1871, when Ellen Swallow, who held an undergraduate degree from Vassar College, requested permission to pursue an MIT advanced degree in chemistry. MIT denied her application for graduate-level work but agreed to let her enter as a "special student" to pursue a second bachelor's degree without charge (a status that gave school leaders a loophole to deny her official presence in case of any objections). The faculty vote of approval described the move as "an experiment" that should not set a broad policy precedent for female admissions. Swallow later recalled being excluded from general classes and literally being closed off to work in separate laboratories.[34]

After Swallow earned her MIT bachelor's degree in 1873, she wed Robert Richards, head of the school's mining engineering program. Although Ellen Swallow Richards hoped to continue working for her doctorate, MIT was not prepared to issue a first

Ph.D. to a woman. Instead, she poured increasing efforts into opening an MIT-based education to other women. She raised funds from sympathetic Boston women (contributing a thousand dollars a year herself) to establish and equip a new Women's Laboratory at MIT in 1876. Volunteering as an unsalaried instructor there, Richards taught chemistry and related subjects. Over the next seven years, approximately five hundred female "special students" studied or worked in the Women's Laboratory. Many were secondary-school teachers who were seeking laboratory experience and advanced training. But Richards and her supporters continued their campaign to win women's admittance to regular courses. To ensure that administrators could not cite inadequate facilities as an excuse to turn away female students, Richards in 1882 raised eight thousand dollars to add a women's lounge and bathrooms to the school's new chemistry building. MIT conceded the point, agreeing to close the Women's Laboratory and start admitting undergraduate students "without distinction of sex." In 1883, eleven women attended MIT along with 432 men; in 1887, there were twenty-five. Meanwhile, MIT appointed Richards to a paid post in the new Sanitation Chemistry Laboratory, where she conducted pioneering work in sanitary and environmental engineering, including sewage treatment, water safety, and food analysis. Reflecting her conviction that women could use science as a tool to improve the family diet, Richards later helped organize the emerging discipline of home economics (often called *domestic science* or, significantly, *domestic engineering*).[35]

Richards had created the first physical places on the MIT campus for women—first, the Women's Laboratory, and second, the Margaret Cheney Women's Lounge, named for a Richards protégée who died prematurely. The Cheney suite served for the next eighty years as "a feminine retreat in the midst of a male environment," which women students appreciated as a comforting "refuge" where they could study, relax, and eat lunch. That last function stirred controversy, which reflected deep uncertainty over women's place at an overwhelmingly male school. Some female students felt uncomfortable eating in the main cafeteria amid throngs of men, but the women's governing committee refused to sympathize, "since the girls have deliberately chosen to attend an institution where men students predominate." Those observers saw no reason "why preferential treatment should be accorded the women students, or why Technology should allow them privileges which the men do not enjoy." Defending women's interest in having lunch in their own lounge, Mrs. Frederick Lord, representing MIT alumnae, responded that eating separately "gives a much needed chance for the girls to know one another." Lord did not challenge "the traditional attitude of the Institute that no discrimination be made in the women's official relation to the Institute." But citing "obvious" differences between male and female students, Lord suggested that MIT could create a special position to address women's distinct educational and personal needs. In 1939, the school named architectural library staffer Florence Stiles, herself a 1923 MIT graduate, to fill the semiofficial post of adviser to women students.[36]

None of MIT's earliest female graduates specialized in engineering. The first was apparently Lydia Weld, who initially entered Bryn Mawr College but moved to MIT to complete a naval architecture and marine engineering degree in 1903. Weld began her career by drawing ship plans for a firm in Newport News, Virginia. In the years leading up to World War I, she worked for the United States War Shipping Board in San Francisco, coordinating supplies of essential materials for accelerated ship construction and repair until illness led her to retire in 1917. The first MIT undergraduate degree in chemical engineering went to Marion Rice in 1913. She was the only female graduate with 312 men in that year's commencement, and the *Technology Review* reported her "receiving the hearty cheers of her fellow students as she was handed her diploma." As the daughter of a wealthy inventor, Marion Rice Hart did not need to seek employment. She earned a geology master's degree at Columbia, completed a three-year round-the-world sailing trip, began flying at age fifty-four, made seven solo trans-Atlantic flights, and wrote a guide to celestial navigation and memoirs of her long and colorful life.[37]

The small ranks of female undergraduates who finished studies at MIT before World War II included at least two graduates in civil engineering, three in electrical engineering, three in chemical engineering, two in engineering administration and management, and one each in metallurgy and naval architecture. The lone female aeronautical engineer, Isabel Caroline Ebel, graduated in 1932 but discovered that a woman's MIT degree did not open doors in the aircraft industry. Amelia Earhart's intervention helped persuade New York University's Guggenheim School of Aeronautics to admit Ebel for advanced studies, but after finishing those extra qualifications in 1934, Ebel did not secure a permanent engineering position until 1939 with Grumman. Among the other early generations of MIT women, Edith Paula Chartkoff completed the metallurgy and mining program in 1925. After writing her undergraduate thesis on "the effect of cold work on austenitic steel," Chartkoff Meyer went to work for the Cleveland Wire Works of General Electric. Harriet Whitney Allen wrote a thesis on the "current capacity of rubber-insulated cables" for her 1927 electrical engineering degree. Ruth Pfeiffer MacFarland wrote a 1934 thesis on "dye effects in sail duck" for her naval architecture degree. Hannah Chapin Moodey entered MIT as a junior from Smith College and earned an electrical engineering degree in 1936. After noting that "some industrial corporations still refuse to hire women engineers, on the ground that living conditions in the field are difficult," a 1957 *Atlantic* article on science careers for women cited

Moodey's job designing cathode ray tubes at RCA Victor's color television laboratory as an encouraging sign.[38]

At the graduate-student level, Edith Clarke became MIT's first female graduate in electrical engineering in 1919. She was the first woman chosen to join the American Institute of Electrical Engineers and was one of her era's best-known women engineers. But in 1912 through 1914, MIT's female enrollment stood at zero. The number of female undergraduates and graduate students enrolled at MIT averaged approximately forty-five a year in the 1920s, fifty per year in the 1930s, and sixty-five in the early 1940s, but among undergraduates and especially in engineering, attrition rates remained high. In 1941, most undergraduate women continued to major in biology and public health, chemistry, or architecture. One accounting by Marilynn Bever records a total of only about seventeen undergraduate degrees granted to women in all MIT engineering programs between 1871 and 1941. The program in mechanical engineering, a perennially popular choice among male MIT students, apparently had at least three women enrolled at various points over those decades but recorded none who completed degrees.[39]

Each woman arriving on campus during this era stood out as a curiosity amid a couple thousand men. MIT's newspaper introduced one 1940 entrant as a New York "glamor girl," who was interested in pursuing cancer research, "naturally" loved parties, and won a hundred-dollar bet from fellow debutantes by getting accepted at MIT. Officially, coeds remained invisible. Writing in the Institute's 1941 *Handbook*, president Karl Compton told incoming students, "In choosing MIT, you've taken on a man-size job, and it will take man-size effort to get it done." The *Handbook* did mention the existence of the Association of Women Students (AWS), created two years earlier, but campus traditions embodied masculinity itself. As an official welcome, the institution held a "smoker" for freshmen and their fathers. MIT camp became the site for freshman initiation, featuring water fights against the sophomores, baseball games with faculty, and similar male-bonding rituals.[40]

## STEERED AWAY FROM ENGINEERING AND TOWARD TECHNICAL HOME ECONOMICS

Young women who were interested in engineering often had to pursue that interest over the questions or outright opposition from professors. When Peggy Shultz decided to enter mechanical engineering and pursue airplane design at Iowa State College in the early 1940s, family members supported her decision because they knew of her childhood interest in aviation. But college faculty members were "not so encouraging. Ever since last summer they have tried to persuade her to take something else."[41]

One possible factor in that faculty ambivalence may have been that Iowa State College had an alternative outlet for female students with technical interests, which channeled them away from engineering and toward the more conventional field of home economics. Iowa State had evolved around a land-grant assumption that, just as studies in engineering and agriculture could make young men better farmers and mechanics, so academic work could make women better homemakers. At Iowa State, that philosophy grew into a new form of technical studies that was devised for female students and was linked to but segregated from masculine engineering. In 1924, Eloise Davison, a graduate student in household administration, test-taught a one-quarter class that focused on making women into modern "household engineers" who would be skilled in evaluating, purchasing, using, and maintaining the new types of household equipment that were just appearing on the market. Reflecting a common sentiment of the 1920s, Davison observed that "The whole modern period in which we live is an age of machinery," which meant that the "average homemaker" must "overcome" her lack of experience with technology. Significantly, the new class was a cooperative venture of Iowa State's home economics division and its agricultural engineering department. By 1929, Iowa State had promoted equipment studies to the status of a department, and its initial roster included not only female home-economics faculty but also mechanical engineering professor Herbert Sayre, signifying the perceived connection between engineering and equipment studies. Elements of engineering authority and masculine representation lent legitimacy to the new discipline, yet from the outset it was clear that women would define the field.[42]

For dozens of women each year from the 1920s onward, Iowa State's equipment curriculum embedded lessons in engineering inside culturally acceptable boundaries of woman's knowledge. Majors took three physics courses and several classes in electrical equipment and circuits. Professors insisted that students learn scientific and technical principles to understand how and why ovens and other appliances worked (or did not work). Hands-on experience reinforced theory. In "electrical laboratory," staff members ordered women to overload a circuit and blow a fuse deliberately, replace the worn-out fuse, and read voltmeters and ammeters at each stage. In lab exercises, students dismantled, inspected, and reassembled gas and electric ranges, making notes on broiler design, heat distribution, and construction quality. Later generations of classes learned to apply the laws of thermodynamics to understand refrigerator mechanisms and to apply principles of heat transfer to plan technical specifications for a complete home heating and cooling system. Some graduates parlayed that knowledge into employment with appliance companies, utilities, and publishing companies.[43]

Faculty and graduate students conducted technical research, experimentally testing the efficiency of the latest ranges, mixers, and other kitchen tools. For her master's degree in 1923, Davison measured the comparative efficiency of coal, kerosene, and gas stoves, using a calorimeter that was designed to measure water evaporation rates. Faculty and students published results in experiment station newsletters and leading home economics journals. Adopting the conventions of engineering and science, these papers were filled with diagrams, data, and tables. Program members shared the engineering world's dedication to technical expertise, but home economics was defined by and for women, explicitly addressing females' presumed sphere of interest, domestic life. In that fashion, Iowa State's program created an alternate vision of gendered knowledge, asserting a link between technical mastery and femininity—at least in the kitchen. Other schools, including the University of Minnesota, Purdue, Ohio State, and Washington State University later created their own equipment courses. Home equipment classes undoubtedly thrived because women's appliance studies did not threaten men, but ultimately, such programs transferred a powerful alternate image of technical knowledge beyond engineering schools to the "women's sphere" of education.[44]

The household equipment department at Iowa State and similar programs at other schools encouraged women to assert their interest in applied technology. On other occasions, however, the existence of such home economics fields proved a gender-stereotyped trap, a place where women with mechanical interests could be pigeonholed and kept away from men's engineering domain. Growing up in a rural home, Lenore Sater had been interested in machinery since childhood, helping her father maintain family cars and handle farmstead repairs. When she arrived at college in the 1920s and expressed interest in civil engineering, Sater received a decidedly lukewarm welcome. Sater recalled:

I hoped to be a civil engineer, connected with some railroad or bridge construction company. My professors discouraged the idea as impractical, however, for I was not interested in the office work which they felt would be the only kind of position open to women. I wanted to supervise the actual construction. I did not like routine, inside office work. Out-of-door sports . . . appeal to me. . . . I like to do hand-work. Wood-carving, lathe work, and tooling have always been fascinating to me.[45]

Aware of the personal and professional challenges that a female engineer in the field would confront, administrators and faculty may have thought that it was logical and natural to steer Sater to the home economics program. They may have worried that a female engineering student might get in over her head and, with the best of intentions, intended to do Sater a favor in shielding her from hostility by indignant men who were unaccustomed to a female presence in engineering classes. Sater earned her master's degree in Iowa State's household equipment program, conducting thesis research on how the thickness of sheet aluminum affected utensils' thermal efficiency. Iowa State promoted her to join the equipment program's faculty. She later took a sabbatical to pursue her physics doctorate, then became head of the Housing and Household Equipment Division at the U.S. Department of Agriculture's Bureau of Nutrition and Home Economics.[46]

## CONCLUSION

It is impossible to produce any definitive numbers or even a reliable estimate of how many young women before World War II felt a pull of interest toward engineering, only to be halted by self-doubts or sidetracked into more traditionally feminine fields. Some professors and university administrators actively deterred some women from enrolling in engineering, and other women dropped out or transferred to more traditionally female subjects; again, it is hard to document exactly how many. Isolation sometimes frustrated the one or two at a time who defied those informal barriers. The women who persisted understood the simple reality of their situation: they needed to tolerate the inevitable skepticism of male classmates, professors, employers, and acquaintances and try to overcome doubts about women's ability, simply by working harder than men.

The fact that there were relatively so few female engineering students, ironically, meant that there was relatively little malicious criticism of them, at least publicly. As long as that small cohort attended institutions that were already coeducational and did not press too aggressively for acceptance in all-male honor societies, then their presence did not arouse much substantial outward opposition. There were plenty of jokes about "Slide Rule Sadies," and by her mere existence, a female engineer attracted inescapable curiosity. But behind the invasion rhetoric, observers assumed that engineering studies would remain firmly in masculine territory.

Female engineers themselves insisted that nothing in the inherent nature of the discipline disqualified women, and some optimistically envisioned that their path-breaking example might inspire more young girls at some point in an unspecified future. But there were no sustained discussions about whether, why, or how to shift the gender balance of engineering. Female engineers found that interacting with even a few others was reassuring. As Lillian Gilbreth's marriage quickly evolved into an informal technical apprenticeship, she took heart from seeing Kate Gleason climbing over construction equipment. Gilbreth later expressed her pleasure at meeting a group of engineering women in Great Britain. But female engineers in the United States were too outnumbered, too scattered, too overworked, and perhaps too intimidated to battle the gender environment of their discipline openly and to organize campaigns to encourage the entrance of more women. Throughout the first decades of the twentieth century, the rarity of women in engineering only seemed to confirm that "Man's sacred domains" (as the University of Minnesota commenter put it) ultimately remained secure from invasion.

The factor that revolutionized many assumptions and attitudes about women and engineering, at least temporarily, was World War II, part of that era's broader transformation in labor demographics. After the Japanese attacks on Pearl Harbor, Hawaii, as the international emergency deepened and the U.S. military absorbed available men, both government and business scrambled to engage female workers, even for heavy industrial jobs. Ambitious propaganda drives helped draw more than 6 million new women into paid wartime employment. The number of American wives who were hired outside the home doubled, encouraged by patriotic slogans such as "The more women at work, the sooner we win!" Although large numbers handled secretarial chores or filled similarly female-oriented slots, roughly two million women took on more traditionally male roles in many aspects of wartime scientific research, technical innovation, and weapons production. Some female mathematicians and human "computers" performed weapons calculations on top-secret U.S. Army and U.S. Navy projects. Others, including Grace Hopper, helped develop and run early electronic computers. The so-called "ENIAC girls" used their mathematics expertise to help program and operate the Electronic Numerical Integrator and Computer, a machine being developed at the University of Pennsylvania to calculate artillery trajectory tables. Female oceanographers and meteorologists analyzed combat conditions, female physicists contributed to the Manhattan Project and helped build the first atom bombs, and nutritionists recommended ways to get the most vitamin value out of wartime food resources.[1] Official government wartime public relations campaigns highlighted the women who made direct, hands-on contributions to defense. A Norman Rockwell *Saturday Evening Post* cover depicted the female workers who assembled ship panels in West Coast navy yards, while other women tested guns for the army and handled dangerous explosives in munitions factories. Hollywood films glamorized the women who manned lathes and drill presses, who comprised up to 50 percent of the workforce for companies such as Boeing.[2]

Rosie the Riveter has remained a well-known popular image of World War II and female empowerment, while scholars of women's history and labor history have extensively analyzed the broad patterns of wartime economic and social change. However, most historians have overlooked the fact that the wartime acceleration of production created manpower shortages not only on the assembly line but also in engineering departments. Employers for the first time turned over essential manufacturing positions to draft-exempt women, and they also raced to hire women for equally nontraditional engineering jobs. As detailed below, aviation companies such as Curtiss-Wright urgently needed female draftsmen and designers to keep up with military orders and help produce new airplanes. The War Department used women to draw army topographical charts, conduct water-seepage experiments to help design drainage wells, and collect river-navigation data for dredging operations. But employers seeking female engineers immediately encountered a complication: there were not enough women available who had already completed full formal college training in the field. To address this shortage, World War II triggered a unique outburst of educational innovation that focused on providing women with crash courses in technical areas and preparing them to serve as emergency engineering aides in industry and government.[3]

## MANPOWER CRISIS IN WARTIME ENGINEERING

By the late 1930s, spreading conflict in Europe and Asia raised tense questions about American military preparedness that brought issues of professional manpower to the forefront. Experts attached national weight to scientific and technical knowledge, as foundations for crucial advances in weaponry. By 1940, a tone of national emergency pervaded discussions of engineering. The U.S. Office of Education warned of a higher-education deficit, that colleges were graduating too few engineers to meet unprecedented defense needs. Penn State engineering dean H. P. Hammond commented that "Twice as many wouldn't be too many." John Studebaker, U.S. Commissioner of Education, summoned presidents and deans from leading engineering colleges to meet representatives from the army, navy, and Office of Education. Their conference concluded that the United States faced "a marked shortage in naval architects, ship draftsmen, marine engineers, engineers skilled in airplane structures, airplane power plants and airplane instruments." Although experts estimated that government and industry would need forty or fifty thousand new engineers in 1941, the graduating class of June 1940 contained only about twelve thousand engineering students, of whom four thousand were immediately absorbed by the military. The Society for the Promotion of

Engineering Education (SPEE) worried that shortages in technically trained manpower could compromise essential defense targets, fatally stalling the expansion plans of the army, navy, and industry.[4]

Technical knowledge represented a limited national resource. Lehigh president C. C. Williams wrote that "Colleges have no synthetic chemistry in sight that will make engineers out of air and coal." Because education leaders could not instantly expand engineering graduation rates, they began thinking about ways to supplement regular degree programs with specialized crash training, meant to upgrade existing workers quickly and train new technical staff to relieve critical shortfalls in vital industries. On Studebaker's recommendation, Congress appropriated $9 million in October 1940 to establish an Engineering Defense Training (EDT) program based in the U.S. Office of Education. Legislation defined EDT's mission as "providing short intensive courses on the engineering college level in fields essential to national defense . . . to give specific training for a particular field of war work."[5]

Deans from prominent engineering schools helped shape the new program and immediately created EDT evening courses at their home institutions. The subjects that were most in demand included production engineering and supervision, engineering drawing, materials inspection and testing, metallurgy, tool engineering, and machine design. Government paid all tuition and lab fees, making EDT courses entirely free to qualified students—all men and women who held at least a high school diploma and were able to handle college-level academic standards. The Office of Education pressured colleges to create classes rapidly to respond to local employers' needs, such as accelerated airplane and ship building in California and the Northeast. By June 1941, officials had approved proposals for 2,350 courses at 144 schools, intended to reach 136,618 students. Building on its existing extension education infrastructure, Penn State prepared courses in fifty cities to train over ten thousand draftsmen, junior engineers, and other technical workers.[6]

Educators soon realized that engineering training without science proved too narrow. Students of airplane design needed to understand physics and meteorology, and companies needed production supervisors who were familiar with business methods. Within ten months, Congress added $17 million to expand EDT to include science and management classes. In December 1941, Japan's attack on Pearl Harbor lent new urgency to training efforts. The Office of Education called for "universal war-mindedness" and renamed its program Engineering, Science, and Management War Training (ESMWT). The Office of Production Management told engineering educators, "We need everything you can give us, as of yesterday."[7]

In American colleges, the declaration of war almost instantly suspended any sem-blance of normal campus life. Pearl Harbor occurred just before Christmas vacation, and thousands of male students enlisted instead of returning to school. Over subsequent months, extended mobilization and the draft further reduced civilian men's presence on college campuses. Enrollment plunged by 20 percent on California campuses by February 1943 and by over 15 percent in Texas. Meanwhile, government sent thousands of men back to college in uniform to study engineering and other technical subjects under the Army Specialized Training Program and Navy V-12 Program. Many schools con-densed their calendars and reduced breaks, which allowed students to complete degrees and enter the workforce in less than three years.[8]

Even as patriotic messages prodded young men to join the ranks or take important defense jobs, wartime threw college women into an identity crisis. With industry beg-ging for female workers to assemble airplane turrets and ship hulls, full-time students contributed seemingly little to war efforts beyond knitting or buying defense stamps. In contrast to the iconic figure of Rosie the Riveter, college women seemed frivo-lous. Majors in art, music, and liberal arts appeared particularly irrelevant to immediate national priorities. A study by the American Council on Education claimed that of 100,000 female college graduates in 1942, fewer than 30 percent majored in subjects that were directly applicable to war needs. Many college men refused to take female classmates seriously. A 1943 poll of 150 Purdue males showed that 46 percent believed that women attended university only to find a husband, 34 percent of men said that fe-male students' main purpose was to enjoy the campus atmosphere, and only 10 percent thought that women actually meant to study. [9]

By late 1942, rumors spread across several campuses that colleges would soon turn over housing and classrooms to the military, dismissing all female students and order-ing them to register for war work. Denying such rumors, Iowa State College president Charles Friley issued a message to students, pointedly emphasizing that "the program for women . . . will continue as at present, with the addition of courses in mathematics and certain engineering fields in which women can do effective work."[10]

At Purdue, questions of "How long will women be allowed to stay?" accelerated after its University Committee on the Education of Women required all twelve hun-dred female students to take science aptitude tests. Rumor suggested that women who performed poorly would be forbidden from reregistering. Seeking to dispel such stories, administrators maintained that the special testing simply aimed to help government as-sess "the number of women who can be prepared to replace men, and . . . to scientifi-cally prove or disprove the belief that few women have aptitudes in physical science and mathematics." Purdue officials called all female students to a compulsory convocation, where they encouraged women who had scored well to register for extra courses in engineering, science, and math. Generations earlier, suffragists had asserted "that women were as good and as efficient as the opposite sex, and they could take the place of any man. . . . Now they have the chance to prove" it, one commenter pointedly noted: "Positions are opening in engineering, radio, chemistry, physics. . . . now women are the only ones who can fill these vacancies. It's a challenge to each Purdue coed and every University woman to learn an essential duty and do it well, immediately."[11]

Although wartime statements often encouraged women to consider studying engineering, realistic observers recognized that most did not wish to switch majors overnight. Of the 47,500 women graduating college in 1942, only 2,545 (5 percent) earned degrees in physical science, and just forty-three in engineering. The American Council on Education suggested that the "large numbers of women . . . continuing to major in the arts and . . . humanities" should supplement that regular coursework with short-term technical or science training: "As every able-bodied man is 'destined for the armed forces,' so every able-bodied woman should likewise sense the obligation to enter some form of war service."[12]

By 1942, ESMWT's program extended nationwide, and female participation rose as high as 17 percent of total enrollment. By ESMWT's end in summer 1945, the federal government had paid to train 1,795,716 Americans, including 282,235 women. Many women chose classes in science or management, but a number also entered engineering. At Penn State, one class retrained two female soda-fountain operators as oil industry core analysts, and a course on Petroleum Laboratory Techniques sent four unemployed women, two former secretaries, and one former salesclerk to Pennzoil as badly needed technicians. Penn State architectural engineering major Virginia Reilly took ESMWT courses in ship construction and subsequently became a junior naval architect at the Philadelphia Navy Yard, the first woman who was hired in its design section. Educators praised female students as both dedicated and surprisingly skilled at technical work. Texas instructor Leland Antes said that after teaching radio technology and electronics to women, "it occurs to me that, although in many cases they are slightly overwhelmed with their first view of the engineering field, they nevertheless feel that they are getting a much firmer grip upon this man's world into which they are being forced."[13]

Beyond trying to recruit more women to join engineering classes, ESMWT leaders created special training targeted at women. Most notably, in 1943, the U.S. Office of Education and the U.S. Civil Service developed a joint program for retraining female college graduates to qualify for junior engineer positions. Rutgers, Texas, and thirteen

other colleges offered tuition-free evening courses titled "Engineering Fundamentals for Women." The twenty-seven-week program introduced women to engineering math, drawing theory and practice, mechanics of materials, surveying, and shop processes. To recruit students, the University of Illinois dean of women sent a letter to all alumnae, stressing that even liberal arts majors could qualify for engineering training if they performed well on aptitude tests and personal interviews. Advertisements promised that women who completed the course could secure positions with the U.S. Navy, War Department, Maritime Commission, Bureau of Ordnance, U.S. Engineers Office, Geological Survey, National Advisory Committee for Aeronautics, Civil Aeronautics Administration, U.S. Coast and Geodetic Survey, and other federal agencies.[14]

With the Civil Service drive to train female junior engineers, even all-male Caltech agreed to set up a class, temporarily reversing its exclusion of female students. Thirty women suddenly appeared on campus, and male engineering students nicknamed them "Jennies," short for their future posts as "junior engineers." Caltech's civil engineering instructor reported that women displayed "excellent" interest and were "doing better work than we at the Institute anticipated." One class member who was already familiar with basic technical terminology through her late husband's engineering career said, "The course is hard and challenging but we enjoy it, especially the outdoor part when we go out to learn surveying." She added, "There is some doubt in most of our minds as to just where we are headed in this work." But for that widow, engineering represented a long-term career goal rather than a brief diversion or wartime necessity. "I'd like to be an engineer right out in the mountains," she declared.[15]

The Army Map Service, which was stretched thin in trying to generate more than 4 million maps monthly, persuaded civil engineering departments in at least fifteen colleges to offer female seniors free ESMWT courses in military map making. The army furnished content outlines, materials, and practice exercises. Civil Service authorities promised that graduates could qualify as engineering aides with annual salaries as high as $2,600, which was far higher than most secretarial pay. At Cornell, fifteen female seniors and nine local women signed up for the concentrated course in topographic and planimetric mapping, photo mapping, military grids, and aeronautical charts.[16]

◀ FIGURE 2.1

This brochure advertised one of the many classes that the University of California at Los Angeles offered during World War II under the auspices of the U.S. Office of Education's Engineering, Science, and Management War Training program. Given the shortage of available manpower, UCLA and other schools often sought to recruit women for this emergency technical training. UC System War Training Program, University Archives, Charles E. Young Research Library, UCLA.

Although program leaders admired the women's capacity to handle technical work, they bemoaned persistent difficulties in both recruiting and placing women. Texas civil engineering professor Phil Ferguson commented that assembling a "satisfactory" class of women demanded three to ten times more publicity and letter writing than recruiting male trainees, since "the work is new for most women and they are uncertain of their abilities and to some extent of their interest." Ferguson also complained that despite government's professed desire to hire female trainees, agencies' actual placement record had been "spotted." The Civil Service Commission had engaged the majority of his first class but more recently had been slow to certify female junior engineers and offered few openings unless women were willing to move to Washington, D.C. Ferguson added that Texas ESMWT classes had "very satisfactory" success placing female technical workers in the oil industry and that the state highway department wanted to hire women engineers but could not offer competitive pay.[17]

As defense industries scaled up production, the demand for female draftsmen seemed virtually unlimited. The University of Maryland's engineering dean reported in 1942 that "The aeronautical industry particularly has asked me to secure as many women enrollees as possible for their work in aeronautical drafting and aeronautical design." Walter Rolfe, chair of the University of Texas at Austin architecture department, admitted that "at first, placement was slow because of the existing prejudices against women. Now that many industries are going over to women programs, this prejudice is rapidly disappearing. Women tend to business better than men—many employers have reported."[18]

Advocates viewed feminine character as being well suited for drafting, citing stereotypical traits of patience, neatness, and precision. According to Rolfe, women excelled at "skills which require the coordination of the hands and the mind, which to them is not a new experience." Texas professor Worth Cottingham added that women "develop a good drawing and lettering technique but usually work slower than men. It has been my experience that women around the age of forty and over do not progress very satisfactorily with either the technique or theory of drawing." Educators who focused on gender differences in skill expected that women entering engineering drawing would also experience certain difficulties. The American Society of Mechanical Engineers noted that on aptitude tests, female students ranked significantly lower than men in physics knowledge and spatial awareness, although women outscored men in number sense, computing, estimating, and reasoning. Female students were "handicapped by lack of knowledge of shop practice and terminology, which men usually have," but experts believed that properly designed classes could remedy that defect.[19]

Numerous schools, including Rice, Ohio State, Johns Hopkins, Washington University in St. Louis, and University of Santa Clara, claimed success in attracting women to drafting classes. At the University of Florida, about forty women signed up for engineering drawing at the behest of the Tampa Shipbuilding Company. At the University of California, women comprised 116 out of 132 students who were enrolled in full-time aircraft and ship drafting courses in 1942. Of that total, 109 students found employment within four months, with over 70 percent hired by the Douglas, Ryan, Consolidated, and Vega airplane companies. Illinois Institute of Technology students organized a new club, the Women's Emergency Engineering Drafting Society (WEEDS), which joined the other feminine "alphabet armies," including the Women's Army Corps (WAC), the navy's Women Accepted for Voluntary Emergency Service (WAVES), and the Women Airforce Service Pilots (WASP).[20]

ESMWT executives encouraged colleges to work directly with business to fill specific personnel gaps. When the York Company reported problems keeping drafting projects on pace, Penn State recruited twenty-three women who were just finishing high school for a full-time, 100-hour engineering fundamentals course. Twenty subsequently assumed posts as junior draftsmen at York. Ultimately, Penn State took 1,945 young women and men through introductory engineering, and 1,164 reportedly found work immediately. Typically, "Foundations of Engineering" classes combined freshman-level math and physics, lessons in slide rules and measuring instruments, and basic engineering methods. The course description told prospective students, "Make no mistake. This course isn't a quick, easy way to become an expert engineer. It provides . . . powerful instruments of immediate practical utility to help you do a better job in industry. . . . What you build on it in the future is up to you."[21]

In a more extensive effort, Grumman Aircraft Engineering Corporation set up a long-term cooperative relationship with the U.S. Office of Education, arranging for Columbia University's ESMWT program to train female engineering aides exclusively for Grumman. Initially, Grumman sought to upgrade current male shop employees for advanced work by sending them to ESMWT courses in aircraft structural layout. But in 1942, as men with sufficient education grew scarce, Grumman's personnel department selected a new class of fifty-five women. All were college or professional-school graduates, and one quarter held master's degrees. Training director Wesley Hennessy explained that although Grumman preferred majors in math, architecture, chemistry, physics, or business administration, the company thought that it could effectively retrain almost any "well-adjusted, healthy and intelligent" woman. Hennessy judged personal characteristics to be more relevant than specific academic background, seeking women with "perseverance and a genuine sense of responsibility," plus "adaptability," since engineering aides must "be able to take and follow orders . . . get along well with other people."[22]

Grumman assumed that female candidates would arrive possessing less technical background than men, so classes started with the basics. Columbia drafting professor Frank Lee taught mechanical drawing, referring to Grumman standards for detail and assembly drawings. Other classes focused on descriptive geometry and its applications, applied mathematics, mechanics, and applied strength of materials. After two months of classroom sessions, the women moved to Long Island's Freeport State Aviation School for one month of hands-on lessons in riveting, sheet-metal fabrication, and use of drill presses, bandsaws, micrometers, and other tools. Grumman believed that this exposure to shop mechanics helped a trainee to "become acquainted with aircraft fabrication methods before meeting them in the plant. She learns to call parts by their right names and . . . [gets] accustomed to noise. . . . Foremen under whom she will work will have respect for her ability, and she will be given responsible and interesting jobs when she goes into the plant." After finally entering Grumman plants, women underwent three months' additional training, rotating through thirteen departments to acquire "actual working knowledge of the construction of an airplane." Each trainee chose one department for a month's immersion, "thereby gaining specific knowledge that will later allow her a degree of specialization when she is in the Engineering Department." Meanwhile, these new engineering aides continued taking ESMWT evening courses in aircraft structural layout to master advanced topics in engineering drafting and descriptive geometry.[23]

Grumman portrayed its program as a grand wartime experiment to place women in engineering. Its recruiting brochures spoke directly to female potential candidates:

You probably never thought of yourself in engineering work—that has long been considered a man's forte. Only a few years ago we too considered it so. But every day the improbable becomes possible, and we have discovered that a girl with certain qualifications and educational background can be trained in a relatively short time for a position of responsibility in engineering work. With a dearth of available graduate engineers facing us, we are looking for young women to assist the men we have; women of intelligence and ambition to whom such an opportunity is also a challenge.[24]

Grumman boasted that its program could convert even artists and social workers to engineering. It portrayed recruits as ordinary women, not as strange unfeminine geeks. The company offered as an example Beatrice Senne, a fashion designer who was eager to aid in defense after her husband enlisted: "Bea is glad to discover some similarity between interpreting blueprints and working with dress patterns. . . . Now she is a regular Grumman shop employee . . . in the job she wanted most, detail drafting in production engineering."[25]

Although wartime expansion had prompted Grumman to train engineering aides, the firm declared that it regarded talented women as more than just temporary substitutes and that those who invested time and effort in technical study would be rewarded with long-term advancement and job security. Its booklet, *A Career for You with Grumman Aircraft Engineering Corporation*, read, "For young women who are interested in a career continuing after the war, the aviation industry is decidedly promising." Predicting that the future would witness "unprecedented demand" for both commercial and civilian flight, the company anticipated a postwar boom "that augurs well for Grumman employees, both men and women, who have proved their worth."[26]

Grumman ran the Columbia ESMWT course five times through early 1944 and graduated 251 women. Entering airplane plants, those engineering aides worked on drafting, technical writing, statistical charting, and tabulating flight test data or as assistants in the stress analysis, aerodynamics, materials control, spare parts, and planning departments. Training director Hennessy told ESMWT officials that the company considered the program to be a great success in helping expand its engineering department, with "consequently a proportionate increase in the amount of work accomplished." He reported that female aides often proved "better suited for the types of jobs given them than would be men. Their satisfactory handling of these sub-professional assignments has left our graduate engineers free to concentrate on the more complex problems of aircraft design and production."[27]

## THE CURTISS-WRIGHT CADETTE PROGRAM

Even as Grumman recruited its first female engineering trainees, another corporation was cultivating even bolder plans in collaboration with the American Council of Education and the Society for the Promotion of Engineering Education. Curtiss-Wright Corporation, one of the country's largest airplane manufacturers, could not hire its normal quota of graduating male engineers and thus risked falling behind on its military orders. The company tried retraining shop men beyond draft age to handle drafting and design, but established male employees resented entry-level assignments. Furthermore, the field of aeronautics had changed so rapidly that older men "were not sufficiently flexible in their thinking to absorb present-day concepts and applications of engineering," explained C. Wilson Cole, Curtiss-Wright's supervisor of engineering personnel.

Curtiss-Wright therefore sought to stretch its existing engineering manpower by investing approximately $1 million to train about seven hundred female "engineering Cadettes" to take over relatively basic but essential and time-consuming tasks.[28]

Over one hundred schools expressed interest in teaching this new form of engineering for women, and in late 1942, Curtiss-Wright signed contracts to collaborate with seven—Purdue, Cornell, Penn State, Minnesota, Texas, Iowa State, and Rensselaer Polytechnic Institute. Curtiss-Wright believed that those schools could supply the appropriate faculty and laboratory facilities, although a formal aeronautics curriculum was still evolving. Purdue had created its aeronautical engineering degree only the previous year, when dean A. A. Potter had presciently argued that the school needed to anticipate forthcoming wartime expansion of airplane production. Penn State had just approved the creation of a two-year aeronautical option as part of mechanical engineering.[29]

Given how many other employers were already competing to hire the small pool of female graduates in engineering, math, and science, Curtiss-Wright reached further down to recruit women who still were finishing degrees. Advertisements in college and city newspapers (including the *New York Times* women's employment section) offered women over age eighteen "a unique opportunity to participate in the war effort . . . if you have a desire to study airplane design." Curtiss-Wright offered Cadettes a $10 weekly salary during their tuition-free training, promising that after entering company factories, engineering aides could earn about $140 a month. Brochures told recruits, "This is your war too, and here is a direct service that college women alone can render. You, as a Curtiss-Wright Cadette, will be a co-worker with the Soldier, the Sailor, and the Marine. . . . This is your opportunity to add your name to the roll of individual service, to give our valiant men the weapons to save America!" The brochure hinted that at least some Cadettes could extend Curtiss-Wright employment into peacetime, declaring that "aviation is facing a new postwar era of expansion and service. At that time individuals—women as well as men—will be evaluated upon personal capabilities and performance."[30]

Curtiss-Wright sought women with good academic records through at least college algebra but also emphasized that future Cadettes should already be somewhat familiar with engineering, if only through acquaintance with a working engineer. The American Council on Education sent recruiting letters to five hundred women's colleges and coeducational schools across the East and Midwest. Curtiss-Wright staff fanned out to evaluate almost four thousand sophomores, juniors, and seniors. Interviewing twenty candidates at the University of Texas, company representative Carol Keplinger found "girls who were very much interested in engineering. Some said they had planned to change their major to engineering . . . but [that] would take five years and they wouldn't be able to help in the war effort."[31]

The Curtiss-Wright program appealed to women for multiple reasons, both intellectual and personal. In a 1991 poll of former Cadettes at the University of Minnesota, nineteen recalled that one main reason they joined was the practical lure of free education, room, and board. Martha Eidson Kissling's father "strongly" encouraged her to apply, since with two other sisters entering college, the family had no money for her to pursue graduate training. Eighteen women recalled that the Cadette program "sounded like an adventure," a thrilling opportunity. At least nineteen women were motivated by desire to contribute to defense; Patricia Levin remembered feeling "gung-ho for patriotism and being able to help in the war effort."[32]

Before seeing Curtiss-Wright's recruiting, the vast majority of Cadettes had never planned to work in engineering; most were preparing for teaching or other traditionally female positions. Vera Dalton Eggers recalled that her mother wanted her to find secretarial employment, and "I knew I wanted something more than that." Before entering Curtiss-Wright, only about five out of forty-three Minnesota Cadettes had considered engineering to be a field that was open to women. Lois Stender Rolke called engineering "a forbidden mysterious world. My father forbade me to study chemistry or physics (in high school) as not suitable for girls." Similarly, Lucy Lodge majored in math education because "when I started college, I wanted to study engineering but my family insisted it was not a job for a girl." She pressured her parents to let her become a Cadette and entered Curtiss-Wright's program against their advice.[33]

Curtiss-Wright's program provided a viable avenue into engineering studies for women who previously had been discouraged from that route. A number of Cadettes said that their premed, English, or other majors represented a second choice because they had ruled out engineering as being purely for men. At least fourteen joined Curtiss-Wright because they "had always wanted to study engineering/technology." Some had become intrigued by engineering through fathers or fiancés in the field. Virginia Terlinden Gilbert recalled helping her brother with electrician's work, while other women read car manuals and *Popular Science*, borrowed fathers' tools, or enjoyed watching their fathers build radios and make repairs. Jean Ordung Bundy's father was a salesman who "had wanted to study engineering, so he was most interested, excited, and encouraging." Bundy liked "accomplishing an unfulfilled dream of my father, to be an engineer." Many Cadettes appreciated the program as an instant community with shared interests where women's talents in math and science won approbation rather than puzzlement or scorn. One said, "Before, most of my acquaintances gave me the impression that my desires in these fields were a little odd for a woman." Other Cadettes were attracted to the aviation angle. A few had built model airplanes, a hobby that was popular with both young women and young men in the 1930s, as historian Joseph Corn has noted.[34] Several had fathers or boyfriends who flew. Edna Kling said that her aviator sister's "love of flying 'rubbed off' on her family." Ruth Granstrom had her own pilot's license and previously taught aeronautics to high school boys.[35]

Curtiss-Wright chose 719 women from three hundred colleges and forty-four states; twenty-four signed up from Cornell alone. Several brought past experience with technical studies. Before becoming a Cadette, Barbara Wolf Bowers was a mechanical engineering sophomore at Ohio State, disappointed that school offered no aeronautical engineering option. Jean Kneeland learned mechanical drawing in high school and wanted to study engineering or architecture, but lack of money forced her to fall back on a business major that she found boring. A few other women had also taken engineering drawing, aerodynamics, aeronautics, surveying, or engineering physics. After graduating from Mount Holyoke, Margaret Douglas had spent over two years at MIT taking calculus, construction, drafting, structural mechanics, physics, and materials science. Josephine Garber had studied calculus and basic engineering at the Carnegie Institute of Technology and the Pittsburgh Institute of Aeronautics. Several Cadettes, including MIT architecture major Barbara How, had already been employed doing drafting and tracing.[36]

However, a majority of Cadettes started cold, with work experience as secretaries, home economics teachers, journalists, and even professional models. Judith Arledge recalled that to her, "engineering was mysterious and remote." Another wrote that technical work had been "as familiar to most of us as the other side of the moon. . . . Two years of psychology or an ability to quote the first eighteen lines of Chaucer don't come in very handy when you want to take a cube root or draw a quarter section of a gear blank. The closest we'd ever come to things mechanical was to stand by somewhat helplessly while some big handsome man changed the flat tire."[37]

The announcement of Curtiss-Wright's extensive, innovative program created a stir on each campus that was involved. Iowa State's newspaper carried a front-page, boldface headline: "Train Women Here: College Will Train Coeds for Work in Plane Engineering." Other than RPI, the schools involved with Cadette training had long been coeducational. But aside from the occasional appearance of one or two women studying engineering, most female undergraduates majored in home economics, teaching, liberal arts, or other traditionally gendered fields. Inevitably, it commanded attention when one hundred female engineering students appeared all at once in February 1943.

Cadettes' arrival at Iowa State instantly almost doubled total enrollment in the aeronautical engineering department.[38]

It was difficult to see how Cadettes would fit into established campus hierarchies. As short-term visitors who were pursuing engineering, they did not have much in common with other female students. Although wartime brought many army and navy men to campus for temporary classes, Cadettes, as females, represented something different. Purdue's newspaper sent a male reporter to inspect this strange species, commenting, "These fresh new faces . . . are looked over carefully by a campus . . . awaiting them for the past few weeks. Are they pretty or . . . goons? Will they offer the [regular] coeds competition? Are they mannish girls with hair clipped short? They are not." Cadettes fascinated male engineering students, who rushed to help newcomers become oriented and enthusiastically volunteered tutoring services. Under the predictable headline, "Cadettes Invade Campus," Purdue's journalists noted that the unusual female presence literally turned men's heads: "The advent of skirts, light footfalls, and the lilt of soprano voices into the heretofore masculine environment of . . . Aero Building and the Mechanical Lab caused many a head to rotate through the angle *theta* and many a neck to exceed all previously known elasticity constants."[39]

The Cadettes' arrival caused an even greater sensation at Rensselaer Polytechnic Institute, which had been all-male for generations since its founding in 1826. The sudden arrival of ninety-seven Cadettes signified "tradition now broken beyond repair." Above a photograph that showed rows of desks occupied by smiling women, the newspaper ran the headline, "Women Invade RPI Campus." A student humorist wrote:

Well . . . we were finally invaded—our forces were temporarily outmaneuvered and thrown into panic. . . . RPI *Plus* Women! At first, our "fightin' engineers" were wary over this feminine intrusion on the last of the male professions; they pictured themselves taking courses in home-economics. . . . But then came the fateful announcement from the front-office: "Curtiss-Wright Cadettes will be of great service to you—they are merely to be assistants to you male engineers, so as to free others for more important work!" And, thus . . . they were heartily greeted by our handsome, intelligent and upright RPI students. . . . Day to day, the gentlemen got a bit more accustomed to the sight of brunettes, red-headed, and, what's even worse, blond engineers. Day to day, the Cadettes got used to the sight of drooling RPI "men," and got a splendid education in cat-calls, whistles, and stares.[40]

War programs strained university facilities to their limit, and a midyear influx of female crowds caused chaos. Iowa State, Penn State, and Texas reshuffled men's housing, converting male dormitories to accommodate Cadettes. Most women found arrangements satisfactory, but those at Texas repeatedly complained about bedbugs and other

FIGURE 2.2
At Iowa State College and the other six schools that trained young women for the Curtiss-Wright
Corporation, Cadettes studied flight theory, aircraft design, structural analysis, and other topics
in aeronautical engineering. Iowa State University Library, Special Collections Department.

infestations, insubstantial meals, lack of privacy, overcrowding, and noisy study halls.
The resulting low morale impeded the women's academic performance, according to
the aeronautical engineering chair, who worried that Cadettes' grumbling might undermine his department's reputation.[41]

Cadettes underwent a ten-month immersion in aviation technology and science for
thirty hours each week. The first term covered engineering mathematics, job terminology and specifications, elements of aircraft drawing and standards, elementary engineering mechanics, and the properties and processing of aircraft materials. The second
term included more engineering mathematics plus theory of flight, aircraft drawing and
design, strength of materials, aircraft structural analysis, and aircraft materials and testing.
Curtiss-Wright instruction was not simplified for women; it involved genuine technical substance. Flight-theory instructors taught fundamental aerodynamics, with lessons
in lift and drag characteristics, airfoil and wing design, load factors, flight conditions,

FIGURE 2.3

During their ten months of accelerated classes, Curtiss-Wright Cadettes devoted substantial time to drafting and engineering design, practicing work with the company's actual airplane parts and blueprints. Iowa State University Library, Special Collections Department.

and airplane stability. Through wind-tunnel testing, Cadettes studied how different airplane models performed under varied conditions. Classes on materials testing and aircraft structural analysis investigated the physical properties of aluminum and magnesium alloys, iron, steel, wood, and plastics, plus the nature of stresses, buckling, and materials failure. Schools entrusted the Curtiss-Wright classes to some of their best engineering faculty, who often assigned the same textbooks that were given to degree-seeking undergraduates in normal peacetime classes. RPI's coordinator commented that for engineering mechanics, "the text selected is a good one, and by no means one of the easier books." Purdue men noticed that the women's class in applied mechanics was "almost identical" to theirs. Their "Aircraft Materials Testing is the Cadettes' equivalent of [our] 'busting lab' . . . so familiar to Purdue engineers."[42]

Because Cadettes trained for immediate industry needs, their concentrated curriculum remained less broad and more applied than the standard curriculum for a four-year

● **By making simple projects of dural, they become familiar with the working properties of this aircraft material. Left to right are Harriet Talmage, Lelaroy Williams, Lois Stender, Instructor Wayne Hay, Glenna Willingham, Jean Mandt, Barbara Sanford.**

FIGURE 2.4

Cadettes in training spent many hours in machine shops and workrooms to become familiar with essential tools and machining operations, production engineering conditions, and the physical properties of materials. These Cadettes at the University of Minnesota built projects using "Dural" (short for "Duralumin"), a high-strength aluminum alloy that was used in aircraft construction. *The Minnesota Technolog* (April 1943): 197. Image courtesy of the University of Minnesota archives, University of Minnesota–Twin Cities.

engineering degree. Over previous decades, engineering schools had teamed with business to offer cooperative education, but wartime experiments dramatically extended such collaboration. Engineering deans and professors planned Cadette lessons in close consultation with Curtiss-Wright engineering staff. Company representatives defined the precise functions that they wanted women to fill, while educators focused their classes accordingly. As Cadette programs ran, Curtiss-Wright frequently sent engineers and other staff to each school to deliver special lectures that were aimed at helping Cadettes connect abstract theory to practical plant operations.[43]

Because the firm needed many women to work in drafting and design, those classes were particularly tailored to company function. For their textbook, Cadettes used Curtiss-Wright's official engineering manual. The firm supplied typical blueprints so that Cadettes could practice detailing authentic airplane parts on standard Curtiss-Wright paper, following official company and navy lettering standards. Midway through training, the firm asked Cornell to give the women additional exercises in ink tracing because "we are faced with a real emergency in getting all of the C-46 drawings traced so . . . plants . . . can start production on this airplane as soon as possible."[44]

Although Curtiss-Wright desperately needed Cadettes to take over drafting work, officials worried that the brightest might become frustrated by monotonous assignments. "Previous efforts by industry . . . indicated that although women might be good draftsmen, they would seldom be proficient in design," Cole wrote, but Curtiss-Wright aimed to help women make greater progress. The firm hoped that many Cadettes could soon "develop into first class layout and . . . stress analysts" or even help with advanced research and testing in mechanical, structural, and aerodynamics labs. To give Cadettes "comprehensive" training for "advancement," Curtiss-Wright tried to "indoctrinate" the best with challenging ideas in engineering science. Cole told the *New York Times* that after starting on basic technical duties, "undoubtedly individual women will break through. . . . We may have a woman chief engineer before many years go by."[45]

Although some Cadettes had previous drafting experience, Curtiss-Wright understood that factory environments would seem alien to most. To familiarize women with industry procedures, Curtiss-Wright created a class in job terminology and production methods, emphasizing how its engineering department functioned. To prepare Cadettes to serve as liaisons between engineers and production lines, women entered school machine shops four hours per week, learning to weld, solder, and operate machine tools. Cadettes built model gliders and airplanes, carving wood to fit precisely calculated wing ratios and balancing tail weights with paper clips and hairpins. Iowa State Cadettes set up a small assembly line, making models of the same A-25 bombers that were manufactured at the Curtiss-Wright St. Louis plant where they were heading.[46]

Before the "educational experiment" began, faculty tried to foresee the challenges that they might encounter when facing students who were very different from the masculine engineering norm. They anticipated that women would be "significantly less informed on physics and mechanics" and "inferior" in spatial ability and mathematics: "this applies [even] to [female] math majors for they forget the placement of decimal points." Some professors worried that Cadettes would prove inept in machine shop and averse to running dirty equipment. Others contended that women were not inherently incompetent but merely novices who needed gradual introduction to tool use. Fred Ocvirk, head of Cornell's Curtiss-Wright program, wrote, "Cadettes . . . are handicapped to the extent that by tradition their experiences have been womanly, and little, if at all, technical. They have not had the advantage of playing with Erector Sets and tinkering with Model T's. They have the rather tough job of catching up on things mechanical . . . in a relatively brief space of time."[47]

Cadette training was innovative in abbreviating aeronautical coursework into a ten-month format, but the real uniqueness came in teaching seven hundred women at once. Engineering faculty felt the strain of adjusting. Minnesota Cadettes remembered a "reputedly tough professor who strode into his first class and suddenly burst into uncontrollable laughter, eventually recovering to admit that he had never before faced 25 females wielding slide rules." But Cadettes could claim to be doing their part for the war and, on those patriotic terms, were welcomed.[48]

As the first semester proceeded, many Curtiss-Wright and academic staff remarked, with degrees of surprise, on how much the Cadettes' performance delighted them. One Iowa State teacher declared, "In comparison to men engineers in the same length of time, the girls are stacking up pretty well." As observers had anticipated, Cadettes impressed professors as "superior" to first-year male engineers in drafting, especially in "neatness and accuracy." At Penn State, roughly one third of Cadettes received grades high enough to qualify for the dean's list. Assistant dean G. M. Gerhardt commented, "These girls could absorb and apply much more engineering training than anyone had anticipated." Those worried about women's performance in shop found that, while initially "clumsy" and "slower than male students," Cadettes "mastered" tools with "gusto" and "a great deal of proficiency."[49]

Cadettes noticed the attitude shift, as faculty regarded female students first "with reservations, then as a challenge, then as educable." Purdue electrical engineer George Cooper admitted that the Cadettes' accomplishments reversed his earlier opinion that women had no role in engineering. Some Cadettes actively sought extra intellectual stimulus and higher levels of technical challenge. Several Minnesota women joined a regular aerospace engineering class for extra credit and helped with laboratory research into simulations of high-altitude flight. Jean Kneeland recalled that "girls learned more and faster than anyone had anticipated and the professors, fascinated, just kept pouring it on."[50]

Faculty members modified their classroom approaches to serve female students, whom they assumed had less exposure to technical thinking than typical men did. Noticing that Cadettes often felt uncomfortable with problems involving projections and three dimensions, faculty incorporated extra practice visualizing the rotation of shapes in space. Mechanics teachers supplemented lectures with extra demonstrations to illustrate abstract physics concepts. In being forced to adapt teaching to women, professors found that rethinking old instruction habits led them to devise improvements that would benefit all students. One said, "I discovered that many things were not instinctively obvious which I had previously taken for granted. . . . Now I throw emphasis on really basic and difficult points. . . . My stock of practical examples . . . is appreciably increased."[51]

Several teachers observed with pleasure that Cadettes were "working longer hours in class than our regular engineering students" and with more positive attitudes. One Penn State professor commented, "They're a lot different from the boys. . . . They don't shuffle their feet ten minutes before the end of class period, and they'd stay for hours afterward, if I'd let them." However, authorities felt disconcerted by many women's habit of knitting during lectures; worried by the distraction, Curtiss-Wright asked teachers to kick knitters out of class.[52]

Cadettes at all seven schools stood out from men in the engineering buildings they shared, but the culture shock proved especially strong at formerly single-sex Rensselaer, where ninety-seven Cadettes joined three thousand men. Coming from all-female Vassar, Elaine Garrabrant found walking to class through "halls lined with boys" stressful. As Cadettes entered the cafeteria, "the boys cheered," and under their stares, Garrabrant tripped on the stairs. Decades later, she still called it a "harrowing experience." But soon "many of us had our date-books filled for weeks in advance," with "coke dates, movie dates, skating dates, dances or just walking home from school with some one carrying one's books." Garrabrant wrote about going "to the most divine place in Albany for dinner . . . oysters, steak and all the trimmings," after which her date "did my electricity homework for me." RPI Cadettes lived adjacent to one fraternity, and residents soon discovered that they could communicate by hanging over the roof. By the time that the Curtiss-Wright training ended, at least nine Cadettes had wed or gotten engaged to RPI men, including one metallurgy instructor.[53]

Cadettes soon established themselves in RPI life, making the cheerleading squad coed and filling the drama troupe, which previously had to import actresses from nearby schools. Despite heavy class loads and homework, Cadettes at all seven institutions became involved with normal student activities, including band, orchestra, and choir. Iowa State singers formed their own "Cadette Sextette." In addition to producing an interschool Cadette newsletter, women wrote for regular campus newspapers, magazines, and yearbooks. The University of Texas invited Cadettes to nominate two "bluebonnet belles" to join regular female undergraduates in the homecoming parade.[54]

Cadettes stood apart from other coeds, instantly conspicuous at school because they wore pants. Cadettes pointed out the stupidity of wearing nice clothes to dirty machine shops. One said, "We find it more practical to dress for work instead of style . . . and since we're here to do a job, we're not interested in looking our prettiest all the time." Jeans represented a distinctive badge of pride for some women, highlighting their patriotic purpose and symbolically linking them to men's uniformed military units on campus. Cadette newsletters featured a sketch of a woman wearing rolled-up trousers, protective gloves, and a welding mask, headed "Fashion Notes by Katy Kadette, the latest in summer attire for Texas." Cadettes made their own fashion rules. Male engineers were easily identified by slide rules hanging from their belts, but most Minnesota Cadettes found that "too masculine" and attached slide rules to their notebooks instead.[55]

Cadettes faced a backlash from female students, who generally wore skirts to class and found their odd, often grubby outfits disrespectful. One Purdue Cadette explained that it would be irrational to change from "greasy overalls" into clean garments just to walk from the machine shop and back to the dorm: "We're still . . . feminine, but a rather . . . unusual schedule precludes our showing it right off unless you choose to look beneath the grimy surface. . . . Maybe those dirty cords . . . *were* the shop clothes of the men. Could be, coeds, that our matching uncouthness has been for the same reasons." A few Minnesota Cadettes reportedly went to a more extreme length to fit in with male-dominated engineering culture. After noticing that nearly all undergraduate engineers smoked pipes, some women also adopted pipes, even in class—an in-joke but also a peculiarity further separating them from other women.[56]

At Iowa State, some men were the ones upset by the Cadettes' appearance. One ranted against pants-wearing women as "a national menace." Reminiscing about prewar "trim coeds in bright wool skirts . . . I reflected on the droves of Iowa State women in slacks rushing to milk cows or strip down engines," who "look like the devil. . . . It's a sacrifice of femininity, which any man in uniform resents."[57]

Accompanying media images of Rosie the Riveter striding into factories in overalls, Cadette publicity often highlighted pants-wearing young women working in machine shops, stepping outside conventional femininity in both fashion and function.

FIGURE 2.5

Photographs featuring Cadettes doing machine-shop work provided wonderful publicity for the war effort. However, the unusual sight of women wearing pants, jeans, overalls, and bandanas on campus created controversy among other female and male students, who considered it a rude violation of traditional feminine standards. Iowa State University Library, Special Collections Department.

Curtiss-Wright's educational program made irresistible public relations in a nation where government campaigns and media efforts were energetically promoting the war effort. Purdue, Cornell, Penn State, and other schools featured Cadettes in front-cover stories for alumni publications and engineering magazines. Pretty girls performing unusual activities in the name of patriotism proved temptingly photogenic. In May 1943, *Life* magazine published a special feature with illustrations of RPI women as they sketched an engine mount, climbed a weather tower to adjust the wind-direction indicator, studied model planes in wind-tunnel tests, and practiced welding. When war heroes and military celebrities visited the University of Texas, they ate lunch with Cadettes.

Lola Stranz
University of Texas

Lieutenant Commander Jack Dempsey, the famous boxer, reportedly said, "I think it is a great thing that these girls are taking up what they are." Ace pilot Lieutenant Colonel Hewitt Wheless added, "If they are as conscientious about their work as they are beautiful and charming, we need not worry about our engineering."[58]

Patriotic imagery surrounded the Curtiss-Wright initiative and provided justification for women taking a radical step into a man's field. One Cadette published an essay titled "With One Purpose—Victory," which another illustrated by drawing a young woman posed before an enormous American flag, with airplanes in the sky behind her, and engineering books, blueprints, and drafting equipment piled at her feet. A Purdue bookstore advertisement depicted a Cadette literally walking arm-in-arm with Uncle Sam. Conveying loyalty to the war effort, Cadettes wrote songs such as the following, to the tune of "Anchors Away" (the fight song of the U.S. Naval Academy):

We are the engineers. . . .
Women pioneers are we,
In fields not dared before:
Aircraft drawing and designing, mathematics, shop,
Gosh, what more!
If we can prove our worth,
For the job ahead,
Columbus [factory] bound we soon will be
To speed the allies on to Victory.

Cornell Cadettes adapted "The U.S. Field Artillery" (changing the original wording, which started out "Over hill, over dale, we will hit the dusty trail, and those caissons go rolling along"):

Curtiss-Wright, Curtiss-Wright,
We're the girls of Curtiss-Wright,
Making airplanes will be our delight.
Over lathe, drawing board,
Getting knowledge we will hoard

◀ FIGURE 2.6
Conscious of being groomed to take part in an essential military industry, Curtiss-Wright Cadettes described their pride in joining male counterparts in "the call to duty." University of Texas trainee Lola Stranz drew this cartoon showing a Cadette who was practically wrapping herself in the American flag with books and technical equipment nearby. Iowa State University Library, Special Collections Department.

Doing our part to help win this fight. . . .
In our fight for freedom and right
(We'll lick the axis).

And new lyrics to a Cole Porter tune at Iowa State read:

Night and Day, we are the ones,
We're the girls behind the men behind the guns.
From the mad assembly line
To the fundamental proposed design,
We'll slave for you,
Night and day.[59]

Cadettes occasionally teased each other about contrasts between engineering life and classical femininity. They kidded about making math mistakes, about building model airplanes lacquered with nail polish, and about asking professors, "What's that do-gig that hooks onto the whoos-i-maginty that turns the gadget on the other side?" Inside jokes bound Cadettes together and reinforced their sense of remaining female in a man's world. But self-deprecating put-downs faded into irrelevance as Cadettes gained confidence and pride. They looked forward to demonstrating slide-rule skills to older brothers. At Iowa State's spring festival, Cadettes displayed drafting samples and machine-shop work along with other engineering college exhibits. Women knew many skeptics had doubted them; RPI's 1944 yearbook contained a photo of a large Cadette group with a one-word question: "Engineers?" One Cadette wrote, "It isn't that a man's vanity is injured by the thought that a woman could do his work; he sincerely believes that the thing is impossible. Classic examples like Curie . . . are only scientific mistakes, like Siamese twins."[60]

Significantly, Cadettes referred to each other as "real aero-engineers." They nick-named themselves "Slide Rule Susies," "Lathe Dog Dianas," and "Slipstick Packin' Mamas." Cadette songs, to the tune of Georgia Tech's "Ramblin' Wreck," referenced their engineering identity:

◀ FIGURE 2.7
In World War II, Rosie the Riveter symbolized the idea that women could contribute to the war effort through activities that temporarily stretched beyond traditional boundaries of femininity. Such rhetoric lent a patriotic aura to the idea of women studying engineering, as seen in this advertisement from Purdue's bookstore that welcomed Cadettes to campus and offered them "best wishes." *Purdue Exponent*, February 12, 1943. Purdue University Libraries.

I'm a rambling Cadette Penn State just met
And a helluva engineer,
And if I had a daughter fair
I'd start her from babyhood to train for Curtiss-Wright.
But if I had a son, sir, I'll tell you what I'd do,
He'd fly the planes his mamma made,
And work for Curtiss too.

And with alternate wording:

I am, I am, I am, I am,
A Female Aero E.
A perfect isometric of
What I really ought to be.
It makes no potential difference
What advice you give to me
I'll work and study and scheme until
I get me an Aero E.[61]

Purdue and several other Curtiss-Wright schools officially classified Cadettes as special students in aeronautical engineering. Cadettes expressed genuine interest in the field, and several who had pilot's licenses proclaimed the joy of flying. Women visited local airfields to inspect planes, examine cockpits, and tour repair shops. They voluntarily wrote articles that explored technical aviation issues, such as a lengthy comparison of Curtiss and Hamilton propeller designs. One Cadette concluded, "We've developed 'drafting board droop' and 'slide-rule squint'. . . and strangest of all, we love it! We're becoming mechanical and air-minded. . . . We crane our necks to see the planes going over and mumble . . . 'cantilever sweepback.'"[62]

Intensifying the Cadettes' identification as real engineering students, the program broke barriers to female participation in campus engineering culture. Because Curtiss-Wright's training represented specially designed wartime education, Cadettes did not mix with male engineering students in classrooms. However, Cadettes were able to enter the male student professional network through engineering society meetings. Iowa State Cadettes were formally enrolled in aeronautical engineering and thus were eligible to join the local chapter of the Institute of Aeronautical Science. Men "seemed glad to welcome girls into their organization, perhaps for reasons other than merely insuring a big membership." Cadettes positioned themselves as serious "*fellow* members of the I.Ae.S." and voluntarily attended technical lectures. Ninety-four came to one I.Ae.S. talk on aircraft identification where men in the audience got "busy establishing themselves in the good graces of certain Cadettes" over cocoa and doughnuts. Ulterior motives aside, it marked a milestone as the first time that college engineering societies included sizable female components. At Penn State, Curtiss-Wright delegates gained official representation in engineering student government, the first time that women ever served there.[63]

More than that, Curtiss-Wright's program provided the first occasion that enough women were studying engineering to form their own sizable organizations. Trainees at each school created a Cadette Engineering Society whose meetings featured movies about aviation and discussions of topics such as high-altitude flying. Cultivating the women's identification with the technical profession, guest lecturer Lillian Gilbreth encouraged Cadettes to recognize past, present, and future challenges that female engineers faced. One personnel staffer at Penn State counseled Cadettes on strategies for success after they finished training and entered company plants. Hinting at stereotypes that denigrated women as being too emotional to survive in industry, he advised women to "remain impersonal" in business dealings and try to "live up to the man's idea of how a man's world should be run."[64]

The Cadette program had impressive completion rates: 94 out of 102 women who started at Minnesota finished, 80 out of 96 at Texas, 83 out of 100 at Purdue, and 96 out of 116 at Cornell. A few women left due to health or family difficulties, and others felt unsuited to the work. Curtiss-Wright chose not to dismiss women who had academic trouble, hoping that any who had disappointing grades in advanced mechanics might still make adequate drafting workers. After deliberation, several schools decided to offer Cadettes college credits. Cornell awarded them the equivalent of one and a half years' worth of regular engineering education. At informal ceremonies in December 1943, Cadettes received certificates to testify that they had "satisfactorily completed the ten-month course of study in aeronautical engineering." RPI's commencement speaker congratulated the eighty-four graduates (out of ninety-seven originally enrolled) for disproving assumptions that women could not handle "men's subjects on a he-man's campus." Meanwhile, Curtiss-Wright had already planned more training, arranging a six-month curriculum at Purdue that by March 1945 produced 175 more Cadettes.[65]

By January 1944, the first Cadettes arrived at Curtiss-Wright's airframe plants in Buffalo, Columbus, and St. Louis and at its New Jersey propeller division. Many worked on drafting and detailing for the latest Helldiver bombers and new experimental airplanes, checking wiring diagrams and correcting wing blueprints and ensuring that rivets were proper lengths and that fuselage drawings fit proper dimensions. Others worked in engineering liaison, writing orders to incorporate shop floor changes into designs and ensuring that subcontractors met Curtiss-Wright's technical standards. Cadettes in service engineering prepared field manuals for new airplane models and issued bulletins about technical updates. Cadettes in experimental design created equipment brackets, rudder trim tabs, and other parts for new planes. Mary Ann Casey Bates recalled that because she could sew, supervisors asked her to design a cockpit-equipment cover. Years later, she still described seeing her name on blueprints as "MOST EXCITING!!!"[66]

Overall, Cadettes felt well prepared for their work, and several soon became eager to advance beyond drafting to more challenging tasks. Engineering aides helped with computation, graphing, and experimental research in Curtiss-Wright laboratories, studying stress analysis, aerodynamics, vibration, flight tests, and materials testing. Jean Kneeland remembered that when she transferred to the structures lab, the department initially classified her as a secretary because it had no formal category for a female engineering aide. Kneeland recognized immediately that despite the short span of Cadette training, she had an intellectual advantage in the rapidly changing world of aeronautics. Minnesota professors had taught her the latest shear-flow techniques of stress analysis, which were still unfamiliar to most regular company engineers.[67]

Most Cadettes praised Curtiss-Wright's atmosphere as supportive and enjoyed rewarding teamwork with bosses and colleagues who seemed to regard the women as peers and respect their work quality. Others said that male coworkers treated them like daughters or sisters. A minority did sense discrimination and resentment. In the drafting department, Vera Dalton Eggers supervised men who were furious at having to take directions from a woman and went over her head when she corrected their drawings. Virginia Henry Bailey, who stayed at Curtiss-Wright until 1946, said one reason that she finally left was that "if any mistakes showed up . . . the fellows blamed the two girls in the department even though we had not seen the jobs." Several women recalled being teased. Barbara Wolf Bowers disliked having the chief engineer refer to her as "the beautiful one," since she "always felt this negated my mental capabilities." But only a few reported encountering what later decades considered sexual harassment. Jean Kneeland actually worried that she was "cramping [the men's] style and "wished that they would loosen up a bit." She was relieved when a foreman finally told her a dirty joke because she took it as a sign she had been "initiated."[68]

Most Cadettes accepted their Curtiss-Wright positions as temporary and realized that "when the men returned from war, the women would be expected to leave or to be *super* competent in order to compete." Martha Eidson Kissling left Curtiss-Wright in April 1945, feeling that she "did not want to keep a returning GI from his rightful job." Most Cadettes had no desire to stay after peacetime came and were eager to find different employment, return to college, or marry. But a few were bothered that Curtiss-Wright offered little prospect for promotion. Although the business praised the Cadettes' performance, the women sensed that engineering remained a man's field where they faced the additional disadvantage of lacking full engineering degrees.[69]

Wartime was chaotic, and many Cadettes transferred frequently between various Curtiss-Wright departments and plants. Out of about two hundred Cadettes who were trained at RPI and Iowa State, at least seventy-five still worked for Curtiss-Wright in late spring 1945. Lila Clarfield told friends, "After three changes I've finally come to roost in the Propeller Test Unit. The work is interesting and varied. I am the only girl in this group and wouldn't switch for the world." After the war, Curtiss-Wright donated its Buffalo plant to Cornell, and at least four former Cadettes continued working at the new Cornell Aeronautical Laboratory on secret projects, including guided-missile research and developing monitoring instruments for atom-bomb tests.[70]

For at least thirty women who left Curtiss-Wright right before or after the war ended, Cadette training served as a gateway to technical employment with other firms. These women appreciated the tangible rewards of performing typically male work. As one commented, "I found engineering offered women higher salaries than were paid in home economics." Staying in the airplane industry, a number of former Cadettes joined North American Aviation, Fairchild Engine and Airplane Corporation, McDonnell Aircraft Corporation, Bendix Corporation, Beechcraft, Boeing Airplane Company, and Pratt & Whitney Aircraft Company, working as drafting workers, designers, engineering aides, and researchers. Starting as a statistician at McDonnell, Mary Jane Blacet won promotion to project planning engineer in August 1945. Marjorie Allen married an Iowa State graduate in aeronautical engineering, and the couple joined the National Advisory Committee for Aeronautics (precursor of the National Aeronautics and Space Administration, NASA) to do design work, wind-tunnel studies, and classified research on experimental aircraft at California's Ames Aeronautical Laboratory. Still other former Cadettes entered technical work or drafting at Goodyear Aircraft Company, Bell Telephone Laboratories, General Electric, General Motors Company, and smaller firms, plus the army, state government, and various federal agencies. Several, like Elizabeth Heckman, continued working to put husbands through graduate school; she acquired further training in electronics and worked for AT&T writing engineering specifications. A few Cadettes returned to college to enroll in regular engineering classes.[71]

Over the long run, most Curtiss-Wright trainees found employment outside engineering or temporarily abandoned the workforce for motherhood. At least three married

Curtiss engineers, and one couple started their own company to design and build special industrial equipment. Many women resumed their original college majors, pursuing degrees in biology, chemistry, sociology, psychology, home economics, teaching, and architecture. Several had sons and daughters who joined the next generation of engineers.

But a significant share of former Cadettes parlayed Curtiss experience into long-term engineering employment with companies including Motorola and General Dynamics. Several pursued careers as computer analysts. Marie Stevens became the "first lady engineer" at the Florida Power and Light Company. June Padget had a thirty-three-year civil engineering career as Washington state's first female professional land surveyor. Mary Ann Trimble Lees designed locomotives and Amoco service stations and handled drafting for Carnegie Tech's cyclotron. Fern Lillian Henderson earned two patents as product design engineer with Cameron Iron Works. After divorcing and selling real estate, Norma Jean Dodge embarked on a twenty-three-year aeronautical engineering career. Her employer, a naval facility, gave her further training and promoted her to senior technician for all process shops reworking jet engines.[72]

Several credited the Cadette program with launching them on successful technical occupations. Theodora Wilkins described it as "a turning point" that gave her "confidence to go on in . . . an engineering career which was just the right direction for my life." Leaving Curtiss-Wright as the war ended, Wilkins worked on design and development of aircraft, gas turbines, and motors at General Electric, where she felt less welcome than she had been at Curtiss-Wright. After marrying a fellow General Electric engineer, Wilkins joined the Bureau of Reclamation and performed civil engineering work on western irrigation. For Patricia Levin, Cadette experience "made me think I could do things in the technological world I never dreamed of. It expanded my horizons." Levin ended up working at MIT, computing the results of wind-tunnel tests. Opal Bellamy Duthie became a Bechtel piping designer and then a nuclear-plant drafting supervisor, crediting Curtiss-Wright for showing her "technology and skills which have helped propel me quickly into lead positions."[73]

Some Cadettes who wished to extend their technical careers had difficulty locating postwar opportunities, especially those who never returned for full college degrees. Despite a Westinghouse policy to "hire no women," Ethelyne Hendrickson managed to become the only female in engineering there in 1947 after a sample of her work impressed the chief engineer. Others became frustrated by perceived discrimination and low pay. Mary Jane Blacet left several jobs after being denied promotion as "the only female in drafting." She later set up an independent business partnership that handled major drafting, including work for Pittsburgh's Three Rivers stadium.[74]

As Margaret Rossiter has noted, early twentieth-century female scientists who were frequently blocked from corporate research posts often had more luck being hired for gender-stereotyped jobs such as technical librarians.[75] Similarly, a number of former Cadettes became engineering secretaries and administrative assistants and technical writers, editors, and librarians at firms such as Honeywell, Martin Marietta Corporation, and North American Aviation. Such positions often proved more open to women than straight engineering but gave them a chance to take advantage of their technical know-how. Mary Blaschak worked as a librarian at Bendix research labs for over twenty years. Lois Broder acquired a doctorate in educational psychology and worked at the University of Wisconsin as a research associate, professor, and administrator. Broder strove to make a long-term difference in the field's gendered nature. During the 1980s and 1990s, she counseled engineering students, advised Madison's Society of Women Engineers chapter, and headed local efforts to attract more women to engineering.[76]

## OTHER WARTIME TRAINING FOR WOMEN

One factor that facilitated the acceptance of seven hundred female Curtiss-Wright Cadettes in engineering programs was the broader context of wartime upheaval in campus life. Other major corporations and the U.S. government soon followed Curtiss-Wright's lead in creating ambitious initiatives for training women in engineering and often ran their programs simultaneously at some of the same schools. In May 1943, just three months after one hundred Curtiss-Wright Cadettes began work at Purdue, they were joined by eighty-five "engineering Cadettes" who were undertaking forty-four weeks of training in radio engineering and electronics for the Radio Corporation of America. The two programs overlapped in more than name. Like Curtiss-Wright, RCA recruited college women from music, home economics, art, and other majors who had some experience and skill in mathematics. As Curtiss-Wright did, RCA offered female "employees in training" an all-expenses-paid course, plus an immediate allowance of $10 each week. The company planned to send Cadettes into six RCA plants to relieve overloaded male engineers of labor-intensive drafting work and assist with quality control, computations, testing, and research.[77]

RCA Cadettes followed a curriculum that was planned by company officials with assistance from Lillian Gilbreth to prepare them for work on the development, design, and production of radio equipment. Women studied engineering math, electrical circuit theory, electronics lab work, communications engineering, electrical measurements, properties of metal, and vacuum tubes. While Curtiss-Wright Cadettes practiced drafting airplane pieces, RCA Cadettes drew diagrams of resistors and condensers. Like Curtiss-Wright women, RCA trainees wore jeans, overalls, and bandanas during hours of machine shop sessions as they familiarized themselves with lathes, drill presses, and other production equipment.[78]

Forming their own community within a male-dominated profession, RCA women joined their Curtiss-Wright counterparts in Purdue's Cadette Engineering Society. At one gathering, engineering dean A. A. Potter lauded female ambition, saying, "This war will force women to be considered equal to men. . . . Married or not, she should have her own career. A woman can never tell when she'll need to have something to fall back on." Especially as the numbers of civilian male engineering students declined, RCA and Curtiss-Wright Cadettes together significantly tipped the gender balance of campus engineering activity. The audience at a May 1943 meeting of Purdue's American Institute of Electrical Engineers was half female, with both Curtiss-Wright and RCA Cadettes coming to hear a lecture on Alaskan airfield construction. Purdue's newspaper published the inevitable headline, "Girls Invade AIEE Meeting."[79]

RCA trainees joked about blowing fuses and crossing wires but ultimately projected themselves, as the Curtiss-Wright Cadettes did, as serious engineering students who were motivated by patriotic duty to venture into fields that were unusual for women. RCA executives presented Cadettes with army-navy "E" pins denoting "essential" war workers. The company's satisfaction with training results led RCA, like Curtiss-Wright, to run a second, truncated program at Purdue in 1944, ultimately involving 137 women.[80]

Meanwhile, in May 1943, Purdue's Curtiss-Wright and RCA Cadettes were joined by yet another group, forty-seven Aircraft Radio Cadettes who were training to become "under engineers" with the U.S. Signal Corps. These engineering aides were preparing to fill vacancies at Dayton, Ohio's Wright Field Laboratory, which conducted secret research and development for all army airborne radio equipment. The Signal Corps had earlier trained all-male classes for Wright Field work, but as available manpower declined, it reframed the effort as an "Adventure in Success: Aircraft Radio War Training Program for Women." Publicity photographs showed women adjusting sea rescue equipment and making preflight checks on a glider's radio. Like RCA and Curtiss-Wright women, Wright Field Cadettes studied engineering materials, math, and drafting but specialized in electronics and electrical transmission, alternating and direct current theory and machinery, radio principles, and antennas and wave propagation. Cadettes joked about accidentally shocking themselves with 150 volts but also spoke about the joy of mastering slide rules and building their own telephone lines. One commented, "At last, I am an engineer."[81]

Like Curtiss-Wright and RCA Cadettes, the Wright Field Cadettes helped create a new, mixed-gender culture of college wartime engineering. Purdue's 1943 yearbook editorialized:

Tradition, established over half a century ago here at Purdue as well as at other major engineering colleges, seems designed to vanish as the demand for manpower opens careers for women in such fields as electrical, mechanical or radio engineering, heretofore fields practically uninvaded by the fair sex. The theory that girls were not mechanically inclined is being pretty well blasted by results shown in the present war industries.[82]

Wright Field women picked up invitations to attend meetings of Purdue's Radio Club and the campus chapter of the American Society of Electrical Engineers.[83]

Impatient to produce as many female engineering assistants as possible, the Signal Corps spread training across several universities in 1943 and 1944, including Illinois, Minnesota, and Purdue. Minnesota also hosted Curtiss-Wright Cadettes during that period, while Illinois was one of eight institutions training female engineering aides for the airplane firm Pratt & Whitney. That program's approach varied from the Curtiss-Wright, RCA, and Wright Field model of recruiting women from different schools and sending them elsewhere for special training. Instead, Pratt & Whitney selected twenty female seniors (and recent graduates) who were already at Illinois but majoring outside science. Those Pratt & Whitney fellows stayed registered as regular students at Illinois, but instead of their intended curriculum, they substituted classes in engineering, physics, and chemistry. In exchange, the women received full scholarships that covered room and board, books, laundry, and training stipends. Following graduation, the company retained a one-year option on the women's time and paid them at least $140 per month. In 1943, the University of Illinois repeatedly contacted female students to pass along program announcements from Pratt & Whitney, Curtiss-Wright, RCA, and Wright Field. Other employers, including Goodyear and General Motors, also courted Illinois women. The dean of women combed files to pull out names of undergraduates who had completed enough math courses to qualify for the program. Listing all branches of engineering as genuine options for women, the University of Illinois War Committee wrote, "There is practically no technical field . . . not now open."[84]

Of all American schools, Penn State was among the most active in women's war-time engineering. In addition to welcoming 107 Curtiss-Wright Cadettes and running one of the nation's most extensive ESMWT programs, Penn State also trained ninety-one women in 1943 to work in drafting, design, testing, and engineering research for the Hamilton Standard propellers division of United Aircraft. Women followed either six-month or full-year curricula in engineering math and mechanics, engineering draw-ing, metallurgy, and aerodynamics. In their first semester, almost 30 percent earned places on the dean's list. Cadettes helped Hamilton trainees settle in and hosted parties for "sister aspiring engineers." Penn State's extension service also ran smaller, shorter classes across the state on behalf of other airline manufacturers. In 1944, the Glenn L. Martin Company recruited eighty-seven women for a three-month training program in Pittsburgh, while Allentown ESMWT prepared seventy-five female draftsmen for Con-solidated Vultee Aircraft Corporation. The *Pittsburgh Press* editorialized that "As grim and determined as any soldier, sailor or marine on a fighting front, women students are pitching in, getting . . . their hands dirty, learning technical jobs to further America's war effort. . . . Man's domain in the machine shop has been invaded."[85]

Relaxation of single-sex rules at RPI opened doors to women other than Curtiss-Wright Cadettes. In fall 1942, Rensselaer taught a class in electrical circuits and electro-mechanical techniques to about seventy-one local women at the suggestion of General Electric, preparing them as assistants in General Electric's engineering laboratory. RPI accepted thirteen other women and eleven men for training as junior procurement inspectors of aircraft materiel for the U.S. Army Air Forces. Underlining how drasti-cally wartime had changed the formerly all-male school, Livingston Houston, RPI's secretary-treasurer, commented, "We believe that in the coming year the training of women, probably in short courses of a specialized nature, will become one of our more important functions."[86]

Different programs gave women a foothold in other engineering institutions for-merly off-limits. Although New York University's Washington Square College had been coeducational since it was established in 1914, NYU's pre–World War II School of Engineering had been all-male. But in 1943, the Chance Vought subsidiary of United Aircraft created scholarships to give women intensive eight months of training at the Daniel Guggenheim School of Aeronautical Engineering. Even before the forty Chance Vought scholars completed their program, NYU began training eighty more women as junior engineers for the Vought-Sikorsky Aircraft Company. Female students gained access to the Guggenheim School's impressive technical resources, including the ad-vanced wind tunnel that simulated speeds up to 130 miles per hour.[87]

A wide range of corporations vied to attract promising women. The Vega Aircraft Corporation set up full-time programs to train female draftsmen at the University of California at Los Angeles and 130 female engineering aides at the California Institute of Technology. Boeing's Seattle plant sent at least four women to take ESMWT en-gineering courses arranged at the University of Washington for the company. Eastern

Aircraft sent at least seventy-five women through ESMWT training at Rutgers, the State University of New Jersey. Such initiatives attracted nationwide publicity. The *New York Times* photographed rows of women as they bent over drafting boards that were temporarily set up on Rutgers' basketball court, and *Time* magazine showed Stanford women watching faculty demonstrate slide rules.[88]

Airplane makers were especially active in courting female trainees, and other defense businesses copied the trend. In a 1942 advertisement headlined "Girls, Girls, Girls," General Electric announced that it was "hiring young college women to do work formerly done by male engineers . . . [to] make computations, chart graphs, and calibrate fine instruments for use in the machine-tool industry." General Electric claimed that it already employed forty-four "test women," including the University of Michigan's Virginia Frey, one of twelve women nationwide who earned engineering degrees that year. Other hires were all physics or math majors, and General Electric hoped to find a total of 150 women: "Although no one expects these girls to become full-fledged engineers, most of them will be given the Company's famous 'test' course."[89]

Advocates believed that virtually any intelligent woman with some background in college math or science could acquire valuable engineering knowledge relatively quickly. Putting that assumption into practice, Iowa State College undertook a special initiative to convert home economics majors into wartime engineering aides by drawing on their preexisting technical skills. In 1929, Iowa State created the nation's first degree program that specialized in studies of household equipment, which aimed at giving women science-based expertise in domestic technologies. Majors took three physics courses, three classes in Household Electrical Equipment, and a course in electric circuits and "electrical laboratory." Students learned to understand home wiring diagrams and technical specifications of heating systems, to know how Ohm's law applied to construction of electric ranges. Professors authored and used textbooks that embedded lessons in physics and engineering inside culturally acceptable boundaries of woman's sphere. Iowa State established household-equipment studies as a female technical space, parallel to men's prized concentration in agricultural engineering. In fact, male engineering faculty joined female home economics professors to develop collaborative classes and research. This domestic science discipline did not challange engineering authority, but from the beginning, the field was defined by and for women.

Prewar household-equipment graduates found jobs with utility companies and appliance manufacturers, but World War II made their technical talents more widely valuable. Following the recommendations of recruiters for the Naval Research Laboratory, Iowa State added special electrical engineering classes plus extra algebra, trigonometry,

and calculus classes to accelerate the preparation of home economics majors for emergency employment. Male engineering undergraduates nicknamed the home economics students "WIRES" for "Women Interested in Real Electrical Subjects." Electrical engineering professor Ben Willis originally planned to give "these girls . . . elementary background [as] a gentle transition from biscuit baking." But "the curious and skeptical . . . who expect[ed] to see the girls changing a fuse or repairing a toaster cord [ended up] sadly disappointed. Baby stuff! They learned those things in their own equipment lab when they were freshmen," Willis wrote. "The familiar way in which they spoke of kVA, power factor and leakage reactance silenced the scoffers." WIRES were ready for "more rugged topics," such as magnetic circuits, vector diagrams, transformers, and synchronous motors. Willis said he developed "wholesome respect for the scientific training gained by home economics students. More and more . . . women are proving their ability to fill positions in electrical testing, computing, design and other fields left vacant by engineers who have gone into military service." Nine women finished the initial WIRES training, and eight immediately began technical work for Western Electric, General Motors, and General Electric. A dozen more female home economics and science seniors enrolled in the second round of special industrial electricity courses, and within months, nine out of the forty-two women who were testing radio transmitters, amplifiers, small motors, and circuit breakers in General Electric's aeronautics and marine laboratories were Iowa State graduates.[90]

## CONCLUSION

Although energetic and expanding in scale, emergency crash courses did not and could not singlehandedly revolutionize the position of women in engineering. With a few exceptions, female engineering aides and draftsmen were presented primarily as substitutes for male workers or assistants to full professionals. As temporary employees without full degrees, they were not perceived as a permanent threat to men's control of the profession.

All told, World War II courses for Curtiss-Wright and other aircraft companies trained about 1,670 women as engineering aides. Hundreds more participated in the RCA, General Electric, and Engineering, Science, and Management War Training programs. Most did not make lifetime careers out of engineering, but a significant minority did, and their collective experience left its mark on American education. A Penn State professor later remembered, "We had [two or three] girls in electrical engineering from the time I got here [in 1931] and I guess they had them before. . . . But to have groups

of them like that!"[91] That was the key difference. Before World War II, the one or two female students who occasionally chose to pursue engineering at schools like Penn State were an anomaly and a curiosity. Wartime programs sponsored by government and private companies suddenly brought their ranks up to critical mass. With several dozen or even a hundred at a time studying technical subjects, the women could provide each other with crucial intellectual and psychological support. Cadettes, ESMWT women, and other female trainees proved to often-skeptical educators and professionals that women could handle difficult subjects and that a feminine presence did not automatically destroy engineering. Their successful "invasion" induced government agencies and military programs, plus at least some American companies, engineering faculty and administrators, and male engineering students and professionals to take a second look at the possibility that women could enter engineering in significant numbers. It served as a crucial precedent for multiple other World War II programs that aimed to accomplish exactly that—to inspire more female students to major in engineering.

Although Rosie the Riveter and her engineering counterpart, the Curtiss-Wright Cadette, represented crucial examples of how war at least temporarily and partially transformed perceptions of gender roles, they had no previous functional equivalent during
peacetime. Before World War II, manufacturers had not normally hired vast numbers of
women for heavy and dangerous work and did not aggressively recruit women to play
a vital role in engineering operations. But as the wartime manpower shortage led to innovative programs that gave women short-term technical training, it also led educators,
policy makers, and young women to rethink the gendered assumptions behind regular
undergraduate education. Although no one expected crowds of female students to rush
into engineering, wartime made it seem both possible and desirable (at least temporarily)
that somewhat larger numbers might begin to pursue and earn full engineering degrees.

As military need underlined the importance of engineering in both government
agencies and defense industries, it lent new significance to the relatively few women
who had earned engineering degrees over previous years and were holding important
wartime assignments. During World War II, Alice Goff, a civil engineering graduate
of the University of Michigan, worked for a steel company and handled the design of
reinforced-concrete structures for a bomber plant under construction. At Lockheed,
Mabel Macferren Rockwell supervised a staff of twenty male engineers in researching
innovations in airplane shop-floor production processes. General Electric ran a wartime
advertisement boasting that the company "employs about a dozen women engineers
and some 200 women in sub-professional jobs."[1]

## FEMALE ENGINEERING MAJORS DURING WARTIME

Not coincidentally, even as Engineering, Science, and Management War Training
(ESMWT) leaders and companies such as General Electric and Curtiss-Wright escalated

special engineering training for women, wartime threw a new spotlight on the women who were pursuing full degrees in engineering. In 1942, the University of Minnesota's engineering publication highlighted its five female students, literally pulling them out from the mass of male classmates. A photograph showing incoming engineering students lined up on campus steps contained a prominently placed inset that showed a close-up of just the women.[2]

Government and industry's urgent need for personnel offered women a patriotic excuse for entering a nontraditional field. Their choice remained unusual but suddenly acquired a new veneer of acceptability in many eyes. One month after Pearl Harbor, the University of Texas newspaper ran a photograph showing five women using slide rules under an instructor's watchful eye. Under the headline "They'll Help Play 'Taps' for Japs," the caption explained, "No knitting or other sissy stuff for these five girls—they're doing their bit for national defense in a manly way. The only female students in the College of Engineering, they are pictured being schooled by Dean W. R. Woolrich." Texas, like other schools around the country, publically emphasized its eagerness to support the war effort. Other front-page stories that day described new physical training planned for male students, the university's research on synthetic rubber, and a national poll showing that 92 percent of college students supported the idea of mobilizing the U.S. Air Force to bomb Japanese cities. Women working slide rules symbolized this new readiness. A Penn State magazine editorialized:

Here's to the Victory girl. . . . The girl who wields a slide rule as deftly as a Lord and Taylor creation. . . . She jumbles up all the old theories about this being a man's world. She can tell an engineer . . . to go to hell. She can even talk to . . . them about dynamos and two-way sockets without feeling like a damn fool. . . . Neat.[3]

Schools were besieged by wartime employers who were searching for women who held engineering degrees. The University of Illinois told companies not to bother coming to interview the lone eligible female graduating in February 1943 because she already "had countless offers for positions" but added that several other promising women would be available in June. New requests arrived weekly at MIT, where the placement bureau wearily responded to General Electric's demand for female electrical engineers by noting, "The Dept. of Electrical Engineering says they have not had a woman graduate for a good many years."[4]

Although female engineering majors remained far fewer than men, wartime support for women studying engineering brought extraordinary jumps in their numbers at many schools. By fall 1945, Cornell had thirty-seven women enrolled in engineering, whereas

## They'll Help Play 'Taps' for Japs

No knitting or other sissy stuff for these five girls—they're doing their bit for national defense in a manly way. The only female students in the College of Engineering, they are pictured being schooled by Dean W. R. Woolrich.

From left to rich, the girls are Orissa Stevenson of Houston; Anne Tally, Ketens; Mickie Jo Carleton, Austin; Anna Perry Wood, Austin; and Margaret Ann Magee, Waco.

FIGURE 3.1

Hundreds of young women signed up for Engineering, Science, and Management War Training classes and other accelerated training programs during World War II, and the war also drew attention to female students who were pursuing regular engineering degrees. This photograph ran at the top of the front page of the University of Texas at Austin campus newspaper on January 6, 1942, showing how these young women's nontraditional choice of studies acquired new patriotic value in the weeks following Pearl Harbor. *Daily Texan*, di_08929, the Dolph Briscoe Center for American History, University of Texas at Austin archives.

in previous decades there had generally been no more than about four and in many years none. Overall, in November 1945, colleges and universities reported a total of 48,977 men and 1,801 women enrolled in engineering courses. A number of engineering schools (including Caltech, the Citadel, Clemson, Colorado School of Mines, Georgia Tech, and Stevens Institute of Technology) still retained a formal male–only admissions policy or had no women registered in undergraduate degree curricula (although several included women in ESMWT courses or other special war training). But of institutions that welcomed female engineering students, a number collected double-digit clusters, with eighty-eight at Purdue, fifty at Ohio State, forty-eight at the University of Minnesota, forty-six at Cincinnati, forty-four at Michigan, forty-one at University of Texas, thirty-three at Colorado, thirty-two at Illinois, twenty-eight at Texas Tech, twenty-seven at Wisconsin, and twenty-six at Iowa State. Of the females who had already selected specialties, the largest fraction chose chemical engineering (288 women out of 1,801 total), followed by mechanical engineering with 204 women, electrical (185 women), architectural (156), aeronautical (116), and civil (113). That preference in majors varied slightly from men's, who registered for mechanical and electrical engineering ahead of chemical. Women were represented in all specialties, although only one female student nationwide majored in naval architecture and marine engineering, just three in mining engineering, and five in petroleum engineering. In the 1944–45 academic year, U.S. and Canadian colleges awarded 4,537 engineering undergraduate degrees to men and fifty-six to women. Of those fifty-six female graduates, five came from MIT, four each from Ohio State and Penn State, and three each from Purdue, RPI, and the University of Texas. Fifteen completed curricula in chemical engineering, eight in electrical, seven in aeronautical, six each in architectural and general engineering, five in mechanical, three each in industrial and in unclassified engineering, two in civil, and one in metallurgical.[5]

World War II brought a number of firsts for female students in engineering as their enrollment climbed and academic achievements mounted. In 1944, Iowa State civil engineering major Ruth Best joined thirty-three men who were selected for the Guard of Saint Patrick, one of the main engineering honor societies. The school's engineering magazine reported, "A woman invaded the Guard of Saint Patrick for the first time in the history of Iowa State College," an initiation traditionally celebrated with the Guard's "informal smoker." That step toward integration opened the gate. In the next year, the Guard welcomed two more women, including Maxine Goodson, Iowa State's first woman graduate in chemical engineering. Eloise Heckert became the first female member of Iowa State's chapter of Pi Tau Sigma, the honorary mechanical engineering fraternity. In her freshman year, Heckert claimed the highest grade-point average in the entire engineering division.[6]

Although such significant tributes highlighted the scholarly success of female engineering majors, one prestigious award remained beyond their grasp. The national honorary engineering fraternity Tau Beta Pi barred even the most talented women. In 1923, the society had issued its first women's badge of merit, given to female engineering majors who satisfied scholastic requirements that were identical to those for male initiates. By 1940, Tau Beta Pi had issued only four such women's badges, but the influx of female engineering majors during wartime brought a surge. At Iowa State in 1945, architectural engineering major Eloise James won the twenty-fourth women's badge issued by Tau Beta Pi. The women's badge remained a consolation prize. At the society's 1938 convention, a suggestion about amending the constitution to accept female members sparked heated argument, then received just a single affirmative vote.[7]

Although some engineering honor societies remained closed to women, other campus engineering groups were glad to accept female members, especially after war drew away male undergraduates. At Penn State and other schools, engineering student publications suddenly relied on a largely female staff. In 1945, architectural engineering junior Mary Krumboltz became the first woman who was elected to edit Iowa State's engineering magazine, giving the new female engineering population a public voice. She immediately wrote an editorial declaring, "Slide-rule-pushing girls are no longer a rarity. . . . We see them on our own campus and they are not the problem they were once expected to be. In fact, they are a problem only inasmuch as their fellow students and instructors choose to make them one." Although Krumboltz admitted that it would be inappropriate and probably impossible for female engineers to supervise heavy construction sites, she maintained that for desk jobs, employers found that women were "naturally more accurate than the average male engineer and far outshine him in patience and perseverance." She warned men who feared that women would steal their jobs to be ready to compete based on competence because "no one is entitled to a specially prepared job awaiting the time he is ready to take it." Besides, she added, realistically, persistent bias among engineering employers meant that male job seekers still enjoyed an enormous advantage:

Obviously there is a long struggle ahead for any woman who presumes to enter a "man's field." Men cannot be expected to share the profession voluntarily, and in the dim, distant future when the break does come—and women are accepted rather than tolerated—the concession will be made only as a matter of necessity. Meanwhile we shall continue with our present compromise.[8]

In addition to joining those honor societies that were willing to acknowledge them and the campus chapters of professional groups such as the American Society of Mechanical Engineers, female engineering majors (like their Curtiss-Wright Cadette sisters) began displaying the critical mass and confidence to form their own organizations. Even before the war, female engineering students at several schools, including the University of Michigan and the University of Texas at Austin, created clubs. Such early small efforts often lost momentum, but wartime brought a resurgence of interest at schools such as the University of Oklahoma. In spring 1945, female students at the University of Illinois established the Association of Women Student Architects and Engineers, which was open to all female students in those majors and offered associate membership to women in math, physics, and chemistry. The official intent was "to promote friendship and understanding among women engineering students, the faculty, and our profession." From the outset, however, officers expressed broader ambitions to advance their prospects in the field by seeking advice from established female engineers. The group also sought to multiply women's future representation in the discipline by distributing information that encouraged high school women to explore options for studying engineering and architecture.[9]

Even as wartime education propaganda tried to persuade female students to consider engineering, individual women still encountered resistance. University of Minnesota 1946 graduate Margery Brimi recalled that when she transferred into engineering as a sophomore, "everyone had a nervous breakdown at the University. I took an enormous battery of weird little tests then," which were not obligatory for men, since "they just thought a woman interested in [technology] was neurotic. They told me I would be unhappy." Brimi remembered faculty dropping insults about women during lecture "just because they didn't know how to cope with having a woman in class."[10]

Wartime brought impressive growth in numbers of female engineering majors, but the starting point was so low that absolute totals on any one campus remained small. Even with twenty, thirty, or more women in engineering, their choice remained a curiosity. Media reports and male classmates alike often assessed their beauty ahead of their brains, rating their potential as girlfriends rather than as future coworkers. In 1944, the University of Michigan's engineering magazine framed photographs of four female engineering majors to resemble queens in a card deck, showing them dramatically lit and with flowing hair. Under the headline "There's Glam in Engineering!," the article noted the women's dating preferences and feelings about Michigan men. It quoted sophomore aeronautical engineer Pat Lyons as saying she "wouldn't switch to any other school in the University for anything" and then added, "Although she is seen at all the dances and functions, she swears that she is 'strictly a student.'"[11]

Everyday campus experiences continually reinforced women's sense of historically being outsiders. The frustration of trying to find ladies' rooms reminded them that engineering schools were literally built for the opposite sex. One anonymous author commented:

All young women, who have come to Penn State
Listen to me, and let me relate
The story of one who has learned the hard way
That technical schools are no place to stay.
When you've stood all morning and you've "got to go,"
First, you'll suspect—and then you'll know
That the [engineering] buildings were made for men at Penn State;
I assure you, my dears, they're not for his mate. . . .
So take my advice, and switch to Home Ec
If you don't want to become a physical wreck.[12]

Wartime emergency needs could not entirely shatter the glass ceiling that blocked women's advancement in engineering. After doing aircraft engine work in France, Lidia Manson escaped the Nazi advance and emigrated to the United States in 1941, becoming a research assistant at Penn State's diesel laboratory. In 1943, Manson earned her master's degree in mechanical engineering there but left when denied permission to extend her studies. The institution maintained that wartime curriculum limitations made it impossible for her to seek a doctorate, but Manson rejected that "pretext" as pure "discrimination."[13]

Idealists suggested that women's accomplishments in wartime engineering had proved their permanent right to share the field. One Purdue woman wrote, "Modern industry has opened a new world for [female students]. There is no longer a limit to the fields in which women may choose a career. . . . Women are actually encouraged to seek jobs once held by men." More wary observers worried that peacetime would bring a resurgence of discrimination. Cynics remembered that after World War I, women had been forced to abandon their slots as streetcar conductors and factory laborers despite the protests of many that they had performed well and needed good wages to help support families. In the early 1940s, even as government, industry, and educators pleaded with women to enter industry or professions that formerly were reserved for men, ominous signs anticipated a postwar backlash against female workers. Common public sentiment suggested that after the veterans returned, women should retreat to homemaking or at least the lower-paid, stereotypically feminized jobs that men typically did not want.[14]

As female engineering enrollments grew during World War II, some observers felt disappointed and concerned that totals were not rising more sharply. But making the decision to pursue a four-year engineering degree during wartime was difficult, especially after corporations such as Curtiss-Wright, General Electric, and others began aggressively courting women for free short-term programs that paid an allowance during training and brought women into salaried, patriotic defense employment relatively fast. In addition, Pearl Bernstein, administrator with New York's Board of Higher Education, declared that women were fully aware of past discrimination in engineering. That history explained why women's representation in the field had been low over previous decades, she said; many interested in engineering ultimately opted for alternate careers where employers treated them fairly. If the country now wanted to stimulate more female representation, Bernstein indicated, business leaders should offer "a definitive statement . . . that after the war is over women will not lose their jobs as engineers . . . merely because they are women . . . [but] will be considered for important industrial positions . . . on the basis of their ability."[15]

The danger that postwar conditions would bring a reassertion of antifemale sentiment in engineering concerned Margaret Barnard Pickel, adviser to female graduate students at Columbia University. Skeptical that wartime efforts to draw more women into technical studies would permanently transform the profession, Pickel was appalled that many women were jumping to conclusions that engineering promised them a long-term welcome: "Illusionary and rosy views of the prospects for women are common just now. . . . We know of one woman engineer, . . . [and] our hearts leap up." Young women had become "bedazzled," so wiser mentors must temper that overconfidence, Pickel wrote in a 1944 *New York Times* essay. She called it dishonest for colleges to encourage women to embark on difficult, costly engineering studies, without issuing caveats that "rewards for women are exceedingly rare, and . . . still uphill all the way." She ridiculed optimistic predictions that female engineers might find postwar employment rebuilding devastated lands, savaging the idea that European working men would accept orders from foreign women. Reconstruction might provide jobs for women trained in nursing, teaching, "interior decoration . . . landscape gardening . . . institutional management and . . . dietetics," Pickel wrote, but not engineering. In short, she suggested, the "average" college woman should "avoid those professions in which men are still pre-eminent" and embrace unselfish service roles that capitalized on inherently feminine strengths and "home-making talent." Referring to "fundamental and honorable qualities that differentiate women from men," Pickel concluded, "If it is a long day before there are women Presidents, construction engineers, or bishops, why mourn when there is so much to be done that women can do better than men?"[16]

## WOMEN ENGINEERS AT PURDUE: RISING NUMBERS AND VISIBILITY

Purdue exemplified the impressive transformation of women's presence in engineering schools during World War II. Purdue had issued its first engineering degree to a woman in 1897, a civil engineering major. In fall 1942, Purdue had about ten women enrolled in engineering; that number multiplied to more than thirty by August, 1944, and eighty-eight by November, 1945. A critical mass made life easier. As Purdue aeronautics major Helen Hoskinson remarked: "Now that lady engineers are not a novelty on this campus, people no longer stare at the sight of a girl clutching a slide rule." Dean A. A. Potter asserted that the engineering school treated "the girls just like the men students; they are all students in engineering. . . . In my opinion, the person with training and ability will have equal opportunity, whether a man or woman" in future. Putting a face to that new population, Purdue's alumni magazine featured a cover photo of 1942 graduate Kathleen Lux working on blueprints, the first female officer in the U.S. Navy Department's Civil Engineer Corps.[17]

Citing the pressing need for more engineers to serve defense industry, Purdue actively encouraged women to consider nontraditional majors. A campus vocational guidance conference for women in October 1942 suggested that "coeds who have burned themselves in chem lab, broken fingernails exploring physics, . . . or been entangled in a slide rule—and still are fascinated by the sciences—have been saved a place in the war effort." Referring to manpower shortages created by the draft, advocates declared that among female students, "A yen to build bridges or to know why an engine goes round is even more useful today than yesterday."[18]

Echoing this sudden fascination with women's potential as engineers, the school promoted as role models the relatively few women who had earned prewar engineering degrees. In a 1943 profile of Josephine Webb, who graduated in electrical engineering in 1940, Purdue's engineering magazine emphasized that Webb was performing her share of wartime service as a research engineer helping develop electronic radio tubes for Westinghouse. The profile portrayed Webb's passion for technology as equal to that of men, originating from a childhood ham-radio hobby shared with her brother. Coverage noted both the ability and femininity of the "pretty and brown-eyed woman" who met her future husband through her radio store job. The couple completed Purdue's electrical engineering program together, and both joined Westinghouse. Although female engineers had been relatively scarce in Webb's prewar generation, Purdue writers in 1943 suggested that rising female enrollments in wartime engineering would permanently raise women's role in the profession. Ignoring any lingering difficulties facing women, the article hinted that the engineering field was now fully open and that Webb's success was "typical of the careers which may be predicted for the future women engineering graduates which the war will produce."[19]

Even while female engineering enrollment grew and that of civilian men decreased, American academic culture retained the principle of engineering schools as naturally male places, as shown in continued use of the "invasion" metaphor. Purdue observers still jested about the shock of seeing "a real, live, female chemical engineer here . . . one of the relatively few women who have invaded the realm of the engineers to explode the theory that engineering is solely a man's profession." In fall 1943, when freshman Betty Carlson and "another girl engineer" walked into a Tau Beta Pi dinner to welcome all first-year engineering students, "they literally stopped the show. The fellows grinned a little. Imagine a girl in Tau Beta Pi! Betty showed them, though, for last semester she received the woman's badge."[20]

The wartime context gave Purdue women leverage to assert an entitlement to share the field. Ellen Zeigler, a member of Purdue's 1942 engineering class, wrote that if anything, it made sense for society to place more women in engineering research and development, leaving men to occupations requiring more physical strength. Given that men had greater muscle, "if men and women have equal minds, why don't we educate the women to do the planning and the designing and let the men do the manual labor. We should have more women in our engineering schools learning how to design and operate the machines, buildings, and equipment that are needed so much by our present civilization and which are doubly important during a war."[21]

Purdue's engineering dean actively promoted women's potential in the discipline, but such supportive leadership did not eliminate all pockets of resentment. Purdue's Helen Hoskinson reported that one electrical engineering professor turned and left the classroom after he saw her sitting there the first day: "He was sure someone had made a mistake and didn't think it was he." Other Purdue women complained that despite their presence, professors still spiked lectures with stale sexist jokes. Patronizing male classmates also annoyed Purdue women, one of whom spoke about how much she resented having to prove that she was serious about her studies, in an atmosphere where "there are too many boys who think you [women] take engineering to get dates."[22]

The issue of women's exclusion for full membership in certain engineering societies, such as Tau Beta Pi, still rankled. A 1945 column in Purdue's engineering magazine, possibly written by associate editor Maxine Baker, asked outright, "Why can't the women be made known as tops in their field?" If existing organizations refused to accept a woman as "one of the fellows," the editorial suggested, Purdue's female engineering enrollment was growing so fast and strong that women could start their own honor society. By bringing supportive speakers to campus and uniting female engineering students, such a group might "interest employers in the possibilities of woman-power" and "gain recognition of the difficulties of a woman entering the field of engineering and perhaps encourage a brighter welcome." The editorial reassured readers that an all-women's engineering honor society was not seeking complete control of the field: "We are not asking the men to move over or to give up their places to us. We want only to be accepted as co-workers, and a Purdue engineering honorary would do much as an incentive to help us fill such a position. Let the feminine voice speak loudly."[23]

Purdue's engineering dean A. A. Potter, a leading national supporter of female students, also weighed in against the injustice of women's "very unfortunate" exclusion from honors. With sixty female students then enrolled in Purdue engineering, Potter observed, "the percentage of superior women students is somewhat greater than in the . . . man students." He found it intolerable that the woman who held one of the highest academic ranks in Purdue's chemical engineering program remained ineligible for both Tau Beta Pi and Omega Chi Epsilon, the chemical engineering honor society. Having heard that certain Tau Beta Pi leaders were "definitely opposed to women," Potter asked MIT president Karl Compton for help in overcoming that resistance to win "proper recognition" for "outstanding women students." Compton disclaimed any political power to influence policy change in Tau Beta Pi but agreed that "in principle, . . . there should not be a distinction against women's membership but that election should be on the basis of proven merit and nothing else except good character."[24]

Recognizing that many female students still felt daunted by the challenges that were inherent in majoring in male-dominated disciplines, inventive staff members at Purdue held wartime discussions about women and engineering and created several alternative pathways for drawing women into technical studies. In 1943, Purdue's engineering school teamed up with the home economics school to start a new program named "Housing." Fifteen female students enrolled in the curriculum, which combined home economics studies with physics, math, chemistry, and six general engineering classes, plus specialized work in civil, mechanical, and electrical engineering. Professors gave women technical knowledge of construction and remodeling, suggesting that graduates could find employment as consultants to home buyers or as laboratory technicians conducting research for manufacturers of building products. Based on assumptions that female students were natural authorities on the home, the program seemed to offer a safe middle ground for technical exploration. It reportedly appealed to women who were "glad of the opportunity to get something a little more revolutionary than the traditionally feminine field of home economics and yet not to have to go the extreme of entering the engineering schools that eds insist upon preserving for themselves." The program also hoped to attract some of the increasing ranks of men choosing majors in home economics.[25]

The next year, Purdue created another crossover course that was meant to offer female home economics majors intensive shop training. Nine women signed up to study plumbing, electrical appliances, and metal finishes. Instructors reported having to force women out of the machine shop after hours, as female students practiced using precision measuring instruments, cutting and shaping wood, and filing, soldering, and riveting metal. Applying their new skills to practical ends, the women designed and made book-ends, wastebaskets, dishes, jewelry, ashtrays, and model railroad cars. Professor O. D. Lascoe argued that even housewives needed to understand modern engineering termi-nology and techniques. Again, the course reflected wartime erasure of strict gender lines. One observer commented that in machine shop, the home economics women "don their slacks, pin back their hair and really assume the role of a woman engineer," step-ping into "a field which, heretofore, was practically unheard of in women's circles."[26]

## COEDUCATION AT RPI AND COLUMBIA

Coeducational schools such as Purdue, Cornell, and Iowa State saw real wartime growth in female engineering majors, but more dramatic changes came when a number of for-merly all-male schools decided to become coeducational. In September 1942, as Rens-selaer Polytechnic Institute planned its Curtiss-Wright Cadette program and advertised its ESMWT training for women, leaders also agreed to accept female undergraduates seeking degrees. The town paper carried a front-page headline (in bolder, larger type than news about the Nazis marching on Stalingrad) reading "RPI Opens Doors to Women: Institute Breaks 116 Year Old Rule Due to War Need." Secretary-treasurer Livington Houston declared that he had recently received "a large number" of female applications and that, even as he spoke to the press, three interested candidates waited outside his office. Houston explained that RPI had never denied women's intellectual potential, but a century's worth of tradition had simply defined RPI as a men's school. Now, given requests "by industries in the Troy, Albany and Schenectady area to . . . accept as many [women] as we can without impeding our training of men," Houston announced that RPI would suspend its all-male status "for the time being at least, as a temporary measure brought about by the exigencies of war." Perhaps to assuage con-cerns of conservative observers, Rensselaer's president justified provisional coeducation as a precaution against the "worrisome" financial loss to come if Congress began drafting eighteen-year-olds, which would cut off enrollment of all men but the physically incapacitated: "Who knows but that this latest move to admit women may help to tide the Institute over the doubtful days that will follow this war?"[27]

The trustees' decision in fall 1942 to admit women that very term offered little lead time, so only seven registered that first year, joining an entering class of 512 men. Female matriculation started slowly, but the Curtiss-Wright Cadette program multiplied the number of women appearing on campus. Rensselaer's 1943 yearbook exclaimed, "A slight revolution had been brought about by the war. RPI had some . . . real, live, honest-to-goodness coeds. . . . Why, a man had to be careful when he let out a good healthy cussword!" The yearbook contrasted a "before" photograph of male students working in laboratories with an "after" version including women. The 1944 yearbook foregrounded the change by placing a photograph of a woman sitting at a drafting board on its foreword page. By November 1945, RPI had a total of forty-two women enrolled in engineering, up from zero just four years before.[28]

Wartime demand for women engineers similarly provided both pressure and a justification for policy change at Columbia University, where undergraduate engineering courses had long remained men-only. All other prewar Columbia professional schools already admitted women, partly due to the influence of Virginia Gildersleeve, dean of Barnard College from 1911 until 1947. Gildersleeve had energetically campaigned for expanding personal and intellectual opportunities for women, both nationwide and at her home institution. Gildersleeve helped ensure that women would be accepted from the start when Columbia's school of journalism and its school of library science opened, both in 1912, and at the school of business, started in 1916. In 1917, when Columbia's medical faculty hesitated to accept a gifted Barnard graduate, Gildersleeve helped persuade them by raising $50,000 to provide women's restrooms. By 1927, under continued pressure from Gildersleeve and Barnard faculty, Columbia's law school reluctantly conceded trial admission for a few women.[29]

During those pre-WWII decades, the question of opening Columbia's engineering school to female students did not arise. Columbia did contract to train engineering aides for Grumman and accepted female students in its other special wartime courses. Indeed, demand outpaced capacity. In 1942, ESMWT could accommodate only thirty-four out of 150 women who applied for a metallurgy lab-practice course, and Columbia promised to open additional sections as soon as possible.[30]

Emergency courses remained set apart from regular degree curricula, administratively and philosophically. But in May 1942, Columbia's department of chemical engineering and division of metallurgy both requested permission to let graduate students (men or women) who were registered in science earn master of arts degrees in industrial chemistry and physical metallurgy. As patriotic justification for this change, metallurgy professor Eric Jette cited the need for female research staff in factories supplying metal to armament manufacturers. To ease tensions over letting women take engineering, a call for a faculty vote stipulated, "These . . . are not intended to be 'engineering' courses—the engineering content is relatively light. . . . The M.S. degree carries an engineering significance and is, therefore, thought not to be suitable, whereas the A.M. degree is regarded as quite appropriate." The memo sternly added, "It should be clearly understood that this is not a proposal to open the courses leading to the several engineering degrees to women." Even with that caution, one observer commented, "This has possibilities of a row."[31]

Advocates of coeducation won the dispute, and Columbia announced that it would accept women in the two engineering master's programs starting in September 1942. Graduate dean George Pegram touted the benefits of "prepar[ing] women for the increasingly important place they will occupy in the engineering sciences." Arthur Hixson, chair of the chemical engineering program, declared that emergency conditions had neutralized any employment prejudice against female engineers and that every available woman in the country holding chemical engineering degrees had already gone to work.[32]

Even as Columbia agreed to let female graduate students study metallurgy and chemical engineering, pressure for relaxing all other gender barriers in the discipline accelerated. In August 1942, by a margin of six to one, the engineering school's committee on instruction approved a motion to admit women to all programs alongside men under identical regulations, at least for the duration. Engineering dean Joseph Barker favored the recommendation, but associate dean James Kip Finch cast the lone negative vote and pleaded with fellow professors to veto such "mistaken and useless action." Finch acknowledged that female technicians and draftsmen could be equal or superior to men and, as an alternate proposal, favored widening Columbia's ESMWT training for women (which would not cost the school anything or impose conditions on regular teaching).[33]

Finch drew the line at admitting female undergraduates, ridiculing such a step as a publicity-seeking gesture. Patriotic claims that coeducation could make a tangible contribution to winning the war were merely "wishful thinking," he said, since advocates conceded that "they would be surprised if a dozen women could be admitted . . . next year." Finch pointed out that under existing procedures, engineering majors took freshman and sophomore preengineering courses through all-male Columbia College. Since neither Barnard nor any other women's college offered equivalent teaching in physics, surveying, or other requirements, "there are thus no women who are prepared today to enter the junior year of the Columbia engineering course." Even if the school could cobble together "one or two, or even five or ten" female engineering students, Finch warned, such a "radical departure" would upset loyal alumni and currently enrolled male students, undermine the school's honor system, force expensive bathroom renovations in engineering buildings, and complicate Columbia's surveying-camp operations: "Unless we can hope to get larger numbers, say fifty or more women undergraduate students, we will simply be getting ourselves into difficulties for no reason at all."[34]

Finch anticipated zero chance for attracting such a sizeable class, given that "engineering has not been 'a woman's profession' and there is no reason to believe that vital changes in this situation are imminent." War had temporarily created a "forced" demand for female engineers, but peacetime would bring "ample manpower" in which employers, "given the choice of a man or a woman," would certainly "follow earlier practice and take the man." He repeatedly urged colleagues to reject the "nonsense" of coeducation, writing that "We must hold to some sense of balance, even in the national crisis which we are now facing. Nor should we be stampeded into a decision which we may long regret. It is easy to grant this so-called right to women, but it will probably be impossible to withdraw it should we desire to do so at a later date."[35]

Disregarding such ominous words, an alumni and student committee seconded the motion for change, and late in 1942, engineering followed Columbia's other professional schools in admitting women. As an honest administrator, Finch vowed that with the vote to "approve this asinine innovation, I will, of course, do my d____est as Associate Dean to make the thing work." Making good on his word, Finch subsequently gave Columbia's School of Mines formal approval to hire a female lab assistant:

I see no reason why women should not also be employed for teaching duties—especially in those classes which are planned primarily for women students. It was obviously the intention of our faculty that there would hereafter be no discrimination as to sex in the school of engineering. I do not favor the idea and have said so, but the majority of our staff does and we must be prepared to accept in full our new responsibilities.[36]

Over at Barnard, Gildersleeve helped proponents advance the case for admitting women to Columbia engineering and repeatedly urged her students to seize the unprecedented wartime opportunity. As it turned out, Columbia's first female engineering graduate did not come from Barnard. Gloria Brooks transferred from Cooper Union after finishing her sophomore year in 1943, when that school curtailed regular undergraduate education in favor of army training. Given that a majority of Columbia's engineering professors had actively embraced coeducation, it is not surprising that Brooks felt accepted by faculty. But "some of my classmates were immature and pulled tricks to get attention. It was silly stuff, but it was annoying," she recalled. Brooks long resented Columbia's failure to give her the free football tickets that were given to all male students under the rationale that they were possible recruits: "It seemed ridiculous since I was one of the few people who wanted to watch a team that had never won a game." Under Columbia's accelerated war calendar, Brooks completed her B.S. degree in electrical engineering in February 1945 and then worked at Bell Labs and on Sperry Gyroscope's radar-system development. Brooks later returned to Columbia for her doctorate (alongside an undergraduate daughter studying engineering there) and chaired the electrical engineering department and bioengineering program at Fairleigh Dickinson University.[37]

Only a handful of women immediately followed Brooks. Eleanor Leland, originally a Cornell chemistry major, transferred to Barnard when she heard that Columbia had begun accepting female engineering students. She described a "wonderful" experience: "my classmates had someone to tease, but I loved to get all the attention." After becoming the first woman to earn a Columbia chemical engineering B.S. in June 1945, Leland joined General Electric. Although Leland had no female classmates, 1948 graduate Carol Schreiber Perrin had two other women in her class. Unlike Brooks, Perrin recalled faculty hostility: "The first day of class in mechanical engineering, I walked into class and the professor looked at me, smashed his fist on the desk and said, '(expletive) you don't mean to tell me I'm going to have a woman in my class!' And I said, 'I'm not going to cramp your style; sometimes I do a little swearing myself.'"[38]

## POSTWAR AND COLD WAR TRENDS IN WOMEN'S ENGINEERING EDUCATION

World War II did not magically remove all institutional barriers to women who wished to study engineering. For the duration, the aeronautics school of New York University trained female engineering aides under contract with both Chance Vought and Vought-Sikorsky aircraft companies. Yet those short-term arrangements did not automatically translate to giving women equal access to formal degree programs. NYU would not officially begin admitting female students to its regular undergraduate engineering college until 1959.[39]

Even in departments that accepted female majors, war did not convert engineering into a feminist paradise. Many skeptics still refused to take women seriously as students who genuinely sought education on equal terms with male classmates. A headline in the 1944 *Cornell Engineer* read "WOES (Women of Engineering Schools) Are Here." The piece said, "Rumor has it that 17 woman engineers are at Cornell. Do they build up morale or do they provide distraction? Are they taking advantage of the boy-girl ratio in engineering, are they just trying to help the war effort, or do they want engineering careers?"[40]

Toward the war's end, Cornell started to worry that women were taking up too much room on campus as returning veterans poured in. Housing grew scarce, and officials feared overcrowded classrooms. For both practical reasons and emotional symbolism, male veterans assumed priority, pushing female students to the margins. At Cornell, harried administrators imposed an artificial cap, ordering all departments except home economics to block the admission of any new female undergraduates for the spring 1945 term. Cornell's engineering dean had already approved admission of nineteen first-year women; combined with the eighteen women who already were enrolled, the engineering college thus exceeded its quota of twenty-five women by 50 percent. Cornell's vice president scolded the engineering school for carelessness and stated that absolutely no more female students would be accepted in engineering that semester.[41]

Yet wartime had generated some permanent changes in opening new opportunities to female engineering students. When peace arrived, RPI did not return to its former all-male status. That official gesture seemed almost superfluous, since everyone assumed that as a technical school, Rensselaer would always remain male-dominated. In 1946, Lois Graham and Mary Ellen Rathbun became the first women to earn RPI degrees. In yearbook photos, the two women, wearing white blouses, stood out strikingly from all the men in military uniforms. Moving to the Illinois Institute of Technology (IIT), Graham earned her doctorate in mechanical engineering and rose to full professor there. She served as president of the Society of Women Engineers and headed IIT initiatives in the 1970s to induce more high school women to consider engineering.[42]

Some administrators remained unconvinced that women actually belonged at RPI, at least in any significant quantity. In 1958, the secretary of Rensselaer's alumni association directly asked the school for official word whether RPI was a coeducational school. RPI's office of external affairs replied that "we are coeducational even though we don't stress the fact because we don't want too many gals around."[43]

One reason that RPI did not wish to cultivate female applications was that it had no official women's residence. Although RPI received as many as a hundred inquiries annually from interested women around the country, in practice it could accept only those who lived in the immediate vicinity. After wartime enthusiasm wore off, female enrollment fluctuated drastically. There were twenty-three women at RPI in 1952, but other years saw no more than six female undergraduates on campus. There were a total of nine in 1960, which RPI's vice president of student affairs praised as a sign of rising female interest. But only one new woman entered Rensselaer in 1960 in a class with 854 men.[44]

To open space for female students, in 1961 Rensselaer initiated a plan for women to live in dormitories at all-female Russell Sage College while pursuing RPI degrees in engineering, architecture, or science. As justification, Rensselaer's president cited America's cold war need for such experts and declared that "many attractive careers" in those areas "lie open to women." Dean of students Ira Harrod commented, "The image of the engineer in field boots should be abandoned." The administration indicated interest in bringing female enrollment up to as high as 20 percent.[45]

That official endorsement failed to convince conservatives such as undergraduate Myles Brand, editor of RPI's engineering magazine, who prominently denounced Rensselaer coeducation as violating both logic and custom. In Brand's opinion, women's presence could bring no conceivable benefit, only a danger of "major distractions" for talented men. Because engineering and science represented the most challenging of all possible studies, he wrote, RPI should not let brilliant men, as future professionals, be "hinder[ed]" by anything inducing "mind-wandering during lectures." Brand warned that coeducation would "inhibit" what he considered RPI's desirable "informality of the classroom," destroying the ideal "friendly relationship" between male faculty and male undergraduates. Brand did not suggest that women were inherently incapable of mastering engineering. Indeed, he believed that creating a new women's college offering technical education would be valuable for yielding a larger pool of professionals. Such separate but parallel training would allow RPI to preserve the all-male status that for Brand defined its excellence: "the only way . . . [to] retain its dignity and tradition."[46]

Brand faced no immediate risk of being sidetracked by throngs of female classmates upsetting his concentration. With RPI's male-female ratio of two hundred to one, outsiders and male classmates alike often assumed that women chose Rensselaer not for its academic benefits but as a prime setup for marriage. Scoffing at such notions, 1963 student Sherry Pelson asked, "What girl is going to spend $1600 for Rensselaer education just to find a husband?" RPI women also mocked speculation that in mingling with RPI men every day, they gained advantages over Sage women in securing dates. RPI women pointed out that the classroom atmosphere discouraged romance, as did their status as engineering and science students. RPI men mentally categorized female classmates as separate from eligible partners, said Jacky Montross: "Sage girls are thought of as girls; we are thought of as tools [nerds]!"[47]

RPI treated all students as male by default. Female undergraduates remarked on the oddity of getting official correspondence addressed to "Mr. Rita Bauer." Although form letters by definition were impersonal, Beverly Dahmer noticed that even most professors addressed her as "Mr. Dahmer." More poignantly, Jacky Montross said about being at Rensselaer, "I always feel out of place." Nevertheless, trends favored women at RPI. The Sage housing scheme attracted nationwide publicity, raising general awareness that Rensselaer accepted women. By 1965, Rensselaer had fifty-five female undergraduates, and the following year, RPI set aside one wing for women in a newly constructed dormitory. Twenty-seven new women enrolled, comprising 2.7 percent of a freshman class of a thousand. Although far short of any 20 percent goal for females, coeducation had become a permanent part of RPI's engineering and science identity.[48]

Like Rensselaer, Columbia's engineering school never revoked its vote for wartime coeducation. Columbia did not see a flood of female majors; instead, they trickled in during the late 1940s and 1950s. Anna Kazanjian Longobardo was one of two women, out of four in her class, to complete her engineering degree in 1949. She recalled "wonderful" treatment from the veterans who then dominated the school and regarded her as a little sister. Longobardo later became a Columbia University trustee, chair of the dean's engineering council, and president of the Engineering School Alumni Association. After starting at Columbia, Klara Salamon Samuels "realized it was up to me to make the boys comfortable and so I became one of them and went to beer parties. For labs, I would change in the ladies room and the boys would change in the hall. I would yell out, 'coming through' and they would say 'OK' and there was no problem."[49]

Women's continued but difficult existence at RPI and Columbia echoed broader trends at the national level of engineering enrollment. Postwar numbers of female engineering students plunged steeply, but they never entirely vanished. After reaching new lows in the early 1950s, female ranks again climbed, finally reapproaching World War II levels by 1957.

In November 1945, a total of 1,801 women and 48,977 men were studying engineering in all American colleges and universities combined. But just as Pearl Harbor had abruptly revolutionized the intellectual, economic, and social character of American higher education, peacetime brought an equally dramatic transformation. The Serviceman's Readjustment Act of 1944 (known popularly as the G.I. Bill) offered most veterans at least a year's worth of education funding, plus stipends for fees, books, and living costs. To make room for an anticipated flood of incoming veterans, Cornell temporarily barred any new women from entering its engineering program. Other institutions, including the University of Michigan, imposed tight overall female quotas, cutting the

TABLE 3.1

Total female engineering enrollment in U.S. higher education institutions, 1949–1959

| Year | Female enrollment | Male enrollment | Percent female |
|------|-------------------|-----------------|----------------|
| 1949 | 763 women | 218,949 men | 0.35% |
| 1950 | 683 | 179,579 | .38% |
| 1951 | 625 | 165,012 | .38% |
| 1952 | 696 | 175,853 | .39% |
| 1953 | 926 | 192,407 | .48% |
| 1954 | 1,161 | 213,253 | .54& |
| 1955 | 1,167 | 242,223 | .48% |
| 1956 | 1,484 | 275,568 | .54% |
| 1957 | 1,783 | 295,294 | .60% |
| 1958 | 1,718 | 287,962 | .59% |
| 1959 | 1,662 | 276,686 | .60% |

*Source:* Ann McGreaham, "The Opposite Sex in Engineering," *Purdue Engineer* (May 1963): 20–24.

number of total first-year women in 1946 by about 33 percent. Even at schools that did not officially cap female enrollment, the postwar G.I. influx dominated headlines and commanded the bulk of attention from overwhelmed faculty and administrators. Between 1944 and 1954, the G.I. Bill helped put about 2.2 million veterans through college. They were mostly (but not exclusively) men because many women who served either did not realize that they qualified or chose to marry right after the war rather than enroll.[50]

The G.I. Bill did not permanently frighten away or block American women from entering college. As Linda Eisenmann has documented in her history of women's higher education, female college enrollment in the United States more than doubled over the postwar era, from almost 700,000 total female undergraduates in 1948 to almost 1.7 million in 1963. The number of women students nationwide rose each year in that period, except for relatively small decreases in 1950 and 1951. Yet that growth pattern camouflaged important distinctions. Female enrollment at large universities was proportionately far smaller than their representation in teachers' colleges, liberal arts colleges, and junior colleges. Women's education was not a high priority for many big institutions, especially the prominent research universities that were increasingly obsessed with chasing after federal funding and industrial contracts to subsidize expanded work in science and engineering. In short, Eisenmann concludes, much of the postwar-era higher-education establishment in general treated women as "incidental students" and auxiliaries to men.[51]

Predictably, that description of women as postwar "incidental students" applied many times over to their position in engineering schools and programs. Women enrolled in engineering were doubly incidental. As women, they comprised a tiny sliver of the engineering student population, and as engineering majors, they comprised a minute fraction of all female college students. In the 1950s, men who earned degrees in engineering represented roughly 10 to 15 percent of all male college graduates. By contrast, women completing engineering degrees amounted to 0.20 percent or less of all female college graduates. Although different accounts provide different total enrollment figures, depending on which schools were included in each survey, which reported numbers, and what was defined as an engineering curriculum, the broad trends remain clear. After the immediate post–World War II decline in women's engineering enrollment, their numbers remained comparatively small through 1952, hitting particular lows in 1950 and 1951, the same years that brought an overall drop in female college enrollment. But 1953 began a new upward trend, and the national total of women's engineering enrollment more than doubled between 1949 and 1959. Because men's engineering enrollment also rose over that period, female students remained well below 1 percent of the total population. Their presence at some institutions still reached well into double digits. For instance, in 1952, seventy-one women were taking engineering courses at the Illinois Institute of Technology, and twenty-eight at the City College of New York.[52]

Anecdotal discussions from this period often conveyed an impression that dropout rates for the relatively few women who enrolled in engineering were astronomical. Such generalizations both reflected and reinforced convictions that women generally could not handle the challenging curriculum and did not want to compete on male territory. But graduation numbers offer a more tempered picture. Data from the U.S. Department of Education showed that in academic year 1958–59, a total of 121 women received undergraduate degrees in engineering. With 38,013 men earning degrees, women constituted 0.32 percent of the total. Although smaller than the female percentage of overall engineering enrollment at that time, such a figure was not embarrassingly low, considering how many women still reported facing discrimination or more subtle but still substantial discouragement.[53]

Even though female undergraduates remained a tiny minority of the total engineering school population, a number of institutions during this period explicitly acknowledged their presence and occasionally even sought to recruit more, albeit in a low-key manner. In 1949, the annual bulletin promoting the University of Illinois engineering school to potential students depicted engineering as entirely masculine territory. Although several female students were enrolled, photographs showed just men working in classrooms and laboratories. The only pictures of women appeared in the section headed "Developing your personality at Illinois," which featured images of men dancing with their dates at the engineering students' St. Pat's Ball. By contrast, the 1952 version of the same booklet carried two images of female students working alongside men in foundry laboratories and machine shop. The 1954 booklet text continued to refer to engineers as male throughout, but it contained one paragraph encouraging high school women with good grades in math and science to "seriously consider the possibilities of engineering careers." The University of Illinois promised "excellent opportunities for women in the engineering profession" and added that "Over the past several years there has been a steady increase in the ratio of women to men students enrolled in engineering." That same year, a committee from the Illinois engineering school prepared alumni who interacted with high school teachers, administrators, and (indirectly) students to field the question, "Is there a place for women in engineering?" The officially sanctioned answer was yes, that Illinois generally enrolled approximately ten female engineering students at any one time, and women had completed degrees in almost every engineering major. The manual explicitly defined an engineer as "a creative professional man or woman."[54]

Those who fought to advance women's status in engineering, both as college students and in the larger profession, gained some ammunition as the cold war renewed public concern about manpower trends in science and engineering.[55] Alarmists warned that the United States was losing ground in the race for superior technical training, and scare stories suggested that Soviet schoolchildren were already mastering calculus in eighth grade. Reports of the Soviet mass production of engineers and scientists led some educators, politicians, and government experts to argue that in order to keep up, Americans needed to broaden their gendered vision of who could and should enter the technical disciplines. In 1952, Arthur Flemming, manpower chief in the U.S. Office of Defense Mobilization, wrote, "[W]e haven't got a chance in the world of taking care of that deficit of engineers . . . unless we get women headed in the direction of engineering schools." Flemming wanted to revive the patriotic atmosphere of World War II, which had invoked a sense of educational emergency in pushing women toward engineering. The cold war posed such a grave threat, he indicated, that "we must convince" women that technical training "will give them an unusual opportunity to serve the nation during a very critical period." Flemming warned that the United States faced a stark, urgent choice because "Russia isn't making this mistake" of squandering half its brainpower.[56]

In 1958, the American Society for Engineering Education (ASEE) and the National Science Foundation sent eight engineering educators from the United States to tour Soviet engineering colleges. One delegation member, University of Illinois dean William Everitt, reported that the Soviet system respected and rewarded technical accomplishment and strongly motivated both boys and girls to excel in such studies. According to Everitt's account, Soviet classes in engineering technology appeared to be roughly one third female.[57]

After the Soviets' *Sputnik* satellite "beat" American rockets into space, the resulting political frenzy whipped up additional pressure to expand fears about inadequate U.S. science and engineering education. In November 1957, President Dwight D. Eisenhower warned that the USSR's pool of scientists and engineers outnumbered those in the United States and that the Soviets were graduating new professionals more rapidly, who matched America's in intellectual quality. Congress poured additional funding into primary and secondary teaching programs in science and engineering, plus scholarships and other incentives to lure more of the country's best students to those fields. Yet the President's Committee on Scientists and Engineers, an action group created by Eisenhower in 1956 to stimulate manpower production, reported that demographic trends threatened to compound the difference between American and Soviet science and engineering enrollment. America would soon face a shrinking labor force of men between ages twenty-five and thirty-four due to reduced birth rates during the Depression and World War II, the committee's final report warned. However, statistics predicted that the numbers of women in the same age groups would slightly rise by 1965.[58]

In 1952, the ASEE's *Journal of Engineering Education* published what author Fred C. Morris, of Virginia Polytechnic Institute, billed as a "logical and intelligent" cold war plan for training women in engineering. Morris praised the women who had ventured into technical jobs during World War II, some of whom were still "doing very well" in related peacetime work. Like wartime engineering educators, he admired what he considered an inherently female eye for painstaking detail and manual dexterity. In a backhanded compliment, Morris emphasized that "quite important, [women] are not adverse to trying something new. Witness, for example, their proclivity to change the furniture around in the house about every three days to see if they can find a more efficient arrangement. This is exactly the procedure that our research scientists use; . . . if you don't know if something will work or not, try it." Morris warned that unless the United States began capitalizing on its "big untapped" supply of female talent, it risked ultimate political subjugation and "virtual slavery."[59]

Morris's support for female engineering training had a catch, however. Although he extolled the "many bright young women in college with inherent technical ability and many more graduating from high school each year," he did not recommend persuading more of them to earn engineering degrees alongside men. Instead, he visualized a cold war equivalent to World War II's Cadette programs and other short-term training, where colleges would offer female science and home economics majors a year's worth of drafting, computation, and other engineering studies. After entering industry, those engineering aides could save important men from wasting time on routine tasks, thereby stretching manpower and multiplying productive efficiency. Morris suggested that feminine character was intrinsically unsuited to the "rough" pressures of pursuing deeper technical ambitions or real scientific excellence: "Girls, being more sensitive and nervous than boys, sometimes become emotionally disturbed by overwork and the fear of failure." Such personal instability barred women from handling serious professional responsibilities, Morris added, but wise employers would understand how "strategic use of a few kind words" could induce "girls" to "work their hearts out for you" as engineering aides.[60]

So despite his anti-Communist anxieties, the blinders of gender stereotyping stopped Morris from envisioning a future when society would encourage more interested and able women to pursue engineering beside men rather than beneath them. The same inability to appreciate women's potential as genuine engineering students afflicted Eric Walker, dean of engineering at Penn State. Despite the fact that a handful of women had successfully completed engineering degrees there over the years, an unimpressed Walker refused to extrapolate their potential to their sex as a whole. In 1955, Walker wrote an article for his school's engineering magazine titled "Women Are NOT for Engineering." He declared that despite the success of "unusual women" such as Lillian Gilbreth and Edith Clarke, most women did not have the "basic capabilities" needed to handle technical work. Moreover, investing time and effort to teach female students did not make sense because "[t]he most evident ambition of many women is to get married and raise a family," Walker concluded: "Few companies are willing to risk $10,000 on a beautiful blonde engineer, no matter how good she may be at mathematics."[61]

In a pointed rebuttal, two female engineering students at Florida State jumped to defend their sex, insisting that women's technical skills and professional commitment deserved respect. In a response headlined "Women Are for Engineering," Wilma Smith noted that marrying did not automatically preclude women from making professional contributions because an increasing number of wives and mothers wanted to continue their careers. Addressing the broader question of discrimination, Penelope Hester added,

"If someone can do a job well, why should he or she be denied the right to do that job? An all-male concept of engineering is based on prejudices and old-fashioned ideas. . . . A woman can be just as devoted to her job as a man, and maybe even more so."[62]

Female engineering majors of the 1950s were well aware that women comprised just about 1 percent of America's total engineering workforce, less if technicians and engineering aides were excluded. They anticipated a strong likelihood they would encounter substantial hostility in job hunting, but they also remained confident that at least one good employer would give a woman a chance to prove her worth. In a 1951 survey of five hundred employers, conducted by the National Society of Professional Engineers, 65 percent of businesses said that they were willing to hire female engineers if they were available. Purdue student Burke Arehart wrote in 1957 that after conversations with established engineers, she had to conclude that her postgraduation prospects "aren't too good." She quoted one Chrysler man as admitting that he would select a male candidate over a more qualified woman, "unless, of course, the difference was such that I'd be doing the company a great disservice by not hiring her because the man would be nothing but a bungler." But at least in public, many female engineering students preferred to think that such bias was becoming outdated. From talking with an engineering personnel representative from Burroughs, Arehart got the impression that the computer company had "no compunction against hiring women if they are qualified." She believed that as more women demonstrated their skills both in school and the workplace, they would naturally overcome resistance and convince more employers to consider them on a footing equal to male engineers.[63]

One factor contributing to this employment optimism among college women was the consensus among engineering professionals that the field itself was changing, reflecting greater demands for specialists in research, development, design, and administration. Laboratory and desk jobs seemed far more female-friendly than macho construction sites or shop floors, where observers expected that most women engineers might have trouble issuing orders and commanding loyalty. But with more employers looking for intellectual maturity rather than physical muscle and experience with heavy machinery, many felt that the profession would soon evolve toward more opportunities for women. Female students of the 1950s believed that they would have to work harder than male classmates to secure employment and promotion, but they anticipated that patience and dedication would win them respect and rewards.[64]

Even before graduation, many women engineering students felt continually on trial by male classmates who resented and mistrusted their presence. The system tested female undergraduates' resoluteness, but some believed that they could win a certain level of acceptance by demonstrating persistence. One woman at Purdue in 1951 said that in her freshman and sophomore years, men conveyed the attitude, "Well, what are you doing here?," but that she saw such suspicion ebbing as she proved her engineering ability. Again, other female engineering students optimistically believed that over time, as their numbers and success grew, such trends would moderate men's wariness.[65]

A CRITICAL MASS FOR ORGANIZING: THE SOCIETY OF WOMEN ENGINEERS

Female engineering majors at schools nationwide recognized that in a number of respects, their treatment remained unfair. Years later, some still bitterly resented that despite being equally qualified academically, Tau Beta Pi refused to initiate them alongside male classmates. But even inside that closed society, female students' new presence in engineering was apparent. Although Tau Beta Pi issued only four women's badges before 1941, thirty-two female engineering students claimed them between 1941 and 1945. A December 1945 article in the Tau Beta Pi journal suggested that such trends made the question of women's status in the society "far more important than it was even a few years back." Louis Monson (who was long involved with the society and former editor of its magazine) diagnosed a split within Tau Beta Pi that was connected to the size and nature of different chapters. Clearly, growing numbers of women were meeting the society's academic requirements for eligibility, but membership elections also involved less tangible factors of fraternal companionship and extracurricular leadership. Monson believed that small chapters, which selected members primarily based on strict scholastic qualifications and which largely held formal business meetings, would be more likely to favor admitting women. Chapters with a larger pool of candidates, however, often bypassed top students to elect "big men on campus" who starred in student activities and whose gregarious manners were an asset to society parties. With that tone of "masculine camaraderie," such groups "are probably going to resent and resist to the bitter end the intrusion of women into the sacred haunts of the male." Monson emphasized to his readers the courage that it took for female students to defy gendered tradition in choosing a college major: "We are astounded . . . that any woman would elect to pursue an engineering education, knowing that she is encroaching on a field that has always been admittedly the sole domain of males—the more rugged ones, at that; that every barrier will be placed before her that can be politely erected; and that she must expect to represent an alien minority throughout her college career." As the father of two daughters and no sons, Monson's sympathy lay with the excluded sex: "Something more than the present recognition is merited by women who not only tackle the engineering curriculum, but do so well at the tackling that they meet our scholarship requirements." Over subsequent years, a substantial number of delegates at Tau Beta Pi conventions agreed with Monson that the society should rescind the ban on female membership, but the proposed amendment to the society's convention failed to win endorsement by the required 75 percent of chapters. Ultimately, Tau Beta Pi did not welcome women on equal membership terms until 1968.[66]

Meanwhile, female engineering students around the country built on the pattern that had gained momentum during wartime, of women establishing their own groups. In 1946, about twenty female engineering students at Iowa State College created a local organization called the "Society of Women Engineers" (preceding the national group carrying the same name) to assist "in orienting new women students in the division." That same year, female students at Syracuse and Cornell vented their frustration at restricted access to societies such as Tau Beta Pi by creating their own organization. By 1948, their new honorary society, Pi Omicron, had chapters at four schools, including Purdue. Members worked to support incoming female engineering majors on each campus and hosted speakers such as Lillian Gilbreth. The mission was "to encourage and reward scholarship and accomplishment . . . among the women students of engineering . . . ; to promote the advancement and spread of education in . . . engineering among women."[67]

Female engineering students still joined those engineering professional societies that were willing to accept them. Pi Omicron's 1948 Purdue vice president Eleanor Costilow also served as secretary of the local Institute of Aeronautical Sciences. When Anna Hanson became president of not just Purdue's Pi Omicron but also the campus chapter of the American Society of Civil Engineers, the school's engineering magazine commented, "You'd think that the many civil engineers at Purdue would at least be able to find one man to head their professional organization." The article went on to praise Hanson for "showing that female engineers DO have a place in this man's world." At Purdue and other schools, Pi Omicron members showed technical films, held initiations, and hosted other activities that often paralleled those pursued by male-dominated student engineering societies. But Pi Omicron activities assumed extra weight, in providing support for under-served women students and making a public statement about their rising numbers. Social events aimed to build fellowship among the outnumbered women, while professional events sought to bolster their confidence and employment viability. Listed alongside older organizations in descriptions of student engineering activities, Pi Omicron's existence made a statement that women's presence in the field should be taken seriously.[68]

Women who had already established themselves in the field were in the process of developing their own networks. By the late 1940s, twenty-four female members of the Western Society of Engineers (with three thousand total members) had formed their own Women's Council, which aimed "to improve the training and broaden the employment opportunities of professional women engineers." In 1950, female engineers in New York, Boston, Philadelphia, and Washington, D.C., began gathering on a semi-regular basis. The following year, they held a national convention in New York City, began publishing a newsletter, and created the first local section, in Pittsburgh. In 1952, the national group, with about sixty members, was officially incorporated as the Society of Women Engineers (SWE), a professional, nonprofit educational service organization. Under SWE by-laws, senior membership was open to any woman "actively engaged in the profession of engineering" who held an engineering degree and had at least six years of engineering experience, or who held a science degree and had at least eight years' engineering experience, or who had worked in the field at least eleven years while demonstrating "engineering competency and achievement." Women more junior to the discipline could qualify as regular or associate members. Significantly, SWE sought to connect established professionals with those just entering the field. It welcomed as student members any women enrolled in full-time degree programs in engineering or science-related engineering.[69]

SWE was concerned with promoting female graduates' long-term status both inside the field and among employers, aiming "to advance the professional interests of women engineers . . . , to inform the public of the qualifications, abilities and achievements of women engineers . . . , [and] to foster congenial relationships between women engineers and industry." Part of its strategy involved maintaining a visible place in the field and in the public eye; in 1954 and 1955, members of Cleveland's SWE appeared on local television programs as proof of women's small but persistent presence in the engineering profession. As another priority, SWE leaders displayed an ambitious desire to draw new generations of young women into the field—"to encourage all women who show an aptitude for and a desire to study engineering."

As one of its first initiatives, SWE created a Professional Guidance and Education Committee, which poured enormous attention into reaching potential converts. Both formally and informally, SWE members wrote to dozens of high school women, sending information about engineering and replying to questions. Irene Carswell Peden, associate professor of electrical engineering at the University of Washington, wrote, "It is important to think of women engineers as real people doing real jobs which the student could do, too." In 1958, Boston's chapter of SWE published a pamphlet containing biographical sketches of a few "typical" women engineers and explanations of how girls could prepare to enter engineering. SWE's authors concluded, "If this pamphlet shall have inspired one young woman to consider an engineering career . . . and one parent to 'encourage' the daughter's desire to enter the technical field, this pamphlet will then have been a worthwhile venture."[70]

Many of SWE's early leaders received their engineering degrees either just before or during World War II. They knew that despite wartime initiatives to bring more women into the field, the entire notion of female engineers remained strange and off-putting to many ordinary women and men, especially with the postwar reassertion of traditional gender roles. Peden warned that "A girl is not likely to choose a career field disapproved by her parents, teachers, classmates, and friends. All of these people . . . seem to be responding in part to an erroneous but popular image of the woman engineer as a cold, . . . aggressive female who trudges through life in her flat-heeled shoes without a man in sight (away from the job)." At a time when many Americans perceived female engineers as odd, manlike creatures, SWE representatives took pains to offer a presentable feminine image, emphasizing that many of them were married and had children. Peden wrote, "Many women engineers are very attractive; most represent a perfectly normal cross section of femininity. The only way that this image can be brought into line with reality . . . is by . . . personal contact." Advocates believed that simply by seeing female engineers and realizing that they seemed like normal women, other Americans would think more positively about engineering as an option for their sisters and daughters.[71]

Such volunteer guidance reflected some of SWE's primary beliefs—that girls often shied away from technical pursuits because they simply did not realize that women could and did go into engineering and that early intervention by supportive women could give them enough encouragement to persist against the odds. Beginning even more active outreach efforts, members in the mid-1950s volunteered to assist at "Junior Engineer and Scientist Summer Institute" (JESSI) programs, which brought high school students to college to explore science and receive educational guidance. At one JESSI session in Colorado, fifty-three girls listened to a five-woman panel discuss why they had chosen engineering careers. Female engineers led JESSI students on visits to industry and gave the girls (and boys) tours of their laboratories.[72]

From the beginning, SWE provided valuable resources that colleges could draw on to promote engineering as an option for women. In its 1959 communication with potential engineering students, the University of Illinois encouraged high school women to contact SWE's New York office to request a free copy of SWE's 1955 booklet *Women in Engineering*. The bibliography also recommended that female readers consult a 1954 Women's Bureau pamphlet on *Professional Engineering: Employment Opportunities for Women.*[73]

SWE encouraged girls at elementary and secondary schools to consider engineering as a career, but it assigned equal importance to supporting college women and increasing their retention in the field, especially given American manpower needs in the cold war competition against the Soviet Union. By 1957, female engineering students at Drexel, Purdue, Colorado, Missouri, Boston, and City College of New York had founded student sections of SWE, and the parent organization welcomed its junior counterparts. Established SWE members vividly remembered how intimidating it felt to be the sole woman in an engineering class. They knew, as Helen O'Bannon wrote, that "being one of a small group following a path that appears to violate society's norms is lonely." Mildred Dresselhaus argued that such young women deserved support from older mentors who could provide the encouragement necessary "to 'keep going when the going gets rough' or when [a girl] begins to ask, 'Is it worth it?'" Successful role models could give new students a boost in confidence, a chance "to see by example that women can 'make it' in engineering." Older professionals especially sympathized with young women just entering Georgia Tech, and in 1958, Atlanta's SWE chapter sent several members to participate in Georgia Tech's start-of-the-year camp for first-year women. SWE commented, "One must realize that there are this year approximately 1300 freshmen at Georgia Tech and only 19 freshman coeds. There will be numerous problems, and SWE Atlanta Section is proud to play an integral part in the quite difficult assimilation of female engineering students in an almost all-male school."[74]

## WOMEN'S ENGINEERING EDUCATION THROUGH THE EARLY 1960S

The early 1960s extended the trends in women's engineering education that had begun in the 1950s. Although female enrollment did not advance in enormous leaps each year, figures overall followed a distinct upward path. Cases of specific schools often revealed similar patterns. For example, the Engineering College at the University of Illinois had twelve women enrolled in early 1960 but twenty-four in 1963, twenty-six in 1964, thirty-four in fall 1965, and thirty-seven in spring 1966.[75]

Even where institutions reached out to invite women to concentrate in engineering, career prescriptions for students differed by gender. A 1962 engineering recruiting brochure from the University of Illinois read, "Girls who enjoy mathematics and science and who like hard work are welcomed as engineering students." But in discussing potential professional directions that engineering majors might follow, the pamphlet

FIGURE 3.2

As this brochure from the Society of Women Engineers illustrates, female engineers and their male allies invoked cold war manpower concerns and the need to "keep up with the Russians" as justification for encouraging more young women to pursue engineering studies. Society of Women Engineers National Records Collection, Walter P. Reuther Libarary and Archives of Labor and Urban Affairs, Wayne State University.

spoke only of men entering engineering administration and sales positions. By contrast, it specifically highlighted the (relatively low-paid) fields of engineering writing, journalism, and publishing as offering satisfying careers for women. The publication also provided detailed reasons why taking engineering courses could prove advantageous even to women who did not pursue ambitious careers: "Through the study of engineering, women can make valuable contributions to the home and to the community. Even on a part-time basis, technically competent women can be of great value in the decision-making required of school board members, parent-teacher members and officers, planning commissions, zoning boards, and city councils or other levels of local, state, and national government." That statement reflected the reality that a sizable share of married female engineers (like other women with babies) did temporarily leave the full-time workforce in the postwar era. Many women engineering students, who actively considered questions of how they would balance work and family, might be reassured by promises that their credentials could prove valuable outside the business world. At the same time, such comments only underlined differing expectations; no one told male engineering graduates that their technical knowledge would be an asset to the Parent-Teacher Association or the golf club. It reinforced the suspicions already expressed, that women students chose engineering primarily as a means to snag a husband, and it gave some engineering faculty, administrators, employers, and professionals an excuse to take female students less seriously. After all, the post–*Sputnik* United States needed to graduate engineers who could keep American technology ahead of the Soviet Union, not engineers who made the local school board run more smoothly.[76]

Indeed, as the ranks of female engineering majors slowly grew, many still felt annoyed by those male classmates who refused to take them seriously as intellectual equals. Ironically, even as male engineering students and other observers dismissed their female counterparts as nothing more than husband hunters, they also dismissed them as true women. The stereotype of female engineers as asexual and romantically undesirable permeated male engineering culture at schools around the country. That canard infuriated female engineering students such as three who told their male counterparts at the University of Minnesota, "We are tired of being called 'dogs,' tired of being thought of as 200 lb. shot-putters in skirts. We are not ugly and . . . not desperate for male company, and we don't sit at home every night pounding the books 'til 3 a.m. We're *girls*." Besides being personally offended, women engineering majors worried that the vicious image contributed to the many other factors that deterred young girls from considering the field. One University of Illinois student called on high school advisers to ask themselves whether they ever urged the female half of their counseling subjects to think about engineering as a potential direction. She added:

You could also stop the proverbial idea that engineering, science and math majors are a bunch of divinely handsome boys and a few strange girls who wear horn-rimmed glasses, skirts practically to their ankles, horrible stripes and plaids together, and go around with straggly hair squinting at everyone as they accidentally fall over a crack in the sidewalk because they were reading the latest book out on entropy. . . . As I see it, our main problem is the idea that a woman cannot be an engineer and still be feminine.[77]

The insults directed at female engineering majors were made worse by the social climate of engineering schools, which was built around male students and crafted a specific role for the engineer's girlfriend. At engineering programs around the country, the annual social highlight was the formal St. Patrick's Ball, headlined by a contest to choose an engineering queen from among campus beauties nominated by male engineering students. With rare exceptions, those engineering queens were not engineering students but majors in more traditionally female areas such as home economics and literature. A casual sexism permeated engineering schools, focusing extensive attention on women's beauty and bodies rather than brains. To show off the versatility of the University of Illinois' new digital computer system for potential students, parents, dignitaries, and other visitors, an engineering Open House ran a program to plot the 36-24-36 measurements of a bikini-clad woman, a drawing meant to represent "the typical U of I coed."[78]

Female engineering students sometimes felt caught in a no-win scenario in which every woman who encountered academic difficulty or felt unhappy enough to change majors confirmed prejudices that women did not belong in the discipline. Those women who survived were still "treated as intruders" by people who assumed that "she can't possibly be interested in [engineering]; she is just too stubborn to change to a 'more suitable' major," Minnesota women complained. General Electric engineer Betty Lou Bailey recalled that as a nervous freshman and sophomore at Illinois, she was "quite careful not to 'let it slip' that I was an engineering student." She did not speak openly about her major until she reached senior year, after a good summer job experience finally bolstered her confidence that she had a future in the field. Bailey was the only woman to receive an engineering degree from the University of Illinois in 1950 along with 1,055 men.[79]

As in any era, individual women's perceptions differed, even studying in the same engineering program, simultaneously. Many reported encountering both the good and the bad. Others thrived in the academic work that they loved and derived precious encouragement from friendly, respectful male classmates, friends, family, and faculty. Some, like Bailey, felt more accepted after they "learned to become 'one of the boys.'" Although not all women were comfortable playing that role, the strategy worked for many. Male allies provided invaluable support for female engineering students, both in private and publicly. At Purdue, Tau Beta Pi president Jay Pettit expressed his dismay that in 1961, the national society still had not offered women full membership honors. Pettit declared flatly, "Women have just as much right in engineering as men; they should be treated equally!'" Such an endorsement meant a lot to female students like Margaret Fern Lewin, an aeronautical engineering senior in 1963 who reported that "the fellows in my class have been wonderful to me . . . and I hope to find the same ease in working with men in industry." According to Lewin, about half her professors seemed to hold mixed feelings about having women in class, and the rest seemed "almost" glad.[80]

Female engineering students and their older counterparts in SWE often pinpointed the issue as one of sheer numbers. They understood that as long as relatively few women entered engineering, isolation made campus life harder for almost all of them. In study-ing and doing homework, male engineers could generally find a neighbor or friend to consult, but female engineering students rarely shared majors with their roommates. Curfew rules and other restrictions limited their access to men's study groups, even if they felt comfortable enough to join.

SWE's professional and campus chapters worked to improve the atmosphere and support for female engineering students at college but argued that in order to achieve any lasting impact, it was vital to draw more women into the pool. They increasingly voiced concerns that as early as elementary school, girls were indoctrinated with mes-sages that engineering was for men and that teaching was the best outlet for any women with talents in science or mathematics. Junior high school and high school funneled young women into typing and home economics classes rather than technical drawing and shop classes, away from the advanced math and science prerequisites for engineer-ing. In response both informally and through organized programs, women engineers discussed their work with younger women, teachers, and counselors. Optimists believed that such small measures could ultimately prove transformational. They hoped that as high school women chatted with successful women engineers and recognized that they "don't have two heads," as one put it, these students would realize that the field was not a sad reserve of odd, unfeminine creatures. More than that, advocates anticipated that hearing about how much female engineers loved their work would remind younger women of their broad intellectual options. Illinois's Betty Lou Bailey noted that she received invaluable advice during high school from her engineer father and also two brothers-in-law in engineering. She worried the field was losing young women without her family background who might flounder, become discouraged, or never even con-sider turning their math and science talents toward engineering.[81]

Although many female engineers and engineering students devoted extensive ef-forts to such outreach programs, the number of personal contacts that they could make was limited by time and logistics. To connect with a wider audience, the Society of Women Engineers in the 1960s sent out an increasingly large volume of published ma-terial that provided both encouragement and information about women's place in the field. Among other things, its pamphlets documented that female engineers were not abnormal and were not a bunch of old maids. One SWE study of six hundred female engineering graduates found that 80 percent were married and that 43 percent were working full-time and 10 percent part-time (with the rest being mainly young mothers who were not employed outside the home). Similar results appeared in a separate 1966 survey of sixty-six women who completed Purdue engineering degrees between 1933 and 1964, which showed 2 percent pursuing advanced degrees, 43 percent holding full-time engineering positions, and 3 percent holding part-time jobs, primarily employed by aerospace companies, the federal government, or colleges.[82]

SWE continued to form new campus chapters, such as the one at the University of Illinois that started in 1964. That group, like its counterparts at other schools, undertook a wide range of activities that included hosting talks by engineering professors and representatives of engineering-related industries, touring local factories, and organizing picnics and panel discussions about campus life for incoming female students. SWE members served as student delegates on the University's Engineering Council and helped run the annual open-house weekend. In addition to providing information about women in engineering for visitors to campus, SWE's own open-house exhibits often featured technologies presumed to be of special relevance to women, including one display of new electronic ovens and another on synthetic foods.[83]

To serve as their faculty advisers, many college SWE chapters were able to draw on a small but slowly widening pool of female faculty and research staff in engineering colleges. SWE's Illinois chapter was advised from the start by associate professor of general engineering Grace Wilson, who had a twenty-seven-year career at the university specializing in architectural and engineering mathematics. Female staff at Purdue in the 1960s included Violet Haas, an MIT mathematics Ph.D. who worked in the electrical engineering school, and Paula Feuer, who helped create the school's innovative program in engineering science, the first in the nation to receive accreditation. Not all women engineering faculty were deeply involved in nurturing female students. Competing demands on their time were extensive, and not everyone felt equally comfortable in that role. Their existence, however, served as an important indirect role model, reminding male and female students, administrators, and fellow faculty that female engineers could indeed "make it" in engineering.

CONCLUSION

World War II gave female engineering students at places like Iowa State, Penn State, and Cornell a collective identity and a chance to build up the numbers over succeeding years. In 1949, there were 763 female students enrolled in engineering at schools across the United States, and by 1957, that total had more than doubled to peak at 1,783. On balance, women remained less than 1 percent of total engineering enrollment. But at individual institutions, the difference was apparent. For years to come, the campus climate for engineering women remained chilly and in some programs even toxic, discouraging some female students to the point of dropping out. But the small, growing number who defied the easy path of choosing more traditionally female majors and instead earned their engineering degrees stepped forward as proof that women could survive and thrive in modern engineering education. Their critical mass fostered the formation of both informal support on campus and nationally networked groups such as the Society of Women Engineers.

The World War II experiences of welcoming relatively large numbers (compared to previous years) of women into engineering, both through emergency training and full degree programs, impressed many observers and won converts among university faculty members, administrators, and other observers. Female students gained the position and visibility to speak for themselves and through their comments and organization defined and defended their right to belong in the discipline. Wartime requirements removed formal barriers to coeducation in engineering programs such as RPI and Columbia, although female students there continued to confront tensions related to their outsider status and low numbers. But equally significantly, powerful, embedded associations between engineering traditions and masculinity fostered continued resistance to coeducation at other institutions, restricting women's access to the discipline. Some of the country's most prestigious engineering schools, including Caltech and Georgia Tech, remained male-only when World War II ended. But those single-sex bastions could not indefinitely ignore the question of coeducation. Over the postwar decades, these universities separately confronted pressures to consider admitting women, extending the dialog about gender and American engineering education.

The story of women's increasing presence in engineering during World War II—both in special programs such as Curtiss-Wright Cadette training and in regular degree classes at schools such as Iowa State College, Purdue, and Cornell—made little difference to Georgia Tech. Starting in 1917, Tech had admitted female students to its Evening School of Commerce based outside the main campus. In 1919, Annie Teitelbaum Wise completed her bachelor of commercial studies degree, becoming the first female Georgia Tech graduate. The commercial studies department immediately appointed Wise as an instructor for the following year, making her also the first woman to teach at Tech.[1] Women soon established a small but definite presence in Tech's Evening School, and at least twenty-five women earned commercial studies degrees by 1932. To lead their graduation procession, the 1932 Tech Evening Studies class chose Juliet Dowling, a scholarship winner who finished the five-year program six months early with honors. Dowling later recalled, "When I ordered my graduation ring, I ordered a man's ring. . . . I figured I'd done a man-sized job graduating from Tech."[2]

## DERIDING THE IDEA OF WOMEN STUDYING ENGINEERING

Although the issue of opening Georgia Tech's full-time regular undergraduate programs to women arose occasionally before World War II, the school's leaders and community members shied away from such "agitation" and expressed relief when this troublesome suggestion "seems to have died down." Nevertheless, the potential controversy lingered, rooted in the wording of state laws governing gender and academic admissions. The relevant statute declared, "All the branch colleges of the University of Georgia . . . except the Georgia School of Technology at Atlanta and the colleges for Negroes, shall be open to all white female students of proper age and qualifications, with equal rights and privileges as those exercised and enjoyed by the male students of such institutions, under such rules and regulations as may be prescribed by the board of Regents." Some interpreted that code as mandating exclusion of women from Georgia Tech, but Board of Regents chair and lawyer Marion Smith argued in 1937 that this policy simply did not require their inclusion. He interpreted the language as giving the state Board of Regents authority to decide whether Tech should become coeducational. Smith suggested that the board should avoid tackling such a contentious issue as long as possible, telling Tech president M. L. Brittain, "It is pretty generally agreed that it will be best for us to simply let the matter of admitting women students to Tech rest where it is for the present."[3]

Such a statement coming from the chair of Georgia's Board of Regents sent a powerful message, and undergraduates who attended Tech before World War II appeared equally comfortable with institutional stasis. In a 1939 open forum, the campus Debating Society rejected the proposal "Resolved: that Tech should be coed!" As the student newspaper summarized the outcome, "Tech ladies' men lost the debate [and] the woman-haters won." The most persuasive point centered on "the fact that the attendance of women would lower Tech's scholastic standards," an assumption that apparently admitted no effective counterargument. *Technique* reporters editorialized, "Maybe the debate came out right. . . . Nearly every dormitory bull session comes to the same conclusions, even though some individuals differ from the general opinion. Tech is a pretty good place as it is."[4]

In their conviction that coeducation would necessarily lessen academic rigor, the community effectively defined Tech's purpose as inculcating men with a sense of technical excellence. The opinion that it remained improper for women to embark on traditionally masculine paths was not unique to Georgia Tech. In a 1942 national college poll, 83 percent of all male students surveyed (and 79 percent of all female students) judged it generally a bad idea for women to attempt to combine marriage and motherhood with careers outside the home. Only 32 percent of men believed that colleges should educate women like men; 35 percent wanted more schools to train women to become primarily wives and mothers, while 29 percent felt that the choice belonged to the individual female student. But reservations about women venturing beyond conventional gender roles proved particularly potent at Georgia Tech. By the 1930s, nonnegligible numbers of female students had already entered medicine, law, and science, yet engineering remained largely unchanged.[5]

As elsewhere across the country in engineering studies, however, World War II emergency programs raised the prospect of sudden transformation. Although Georgia Tech did not participate in the Curtiss-Wright Cadette program, the U.S. Chemical Warfare Service chose Tech as one of its first sites to train women to supervise inspection processes in weapons plants. The government selected female trainees from across the South based on their previous success in college math and laboratory science and on personal interviews to measure character. In 1942, thirty young women arrived on campus to take classes on the theoretical principles of munitions quality control, techniques of precision measurement, and metallurgical testing. Public comments on these female wartime students emphasized their patriotic seriousness of purpose, while drawing out the oddity of women's presence in an engineering school. Press coverage highlighted both the women's dedication and their femininity, with photographs showing women in pearl necklaces with sweater sleeves rolled up, absorbed in mastering use of gauges. The *Atlanta Constitution* ran a headline, "It's Really War, Boys! Girls Enroll at Tech," while the *Atlanta Journal's* society reporter editorialized, "Southern girls have traveled a long way from the magnolia and crinoline era! Many of them already are practicing Mrs. Roosevelt's advice to 'go into a factory, learn a skill and be useful.'"[6]

The novelty of women's appearance in Tech engineering caused a stir:

Faculty peer over the rims of their double-lensed glasses with the stricken awe of a hen which has just discovered a doorknob in her nest. . . . Delighted members of the student body leap into the air tossing hats and twiddling their feet like Russian dancers, giving vent the while to robust catcalls and whistles. The faculty members, after the first shock, thought hard for a moment, recalled there is a war going on and relapsed pleasantly into their fog of philosophical acceptance that almost anything can be expected anywhere at any time of a war, day or night. The students, watching the 30 coeds drift by to their classes, felt the inspired happiness of a DeSoto catching his first view of the Mississippi.[7]

This sense of radical change was intensified by use of the "invasion" metaphor. The *Technique* commented, "In a complete blitzkrieg, thirty girls, the first coeds in Tech's history, descended on the campus week before last in an invasion that had even ordinarily staid professors casting furtive glances as they filed to class in the ME building." The *Atlanta Constitution* published a photograph of three women touring Tech's machine shop with the caption, "The first batch of [girl] day students ever to invade the Georgia Tech campus . . . [are] learning war jobs [and] they're working hard."[8]

Tension rose when the "beautiful young things" shared space with male undergraduates, taking classes in the mechanical engineering building. As a jest, one student posted a sign "Defense Damsels Welcome" on the engineering lounge, allegedly leading two "damsels" to unwittingly intrude on a male student who was changing his pants: "A beautiful head appeared . . . ; then there was a scream and the slam of a door. It is needless to say that . . . [a new] sign [has been posted] which reads, 'No Women Allowed— Trespassers Will Be Prosecuted.'"[9]

Yet Georgia Tech's reaction to the "invasion" by women was muted by their status as a wartime curiosity. They were not sitting beside undergraduate men in degree-seeking engineering classes. They received substantive short technical training but never planned to enter the engineering profession. Instead, they prepared for wartime service of a type that many "experts" considered particularly appropriate for women, work that demanded patience and meticulousness. Tech's female trainees themselves promoted this characterization: "An attractive Alabama State graduate claimed, 'We can do this work just as well as men, if not better. We will pay more attention to details.' Their claim is borne out by Professor J. O. King, head of the ME [mechanical engineering] department, who stated, 'Women are much better at the routine jobs in industry, since they are more particular.'"[10] The limited nature of the female inspectors' training program meant that these women's presence had limited effect on the long-term question of Tech coeducation. The school's engineering community was free to maintain its assumption that the average college girl had no interest in engineering.

Following the U.S. Chemical Warfare Service program, Georgia Tech hosted several other special wartime efforts to give women limited technical training. The Evening School offered a wide range of scientific and technical courses for vocational education that included female students. Other women came from across the South to take the three-month Engineering, Science, and Management War Training (ESMWT) program in engineering drafting, elementary mechanics, manufacturing processes, and surveying. They prepared to become engineering aides in the U.S. Army Corps of Engineers or industrial technicians with companies such as Douglas Aircraft, which were eager to hire trained women. The U.S. Coast Guard sent two hundred SPARS, members of its women's auxiliary, for training at Georgia Tech, while fifty U.S. Navy WAVES studied to become operators of the Link flight trainer. These military women joined men for three daily meals in the Tech dining hall but lived off-campus at the Biltmore Hotel. Male undergraduates occasionally referred to these temporary female students as "Tech coeds," but school officials rigidly policed the boundaries demarcating campus life. After the band director proposed adding majorettes to performances at athletic events, the team coach "tersely replied, 'No girls.'"[11]

The Georgia Tech community noticed that wartime trends had brought women into engineering at other schools. In a 1945 front-page story (headlined "Women Invading Engineering Field"), the *Technique* remarked on the presence of a freshman female engineering major at New Mexico's College of A&M: "No longer can the fellow who wants to get away from the gals hide behind a slide-rule. . . . Now he's likely to find the gals there before him. In fact, about the only course left to the woman-hater is military science." Indeed, Georgia Tech's own registrar received information requests from interested high school women on almost a weekly basis: "Most of the girls who fancy themselves as potential Rambling Wrecks usually give up and go elsewhere when they are told that Tech is for men only. But a short time ago, a cute little blonde from North Carolina decided that, rules or no rules, she wanted to study engineering—at Tech. She wasted a lot of money and kept the wires hot talking to the Registrar by telephone — and we are still waiting to see her hit the campus."[12]

Women interested in Georgia Tech degrees were wasting their time making inquiries, as long as the school preferred to duck the issue. Even as record numbers of women studied technical subjects in special wartime classes or pursued engineering degrees at coed colleges, Tech's male undergraduates treated the notion of coeducation as a joke. The *Technique's* 1944 April Fool's issue offered a mock announcement that Tech had decided to admit women and would remodel the cafeteria to add dollar-per-hour "private booths for smooching" and "stock up on lipstick and powder puffs, plus many unmentionables." The student newspaper published weekly pinup cartoons that categorized women as desirable sex objects while undermining any notion of them as serious engineering students. One such drawing showed a kneeling girl in a short skirt and tight T-shirt that had "Georgia Tech" written on the curvaceous front; one hand held a slide rule, while the other curled behind her head to accent her breasts. The accompanying poem read, "Figures don't lie; the answer's here! Tech's Coed—I'm an engineer!" The paper encouraged readers to "drool" over another cartoon, under the headline "Things to Come?" that similarly showed a young lady in a tight sweater with "Georgia Tech" lettering highlighting the bosom. An accompanying article discussing possible postwar changes on campus asked, "Are we prepared to sacrifice the privacy of our manly institution?"[13]

Indeed, the postwar Georgia Tech student body defined its male engineering identity partly through ridicule of coeducation. Humorists portrayed women as naturally unable to handle serious technical study. A cover drawing for the Tech humor magazine showed a panicking female student desperately trying to cheat her way through a test by peeking at formulas scribbled on her thigh under a short skirt. The *Yellow Jacket* devoted an entire 1947 issue to mocking the notion of female engineers. Two full pages of cartoons depicted feminine incompetence in the laboratory, showing women wearing tight sweaters and short skirts and getting long hair caught in a shop machine.

Portraying women as flighty flirts who did not take college studies seriously, other cartoons showed female students powdering their noses while chemistry experiments boiled over and using hydraulic testing machines to crack pecans. To emphasize that

women's presence would destroy Tech's all-male atmosphere, one cartoon showed a scrawny skinnydipper in the campus pool exclaiming, "Ye gods! Coeds!" as he spotted women in tight bathing suits sitting at the side. The *Yellow Jacket* defined women strictly in sexual terms. After disparaging blind dates from Wesleyan, Agnes Scott, and the University of Georgia as "zombies" and "monsters," one columnist declared that he would support Tech coeducation only if female entrance standards mandated a thirty-eight-inch bust. An accompanying fake advertisement read:

The Georgia School of Technology is now accepting applications of qualified young women to enroll in the following courses: aeronautical engineering, . . . civil engineering, mechanical engineering, industrial management, millinery design, cosmetic engineering, nylon synthesis, basket weaving, how to win a husband. Minimum requirements for acceptance: algebra 2, trigonometry 1/2, . . . waist 24, foreign language 1, chemistry 1, bust 34, physics 1, hips 33, shop work 1. A full-length picture in bathing suit or form-fitting dress must be submitted with application.[14]

As a publicity stunt for its issue mocking engineering coeducation, *Yellow Jacket* editors arranged for fifty high school women to converge on Tech's engineering-school office and demand applications. "The unheralded assortment of pulchritude," reported the *Atlanta Journal*, "received a stern, 'I'm sorry' from William L. Carmichael, director of admissions." The *Yellow Jacket* then used a photo of this crowd of pretend female applicants to illustrate its poem, "Ten Little Coeds":

Pretty little coeds, signing up at Tech,
Worried Mr. Chapin said, "I'm a nervous wreck."
Ten little coeds, registration line,
One was lacking chemistry, then there were but nine.
Nine little coeds, hunting for a mate,
One grabbed a quarterback—that left only eight. . . .
Seven little coeds; one was in a fix,
Stumbled over algebra—that left only six. . . .
Four little coeds, cute as they could be,
One blew up in chem lab, that left only three.
Three little coeds, one with heart so true,
Married a professor, then there were but two.

◀ FIGURE 4.1
In the years when Georgia Tech remained all-male, its humor magazine enjoyed joking about how ridiculous coeducation would be. Cartoonists assumed that women were inherently incapable of handling engineering studies and suggested that a female presence would undermine the quality of a Tech education. *Georgia Tech Yellow Jacket*, March 1947. Used with permission.

Two little coeds, having lots of fun,
One got drunk, forgot herself—one from two left one.
Single little coed, where there were so many,
Graduated finally, then Tech hadn't any.
Ten more little coeds, asking where and when.
Worried Mr. Chapin says, "You starting *that* again?"[15]

In their humor stressing that women simply did not belong in engineering, Tech men poked fun at their own stereotypes. A 1947 *Technique* column calling coeducation "clearly [an] invitation to disaster," opened, "Coeds at Tech? Never! May the powers that be forbid . . . bring[ing] the pitch of feminine voices into our hallowed classrooms. Why lower our noble institution to the gutters alongside that thing about 70 miles east of Atlanta? With the inroad of petticoats at G.T., we might as well change our name to the Georgia School of Gynecology." The column proceeded to complain that with "invasion of the opposite sex," men would "have to wash our faces every week." More than that, "those silly women would even do their assignments," sending professors into shock and destroying the hopes of men who were accustomed to squeaking through thanks to teachers who graded on the curve. In conclusion, the *Technique* rewrote lyrics to the school song:

Oh, I'm a student coed from Georgia Tech,
And a powder-puff engineer,
A motherly woman, and chastely, maidenly feminine engineer.
Like all other numerous females
I indulge in my one-percent beer,
'Cause I'm a modest coed from Georgia Tech,
and a powder-puff engineer.[16]

But underneath such jokes, postwar Tech undergraduates and, by extension, the Georgia Tech community valued school tradition and high-quality technical education, both of which were judged incompatible with coeducation. A 1947 poll of Tech students showed that while 58 percent felt they were missing a substantial part of college life by attending an all-male school, 63.7 percent opposed admitting women to Georgia Tech. Almost 54 percent said that a female presence in classes would be distracting.[17]

PRESSURE FOR TECH COEDUCATION

But during precisely this time, a direct challenge to Georgia Tech's single-sex status was gathering momentum. In 1943, high school senior Anne Bonds, daughter of an Atlanta electrical engineer, decided that she wanted to follow in her father's footsteps and attend Tech to study engineering. After the all-male school denied her application, Bonds began attending Alabama Polytechnic Institute in Auburn but continued her fight for admission to the only public school that offered engineering in her home state. Bonds turned to local women's groups for support, and the Georgia Federation of Business and Professional Women's Clubs joined her in petitioning the University System's Board of Regents to make Georgia Tech coeducational. Although wartime priorities and bureaucratic procedures delayed consideration, the Board of Regents in 1947 publicly directed its Education Committee to work with Tech officials on reviewing the feasibility of admitting women as regular undergraduates.[18]

Regents chair Marion Smith repeated his prewar opinion that the wording of Georgia education law did not explicitly ban admission of women to Tech, thus giving the regents the discretion to order the school to become fully coeducational. Not all observers agreed with this assessment of the state statutes. Georgia Tech dean Phil B. Narmore maintained that the school's original charter specified that admissions be limited to men, meaning that any move to coeducation would first require that the state legislature amend the charter.[19]

News of the regents' potential openness to coeducation surprised current undergrads, whose identification with Tech was closely tied to its single-sex climate and belief in the masculine nature of engineering. A number expressed concern that "I can't study around 'em [women]." One declared, "If women desire a technical education, I think they have a right to attend Tech. However, it would be hard to study." The *Atlanta Constitution* summarized the reaction as "Co-ed Plan Takes Tech Aback: What! What! Rah! Rah! Woe! Woe!"[20] In what it dubbed the "battle of the coeds," the student newspaper promptly wrote an editorial listing the reasons for "a big, bold, emphatic NO!" It warned that coeducation would create multiple problems, including the need to rearrange classes and dormitories, an obligation to impose strict behavior and speech rules, the risks of academic distraction and "unavoidable professorial favoritism," a trend toward less specialization, and significantly for Tech fans, the disaster of worse football seats. The column concluded, "With the advent of a whooping, light-headed, irresponsible bunch of females go the traditions, glory, and leadership of Georgia Tech. . . . [T]he value of your diploma depends upon the reputation of your school. If these traditions, that glory, and that highly-vaunted leadership are to be set aside . . . you will have suffered a personal loss. Choose now, her personality or your personal loss." The *Technique* published a cartoon showing a man (labeled "Regents") looking with puzzlement at a plug labeled "coeducation at Tech" and three sockets marked "benefits to state," "benefits to school," and "benefits to students," while a young girl asked, "Won't It Fit Any Place, Daddy?"[21]

Meanwhile, Blake Van Leer, former dean of engineering at the University of Florida and at North Carolina State, replaced Brittain in 1944 as Tech's president. As Tech's new leader, Van Leer issued a memo to various school officials, including deans, the athletic director, and the (female) librarian, asking these "gentlemen" to serve as a committee to study the legalities and practicality of coeducation. Dean Narmore indicated that admitting women would require expensive and time-consuming changes to the dorms, bathrooms, athletic programs and swimming pool, and hospital and counseling staff. Moreover, the postwar education boom had left all but six Tech majors already overcrowded.[22]

Regarding the question of whether Tech coeducation was required or even permissible, Georgia's deputy assistant attorney general, Hamilton Lokey, told Tech officials that he could "find no law which limits the student body of the Georgia School of technology to male students, or bars women from the student body." Contrary to the opinion of Narmore and some others, Lokey stated that the original 1885 act establishing Georgia Tech was "silent as to the sex of the student body." Lokey concluded, "It is apparent . . . that, while there has been no legal enactment limiting the student body of Georgia Tech to men, there has developed over the years a strong policy to that effect." His opinion seconded that of regents chair Smith that the Board of Regents had the power to decide whether to open Tech to women.[23]

As it happened, the Georgia Federation of Business and Professional Women's Clubs was continuing to advocate for admission of Anne Bonds. In a June 1947 vote, the regents declined a proposal to make Georgia Tech (and also the two state women's colleges) coed. The women polled each board member to explain their vote. Smith attributed the regents' decision primarily to concerns that Tech had no dormitory housing available for women and was already "fearfully overcrowded," adding that a future board might vote otherwise if it saw that school facilities "did not present such a difficult practical situation." Smith declared that he had voted for coeducation because "it seemed to me that it was more important to establish the principle involved." Other regents confirmed that their objections centered on space issues. William S. Morris wrote that Tech's "crowded conditions" made it "unwise" at present to admit women. Sandy Beaver wrote that he was "definitely in favor at this time of permitting the comparatively few Georgia women who are interested in engineering education to attend classes" at Tech but not to live in the dorms. Other regents agreed that the state should "provide some way for Georgia women to secure an engineering education if they so desire" and even said that "there is no doubt . . . that women will eventually be admitted" to Tech but not "until conditions became a little more normal." H. L. Wingate felt "very much in favor" of making Tech coed but opposed the measure since it also required admitting men to Georgia's female colleges. Regents Frank Spratlin and C. Jay Smith declared their opposition to Tech coeducation, especially "if by so doing Georgia Tech will not be able to take care of all male students."[24]

The 1947 vote against coeducation left Tech men free to continue joking about the ridiculousness of female engineers. A 1948 *Yellow Jacket* columnist wrote that arrival of "little girls all over the place" would be "nothing short of devastating," sending mechanical engineering professors into despair over how to explain to innocent women the difference between male and female pipe connections. The column joked that chemical engineering faculty intended to frighten away women by delivering two-hour lectures on "Crushing and Grinding, Size Separation and Countercurrent Decantation of Titaniferrous Ore," while a prominent math professor "was prepared to transfer to Cal Tech . . . rather than admit a single co-ed to one of his classes" because "compared to women a bunny was a mental giant."[25]

During the postwar years, women continued to attend Tech's evening school, studying subjects that included mechanics and architecture. Photographs showed those "Tech coeds" wearing school hats and waving Tech pennants, implicitly tweaking those who identified school spirit as solely and necessarily male. Meanwhile, in fall 1948, Barbara "Bobbie" Hudson became the only woman to join 285 men at the newly created Georgia Technical Institute, a part of Georgia Tech's engineering extension division that was intended to train draftsmen, inspectors, and other industrial technicians to boost the Southern economy. Hudson reported initially receiving "curious states" from male classmates. "I felt awkward—like a canary in a goldfish bowl," but "everybody's been so nice, . . . I couldn't help being happy here" after a few months. "The boys tease 'Hudson' about everything, especially the big question in their minds as to whether she'll get married before she finishes the Institute." Even her grandmother reportedly wondered whether outsiders would assume that the young woman had signed up for technical studies merely to find a husband. In telling Hudson's story, the *Atlanta Journal* remarked, "If Bobbie's only ambition were to get married, . . . she could certainly find an easier way than via logarithms, Ohm's law, and electronics," adding that after finishing her courses at age eighteen, "she'll have plenty of time later to look for a husband." Much of the ribbing hinted that attending technical classes with men risked turning a girl into a boy. Former high school friends asked Hudson whether she had joined a fraternity, and one date took her to a construction site as a joke. To reassure readers that this female

Tech student remained domestic-minded, the *Atlanta Journal* emphasized that Hudson sang, cooked, and was sewing a dress for her baby sister. The Technical Institute's student paper preferred to focus on her physical femininity, calling her "a 1952 Streamliner Model—a small 4 ft 11 1/2 incher with . . . bright eyes which tune in like radar and . . . dimples in each cheek." At least some teachers accommodated supposed feminine sensibilities by quickly apologizing for using slang in class. Hudson said that being "willing to overlook some things she hears" helped a woman get along well at a male school: "Engineering lingo has been toned down considerably for Hudson's sake. . . . Instructors think she's also a good influence on grades—the men don't want a mere 'bobby-sox Bobbie' to get ahead of them." Building-design instructors valued Hudson's presence as serving to remind the male future draftsmen with "the 'woman's point of view'. . . what a housewife wants." Hudson was saving her money to pursue an architecture degree at an out-of-state engineering college unless "maybe Tech will open its daytime doors to women by then. . . . I don't blame the boys for howling, but if a girl happens to like engineering and hasn't got the money to go out of state, she's stuck!"[26]

Hudson's comment underlined the message of the Anne Bonds case, that Georgia female high school students who were interested in earning engineering degrees did not have any option to receive public in-state training. Bonds transferred to Washington State College for her junior year in 1947, maintaining that she still hoped to enroll at Georgia Tech as a senior. When this remained impossible, Bonds applied to Georgia's regents for reimbursement of the out-of-state fees she had been forced to pay at Auburn and Washington. The board's own analysis admitted that Bonds had spent at least $126.07 in extra expenses but worried that complying with her request could set dangerous precedent: "If this differential is paid, the Board of Regents will receive requests from other women students attending out-of-state institutions taking courses offered by the Georgia Institute of Technology."[27] Georgia Tech's controller supported Bonds, arguing she had "every legal right to collect the difference" in cost by being forced to pursue her education elsewhere. But in opposition, regent Sandy Beaver argued that "insufficient facilities or appropriations" legitimately left the state unable "to furnish instruction to every citizen of Georgia who asks for it." The board ultimately voted to deny Bonds reimbursement.[28]

Deliberations over the Bonds case ensured that the question of Tech coeducation did not vanish. Crucially, Tech's new president proved open to change. Van Leer wrote:

My personal feelings sway me in one direction, whereas my official position influences me another way. . . . I have been associated with co-educational institutions practically all of my life, and I have always felt it was wrong to discriminate against a student because she happened to be a woman. . . . Miss Bonds . . . is obviously a Georgia citizen and a qualified and responsible engineering student; this makes it seem wrong to me for her to be denied an engineering education in her native state simply because she is a woman. On the other hand, Georgia Tech is traditionally a man's school.[29]

Van Leer's personal inclination toward coeducation connected strongly to the women in his life. His wife, Ella, had completed an architecture degree from the University of California at Berkeley and worked both in commercial illustration and wartime army design and drafting. The Van Leers' daughter, Maryly, had just finished high school and would have liked to study engineering at Georgia Tech but taking that as impossible, enrolled at Vanderbilt. The Van Leers' son, Samuel, who entered Tech in 1952 and earned a civil engineering degree in 1956, later recalled, "The main concern for the men was, when you were a Ramblin' Reck from Georgia Tech, it was hard to picture a female. . . . Dad always was progressive. . . . He wasn't concerned about male students and female students, he was concerned about quality students. He could imagine a Ramblin' Reck from Georgia Tech being a female."[30]

Van Leer knew and admitted that "a large portion" of Tech's current students, faculty, and alumni favored keeping their beloved school as single-sex. But he emphasized that the state had an obligation to serve high school women who wanted an engineering degree. Given that it would be impractical to set up a special school just for "the small number of qualified women" who were interested in engineering, he said, these women "must eventually be admitted to Georgia Tech." Van Leer assumed that only a few women were suited to engineering and suggested that this actually represented an advantage, a way to defuse potential conflicts over coeducation: "If women are admitted to Georgia Tech, we will never have more than thirty or forty of them here at any one time, and since they will be in such small numbers, they will present no particular problem." He stressed that Tech should remain an engineering school and not copy the University of Georgia in adding liberal arts or worse, home economics programs, which might bring "a flood of women students which would seriously embarrass us because we would have to provide women's dormitories, gyms, etc."[31]

Van Leer's ruminations reflected his expectation that, although Bonds was about to complete her engineering degree out of state, outside pressure for coeducation would not cease: "I have heard rumors . . . that the women's organizations in Atlanta are raising funds with which to carry through the courts the question of compelling the Regents to admit women to Georgia Tech. . . . The Attorney General . . . tells me that if they do . . . , they are certain to win." Van Leer accordingly began analyzing the practical changes that Tech would need to make to accommodate female admission, starting with the bathroom situation. A tabulation of existing facilities showed that the number of available toilets was already inadequate, especially in older, cramped buildings. Campus engineers estimated that at $5,000 each, adding enough lavatories to accommodate undergrad women in ten of the university's sixty buildings would cost at least $50,000.[32] But Van Leer maintained that Tech could handle such necessary physical-plant adjustments given at least six months' lead time and funds for new construction. Lloyd Chapin, Tech's dean of faculties, agreed and told Van Leer, "Personally, I see no sound objection to our admitting women students, other than an objection—more real than logical—to breaking a historic tradition." Chapin reminded the president, however, that the "attitude of faculty, student body, and alumni toward co-education is a mixed one, and I do not believe that anyone at the present time is accurately informed as to what the reaction would be." Indeed, the school's sports fanatics had even objected to a 1949 student council proposal to add female football cheerleaders (borrowed from nearby coed or women's colleges) to the all-male cheer squad. Tech's athletic director commented, "This is a boys school. Boys play football, boys play in the band, boys lead cheers. I don't see any sense in bringing in girls to help them." The *Atlanta Journal* added, "Old grads have taken pride in the strictly stag character of Tech athletics and . . . are depending on 'The Old Man' to keep the picture clear of swirling skirts."[33]

Meanwhile, Georgia Tech kept advertising technical education classes at its Engineering Evening School as "available to men and women," especially promoting opportunities for employment in the growing cold war defense industry. In 1950, the Georgia Society of Professional Engineers admitted its first female member, Anne Bonds, who had completed her electrical engineering degree from Alabama Polytechnic Institute and was employed as an engineer-in-training with the Georgia Power Company. The *Atlanta Constitution* editorialized, "Engineering with a feminine slant is bound to bring a fresh viewpoint to her fellow members of GSPE, to say nothing of the boost in their beauty status." In 1950, Barbara Hudson also finished her certificate in building construction technology at Tech's auxiliary, the Georgia Technical Institute, commenting, "Only one or two of the teachers made me feel like I had no business there. . . . The rest were wonderful. So were my fellow students. . . . Of course they are always telling me to go out of the room so they can tell dirty jokes." Describing the experiences of the "winsome sultan with a harem numbering 389," the *Atlanta Constitution* reported that Hudson's numerous activities had kept her too busy for dating and unable to reap the benefits of her "male paradise." The newspaper also joked about whether Hudson's studies of concrete mixing would "help her cooking."[34]

In February 1952, promoters of coeducation made their next move. The Women's Chamber of Commerce of Atlanta passed a resolution urging Georgia's University Board of Regents to end the ban on female admission to Tech: "The continued policy . . . constitutes a gross discrimination to the females of Georgia in that it requires any female desiring to matriculate in the fields of engineering and/or architecture to seek facilities out of the State of Georgia for such study, resulting in additional expense . . . in tuition." According to later accounts, Mrs. Van Leer played an influential role behind the scenes in encouraging the Women's Chamber of Commerce to press the matter. The women argued that recent construction at Tech had solved problems of overcrowding, stretching capacity to at least five thousand students. Given an assumption that few women would be interested in or qualified for engineering studies, advocates asserted, newly built bathrooms and apartments could easily absorb that small number. Amber Anderson, who drew up the women's motion, said, "I doubt if there would be one hundred women in Atlanta, or in the whole state of Georgia for that matter, seeking engineering degrees."[35]

The Board of Regents subsequently referred the women's petition to Van Leer, who surveyed current students, faculty, alumni, and school officials. Tech administrators seemed to favor coeducation. They noted that Tech already had many "young and attractive" women working on campus as clerks, stenographers, cashiers, computers, waitresses, technicians, and nurses, with "no unseemly incidents" or risk to morality. Tech leaders also worried about the school's recent steep drop in enrollment, reflecting both the end of the GI Bill veterans' rush and a shrinking population of college-age students. Admitting women could offset that decline, optimists suggested, even given assumptions that relatively few female students would be attracted to or qualified for engineering studies.[36]

Student opinion remained divided but suggested a slight momentum favoring change. Tech's student council, by a thirteen to nine vote, adopted a motion urging the regents to allow women's enrollment "on the same basis as male students" because the school now had room and the ban posed an injustice to female high school students who sought in-state engineering education. The council's decision fanned dissent among students who denounced the vote as misrepresenting the bulk of undergrad feeling. Talk in "a number of 'bull sessions,'" one critic asserted, proved that "a large number of students are against this move," fearing "lowered standards" and a loss in "freedom of classroom relations." He warned that admitting women necessarily meant institutional decline and that any women interested in attending Tech would be less than desirably feminine: "I am against co-eds in any form (or lack of form) whatsoever. . . . For the few women (neuters?) we would get, why mess up Tech? Personally, I want my diploma to be worth something in years to come. I do not believe this value will be retained when co-eds arrive (bags and baggage). Put me down as a charter member of the KKK Anti-Coed Division."[37]

The *Technique*, which in earlier years had mocked the notion of female engineers, now favored their admission, although with lingering concern that coeducation, handled improperly, risked institutional loss of prestige. The editors (like many other observers) considered it unfair that Georgia paid the differential expenses for African Americans who were forced to study engineering at out-of-state schools (due to exclusionary rules in-state) but refused to reimburse women. Making Tech coed would redress such "discrimination," the paper said, while creating "few disadvantages" for current undergraduates: "Such things as having to make a better personal appearance, having to control the vulgarity in our speech, etc., while needing to be considered, carry very little weight." Given that Tech now owned sufficient space, the paper said, the school should add women "IF they are admitted on the same basis as male students, and IF, no 'special' courses are ever instituted for the female students that might lower the standards, or mar the fine reputation that Georgia Tech now enjoys as an all-male institution."[38]

The Tech alumni magazine similarly editorialized in ambivalent fashion in favor of coeducation. The editors acknowledged that Tech "has always been a sort of male sanctuary. . . . Some of the alumni will not favor a break in tradition, but someone always is opposed to change no matter how sound the reasoning." The alumni secretary sought to reassure readers that the school would not be overwhelmed by female would-be engineers: "I do not think that there will be any great stampede on the registrar on the part of women wanting to take the courses offered here at Tech. My thinking is that there will be only a few to begin with; possibly 10–15. The figure of 50–100 is being bandied about, but I doubt if we will get that high any time in the near future." Especially with such low female numbers, he declared, coeducation would not destroy the institution, "as long as the entrance requirements are as tough as they are. Should some [women] . . . come here as a lark, or looking for husbands, they certainly would not last long."[39]

At a faculty meeting, Tech's professors, with just two or three exceptions, voted in favor of coeducation. Noting a similar preponderance of opinion among administrators, Van Leer wrote to Harmon Caldwell, chancellor of Georgia's University System, on March 6, 1952, with his official recommendation that "qualified women students be admitted to the Georgia Institute of Technology on exactly the same basis as male students are now." Van Leer noted that the 276 women on campus as secretaries and staff "presented no problems whatsoever; in fact, they have become almost indispensable

. . . competent, loyal, and industrious. Flirtatious, unseemly conduct by our women employees is practically non-existent." He reminded the regents that almost all other state-owned engineering schools in the country already admitted women. Van Leer reassured the regents that as long as Tech maintained admission requirements of three years of high school math and one year of science, coeducation would remain minimal:

Not many women students are interested in or have an aptitude for these subjects. . . . I seriously doubt if more than fifty women would apply immediately for admission, and I also doubt if the number ever enrolled would exceed one hundred fifty. Hence, in a student body numbering from 3,000 to 5,000, this small number would not present any serious problems.[40]

Finally, Van Leer cited the inconsistency that Georgia had "been much more liberal with the negro students than it has been with the women," giving African Americans financial aid for out-of-state study: "This is a discrimination based entirely on sex, and it should be eliminated as soon as our material and educational resources would permit. I believe that time is now."[41]

To overcome lingering resistance on the board, Van Leer encouraged the Women's Chamber of Commerce to write letters to individual regents. In turn, the Women's Chamber urged the Georgia Association of Women Lawyers, Soroptimist Club, Business and Professional Women's Club, National Secretaries Association, the University Women's Association, the League of Women Voters, and other women's organizations to lobby the regents.[42] As further ammunition, Van Leer asked Dorothy Crosland, Tech's main librarian, to document that Georgia was out of step with other states that already offered engineering training to women. As Crosland later recalled, the president told her, "I now turn the matter over to you—see what you can do to get women into Georgia Tech." Crosland retrieved statistics from the Engineers' Council for Professional Development, showing that out of 147 schools nationwide with accredited engineering courses, only twenty-eight excluded women (and four of those were military schools). The other twenty-four male-only schools were located where women could find engineering instruction at alternative in-state institutions. For instance, California had two all-male schools teaching engineering (Caltech and the University of Santa Clara) but four other schools that allowed female students (Berkeley, UCLA, USC, and Stanford). "Georgia is the only state that does not have a school, either private or state controlled, offering architecture or engineering to women," Crosland wrote. "Should women be barred from contributing to the advancement of science and industry in the State? . . . The State is attempting to equalize education for negroes; should it not provide equal opportunities in education for women?" Crosland told the regents that it was ridiculous for Tech to hire female technicians and researchers, who received degrees elsewhere, to work at its engineering experiment station and aeronautical-engineering wind tunnel, while refusing to admit them to Tech's classes. She connected the issue of Tech coeducation to a broad feminist stance, noting that women comprised a majority of the U.S. population and a majority of voters, owned a majority of the country's wealth, and yet still did not receive fair treatment. Even the workplace had changed, she wrote, with almost one woman at work for every two men: "More and more industries are employing women for jobs formerly held by men. . . . If men and women can work together, the idea that they cannot study together successfully is archaic and unnatural." Crosland mailed the regents copies of articles promoting equal rights for women, especially in science and the professions. She echoed the assumption that Tech should not fear a flood of females taking over: "because of our requirements for entrance, the possibility of Georgia Tech ever having many women students is very slight."[43]

At the request of the regents, Crosland contacted fourteen coed schools to collect further statistics on female enrollment in engineering. Cornell's registrar answered that the school currently had eight women in engineering (amid 1,532 men); the University of Illinois had four female students registered in engineering, Ohio State had five, and the University of Texas had nine. The highest figures came from Iowa State College (nineteen female engineering students) and MIT (with thirty-seven, counting fifteen in architecture as part of its engineering school). Penn State had thirteen female engineering students (including eight in architecture), Alabama Polytechnic Institute had two women in engineering, and the North Carolina State College of Agriculture and Engineering had four. Virginia Polytechnic Institute, the Carnegie Institute of Technology, and Louisiana State University each reported having a lone female enrolled in engineering, while Tulane and the University of Iowa had none at that moment. Crosland concluded, "The picture is the same throughout the country. Few women have the requirements for engineering." Reminding regents that a total 563 women had enrolled in engineering at all schools nationwide in 1950 (versus 142,391 men), Crosland wrote, "If our alumni think that co-education would mean another university like the University of GA, they have only to look at the engineering enrollments."[44]

The Board of Regents' Committee on Education, headed by Rutherford Ellis, accepted these arguments that Georgia should not remain the sole state to offer women no opportunity to study engineering. Yet one final twist remained. The motion that the Education Committee recommended for approval to the entire board stopped short of mandating wholesale coeducation of Georgia Tech admissions. Instead, the proposal called for limiting female enrollment solely to Tech's engineering programs to pursue only those degrees that were otherwise unavailable to women in-state. The motion further specified that women must meet academic entrance requirements that were identical to those for men and that Georgia Tech must not create "any courses designed primarily for women" or otherwise "depart from its engineering curricula."[45]

Even carrying those detailed restrictions, the motion evoked "bitter" responses from opponents, setting off "one of the most heated debates in recent sessions." Ellis fought fiercely to defend the motion, calling coeducation at Tech "practically inevitable" yet also relatively minor in effect, given the consensus that few female students would be interested anyway. Ellis commented, "Don't envision the campus being overrun with women." Regent chair Robert O. Arnold defended the all-male tradition, asserting that he had spoken with "a number of leading Tech alumni" and found none who wanted coeducation. More than that, Arnold warned that coeducation must inevitably alter the quintessential nature of Tech and that a female presence would dilute the focus on engineering excellence: "I'm afraid the moment we get women on the campus they'll be coming in and saying we've got future mothers on our hands and we ought to prepare them for it." Regent Edgar Dunlap expressed that concern more pointedly, maintaining, "Here is where the women get their noses under the tent. . . . We'll have home economics and dressmaking at Tech yet." At its April 9 meeting, the regents voted seven in favor and five against to allow women's admission to a limited set of Georgia Tech's engineering programs.[46]

Praising the regents' "wise" move as a "matter of equity," Van Leer publicly credited local women's groups for helping to end discrimination against Georgia women at tax-supported schools. He also privately thanked Ellis for his "very valiant work" in fighting for change as "the just, fair, equitable thing to do." Van Leer wrote, "I don't want you to think for a moment that I am not aware that problems will arise. We have them with the male students, and we may have more with the women in proportion to their number until a pattern for dealing with them has been worked out." But the president felt confident of finding solutions to any difficulties and making the change ultimately "successful."[47]

Van Leer confirmed that for the next term, Tech would start accepting women who wanted to pursue engineering or architecture degrees and added that he expected no more than twenty-five female students to enter in 1952, given that few high school women took enough math and science courses to qualify. Van Leer dismissed fears that this small female presence would destroy male students with "social frivolities" on campus or that Tech would destroy women students' natural and desirable femininity. Citing the personal experiences that his wife had studying architecture and his daughter had earning a chemical engineering degree, Van Leer declared that "constant association with a majority of male students didn't make roughnecks out of them."[48]

Female students who arrived at Tech in 1952 would not be allowed to earn bachelor's degrees in industrial management, physics, or chemistry, since other in-state schools like the University of Georgia already offered women those options. The regents limited coed enrollment to bachelor's degrees in mechanical engineering, electrical, civil, ceramic engineering, chemical, aeronautical, industrial, and textile engineering. Van Leer cautioned Tech's registrar:

Please follow these instructions literally and specifically, and under no circumstances are any exceptions or deviations to be made. . . . We must be very strict about this matter. If any questions arise about the admission of a young woman, the case should be decided against her. We want to be absolutely sure that they meet all of the entrance requirements that we prescribe for men, and no women are to be admitted with any entrance deficiencies whatsoever.[49]

Student opinion remained divided, and the change caught many current undergraduates by surprise. Such students worried that coeducation would burden them with oppressive new rules, while undermining campus standards and eroding the value of a Georgia Tech degree. Men often held two assumptions that female engineering students would be forced to fight—that women generally were incapable of handling the challenge of engineering class and that women were more interested in husband hunting than serious study. "It's just the normal male attitude (egotism) to think that girls can't make the grade at Tech," the Student Council president said, but "we don't mind having women on the campus. They have a right to earn an engineering degree if they can meet the qualifications." The senior class president added that if women "come here to study student engineers instead of engineering, they won't stay long." The *Technique* wrote, "The students are determined that no tradition be changed for members of the fair sex. When they show they have the ability, then they'll be accepted as one of us."[50]

PARTIAL COEDUCATION: TENSION AND WOMEN'S ADJUSTMENT

The coeducation decision created a field-day for observers to make cracks that both reflected and perpetuated an inability to take seriously the notion of women as good Tech engineering students. The *Rome News Tribune* wrote that "it won't seem right to hear soprano voices singing, 'I'm a ramblin' wreck,'" while the *Griffin Daily-News* proposed new lyrics for that school song: "I'm a ramblin' wreck from Georgia Tech, I keep my lipstick near." Joking about the difficulty of telling which female applicants "want to pursue engineering and which ones want to pursue engineers," that paper continued:

The ladies, bless their sweet hearts, have just about taken over all the citadels of man. . . . I hope that "related work in engineering" will include how to put a new washer in a leaky faucet, how to install a fresh fuse in place of one that's blown and how to remove the appliance plug from the wall receptacle without yanking on the cord.[51]

The *Atlanta Constitution* editorialized in favor of the change, declaring that in modern life, "when skills are so heavily in demand, there is no logical reason why women should be prohibited from training in any vocation in which they are capable of serving." But an accompanying illustration focused on the physical attributes of female would-be engineers, with a drawing of a curvaceous woman who was wearing a tight blouse and high heels and using a transit.[52]

At the time of the regents' decision, Tech had no female applications in hand, but within four days, the first interested woman stepped forward. Reaction highlighted both her intellect and her "golden-haired, blue-eyed" prettiness, and commentators employed the predictable "invasion" analogy. The *Atlanta Journal and Constitution* wrote, "A petite blonde is the first woman to attempt invasion of the home of the 'Ramblin' Wreck' since that male stronghold became coed last week. She is golden-haired Mary Joan Coffee, . . . who makes it plain that she is going to Tech strictly to study and not to look for boy friends." Other press coverage described Coffee as "a blonde sweater girl . . . preparing to brave the no-woman's land." Coffee had intended to attend Vanderbilt, but as both the daughter and niece of Tech men, "weaned on a slide rule and T square," she quickly assembled a Tech application. The press used Coffee's interest to personalize and dramatize the impending revolution in campus culture. The *Atlanta Constitution* ran a photograph showing Coffee walking past Tech's bell tower as a male student swiveled his head all the way around to stare at her. A *Constitution* cartoonist drew a picture showing women's lingerie hanging out to dry on a clothesline slung from Tech's campanile.[53]

Alumni sentiment increasingly (although still often grudgingly) acknowledged that "whether we like it or not," change at Tech had arrived. "You'll have to admit" that the state was discriminating against women, the *Georgia Tech Alumnus* told readers:

The fragile flower of humanity, long represented on our campus only by snapshots in an engineer's wallet, are taking T-squares in hand and heading straight for one of the greatest masculine strongholds in our country today—your Alma Mater. . . . Now that everyone is getting used to the idea, most of the talk has died down. The prevailing sentiment seems to be, "We've got 'em, God bless 'em."[54]

The publication downplayed the threat of radical upheaval, emphasizing that female enrollment was allowed only in limited fields, that the regents stipulated that there would be no change in curriculum or scholarly expectations, and that Tech administrators felt sure that no more than twenty-five women would appear in the near future: "To ward off your worst fears . . . , we hasten to add," that as of May 1952, Tech had received applications from a mere four women.[55]

Those low numbers might reassure skeptics who wanted to defend Tech's male tradition, but Van Leer found them disappointing. Arguing that it was not only possible but desirable to train more female engineers, Van Leer used a July 1952 speech to the Regional Defense Mobilization Committee to warn of an alarming manpower shortage in engineering. Reports suggested that the Soviet Union was producing three times as many engineers as the United States. To protect the nation's future economic expansion and industrial innovation, Van Leer said, American families, educators, and authorities must urge more female students to consider the field: "Of the 148,000 students enrolled in engineering schools, 518 are women, or four tenths of 1 per cent. . . . More women than that have sufficient mathematical ability. I married one and I know. . . . But no one has encouraged them. . . . We should have 4,000 or 5,000 enrolled." Van Leer dismissed fears that females would steal opportunities from returning veterans, saying that "women engineers would likely work four or five years and then get married and give up their jobs—but those four or five years of their service would be of a great benefit."[56]

The first female student to register for fall 1952 was not Coffee but "blue-eyed" Elizabeth Herndon, who had lost her army husband to tropical disease and then delayed applying to college in the fruitless hope that government would provide educational benefits for war widows. In registering at Tech, the former accountant cited the country's "definite need for engineers." Herndon reported that she had received friendly and respectful treatment from male teachers in Tech's evening school, where she completed summer refresher studies in physics and algebra. Male classmates grinned upon first seeing her and teased her. One fellow asked, "You gonna study electrical engineering? Why I bet you can't even tell me what kind of juice is running through this wire, AC or DC," to which Herndon amiably replied, "If I knew the answer to big old questions like that, I wouldn't need to be here." But women had enrolled in Tech's evening school for years. Herndon's first daytime campus visit for registration was greeted with wolf whistles and catcalls such as, "What kind of electricity is running through that wire?" The *Atlanta Journal* reported that Herndon not only "successfully survived" but declared that she found such greetings amusing rather than embarrassing. "I will probably enjoy all of their smart remarks," Herndon said, adding that Tech's men needed to "learn the girls, learn their personalities and treat them as an equal person." Press coverage played up Herndon's domesticity, noting that she designed and sewed her own clothing and that as sole parent of a ten-year-old son, she could not neglect obligations at home. The *Journal* threw in the editorial quip, "Her particular interest in electronics is radar, which seems a natural choice since it's the next best thing to a woman's intuition for showing what's going on beyond her line of vision." Photos showed Herndon's son watching his smiling mother try on Tech's traditional freshman "rat cap," adding, "If the [fellow] wearers of rat caps give her any trouble, she can always call out her troop of Cub Scouts to protect their den mother."[57]

Debate had already arisen about whether first-year females would be required (or indeed permitted) to wear the freshman headgear, which "has come to rival the ivy on the halls in college sentiment." The comment, "Horrors! Girlie Touch Looms for Rat Caps," highlighted the potential threat to tradition. A student committee agreed that like male classmates, female first-year students would be expected to wear rat caps at least until Thanksgiving, when Tech's football team played against the University of Georgia. It was unclear whether student enforcers were prepared to apply to women "the customary punishment of shaving a T-shaped haircut onto the head of any frosh who defied the complex codes surrounding cap-wearing."[58]

The only other woman to join Herndon for Tech's first coed semester was Barbara Michel, an engineer's daughter and "small blonde" valedictorian who loved "math, astronomy, and cars." Michel later explained that as soon as Tech became coed, her father urged her to drop plans to go elsewhere and "get the best engineering education while I was about it. . . . The challenge of being the first to do something intrigued me, as it would any American."[59] With only two women entering, coeducation did not immediately force many changes in campus routine. Both Herndon and Michel had family residences in Atlanta, letting the school avoid the tricky question of where to house women. Campus athletic and military programs did not change; Herndon and Michel simply did not participate. Nor did either fit the dreaded stereotype of flighty girls who were keen to distract boys from serious study. Herndon was a widow and mother, not a bubbly seventeen-year-old, and engineering major Michel was already engaged to a Tech student and therefore romantically unavailable.[60]

But as novelties, Michel and Herndon became instant press magnets, set apart from the almost 3,800 male students. Seeming to document their first steps on campus, a photo showing a woman in pumps descending a staircase was captioned, "The oaken stair

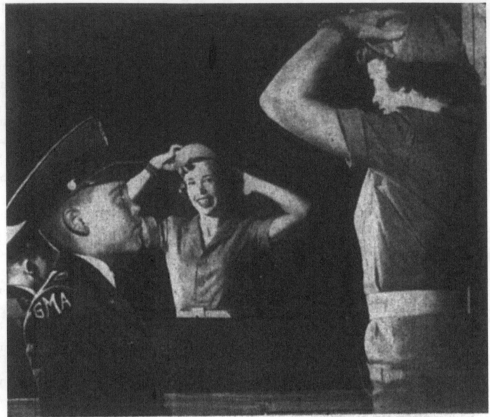

**TECH RAT CAP BECOMING TO FRESHMAN   ELIZABETH HERNDON**
Son Ashley, "No Comment."—Staff Photo by Bill Warren

FIGURE 4.2
In 1952, anyone who was connected with Georgia Tech knew the significance of the freshman "rat cap," and so this image of Elizabeth Herndon, a war widow and single mother, trying on the standard headgear defied all precedent. *Atlanta Journal*, July 28, 1952.

tread in Georgia Tech's main building have been worn deep by scuffing received from generations of men. First enrolled coed to use stairs is Betty Herndon." The *Georgia Tech Alumnus* ran a cover shot of the two women poring over a campus map, with the note, "Time Changes Everything." Publications around the state ran photos showing these "lone co-eds" on campus and trotted out lame jokes about "lady engineers" who had no intention of trying out for football. Although neither Michel nor Herndon fit the image of a flirt, the *Atlanta Constitution* referred to them as the "first 'Betty Coed[s]' to storm

Tech's male stronghold," referring to the prewar song about an alluring woman who conquered every college man's heart. Images sexualized women's presence. A 1952 *Technique* cartoon showed a pudgy, balding, bespectacled Tech alumnus on campus for homecoming, looking aghast at being greeted by a curvaceous coed in a tight sweater.[61]

Yet behind the joking, headlines, and photographs highlighting the novelty of women's presence, doubts remained about the wisdom of allowing coeducation at all. Reassuring alumni and tacitly validating the link between quality engineering and masculinity, the *Technique* vowed that "we admire the standards they [former students] have set for us to follow." Effectively erasing women's presence, editors wrote, "We revel in the brotherhood of Techmen." The *Technique* also continued to print antiwoman "humor." An editorial on the "chemical composition of females" suggested that girls might be made of "sugar and spice and all that's nice" but that college-age women were formed from less desirable ingredients, including lime, "glycerine enough . . . [for] bursting a bomb, . . . sufficient fat to make several pounds of soap, . . . sulphur enough to rid a dog or cat of fleas, and finally a measly quarter pound of sugar, which is utterly insufficient to sweeten and render palatable such a goshawful mixture."[62]

Observers started from a baseline of skepticism about women's ability. One article opened by saying, "Even if the . . . feminine ramblin' recks survive through their rigorous engineering courses and graduate. . . ." The *Technique* celebrated traditional gender roles at a time when over 15 percent of Tech students were married. It praised the postwar custom of distributing "wives' diplomas" that honored the women who put husbands through school while providing home comfort and psychological support: "They have suffered through formulas and theorems almost as much as their husbands have and they probably knew almost as much about engineering before it was over. . . . The Wife's Diploma . . . should be highly honored."[63]

Herndon and Michel were greeted at Tech's Freshman Week with whistles and what they described as a "strained" atmosphere. When the women were introduced, their male classmates clapped, but singling out the women for introductions reinforced their oddity. Herndon later remembered, "I felt that thousands of eyes were following me at every move I made." After the first two quarters, Michel and Herndon reported still being "teased a bit" by classmates, some of whom remained "resentful." Any perceptions that the women might be getting preferential treatment could elicit umbrage. One male undergraduate was quick to complain when he believed that Tech's women were given privileged access to employee parking lots: "We had thought that the co-eds were to be considered no better or worse than ourselves and do not favor granting them special favors that we ourselves do not enjoy. This smacks to us of favoritism, and I, among others, would like to know why." Herndon admitted that she had once parked in an out-of-bounds spot and received a justified ticket: "I do not seek any special privilege, or ask any special favors because of sex. I hope to obey all rules just as all freshmen are required to do."[64]

By midyear, both Herndon and Michel earned satisfactory grades that proved their ability to handle engineering classwork. Herndon declared that she felt "at home" and reported that she averaged thirty weekly hours of homework, which placed her within the Tech community ideal of seriousness. The T-square that she carried marked her as just another technical student. But even as photos showed Herndon studying with a table full of men, photos (in the same articles) of her ironing clothes stressed a difference: she also "does all of her own housework."[65] Herndon's excellent grades met the qualifications for membership in Tech's chapter of the national freshman honor society. But as an all-male organization, the Phi Eta Sigma Scholarship Club could not admit her, and Tech had no chapter of the parallel female honor society. The *Atlanta Constitution* editorialized, "But she had the satisfaction of knowing she had made it. She was later compensated for this disappointment when a fellow [Tech engineering] student . . . sought her out. . . . The meeting led to romance, and one Sunday Betty replaced T-squares and slide rules with wedding rings and church bells." The *Atlanta Journal* ran a photograph of Herndon posed in the kitchen while consulting a slide rule, under the caption, "Slide Rule Approach to a New Recipe: Mrs. Elizabeth Herndon Brushes Up on Cooking."[66]

Making it clear that coeducation had come to stay, two other Atlanta women joined Herndon and Michel in January 1953, doubling female enrollment. Early reports on Patricia Sargent stressed her domesticity and, like Michel and Herndon, the fact that she already had established personal relationships. A former home economics major in Texas, Sargent followed her air force husband to Georgia and announced her intent to study chemical engineering: "Her major outside activity is housekeeping for her and her husband. . . . Since Bill graduated as a petroleum engineer, Pat thinks she will get some much-needed help from him. . . . Asked his opinion about his wife attending Tech . . . he simply answered, 'It doesn't bother me!' He's quite proud of his wife since she is both a good housewife and a good student."[67]

Frances Lillard had no preexisting romantic commitments, but as the press immediately established, a few Tech classmates soon approached her for dates. Within two months, Lillard sparked a minor outrage among some male classmates by daring to poke fun at Tech campus culture. Lillard joked about Techmen who slept in their clothes, with "a mop of uncombed hair" and "sagging shoulders . . . from bending over a study desk half the night." She sarcastically praised the "style and charm" of untucked shirts,

SLIDE RULE APPROACH TO A NEW RECIPE.
Mrs. Elizabeth Herndon Brushes Up on Cooking

*VERY CIVIL ENGINEER*

**Tech's First Coed
To Wed Fellow Student**

FIGURE 4.3

To Atlanta newspapers, the novelty of the first engagement between two Georgia Tech students was newsworthy. In showing Herndon literally balancing a slide rule and cookbook, the *Atlanta Journal* encapsulated all the questions about whether and how women could reconcile traditionally male engineering ambitions with assumptions about women's traditional family responsibilities. *Atlanta Journal*, March 27, 1953.

droopy socks, and dirty jeans: "Who but the well-dressed Tech man would go to the trouble of searching his bureau drawer each morning for a nice, wrinkled shirt . . . and stretch the neck? Nobody likes . . . a well-fitting T-shirt. . . . Who wants to show off mangy old muscles, especially since the distaff side has been added to our campus?" Referring to the women's presence, Lillard wrote, "Coeds have not been here long enough, nor are there enough of them, to have set styles of their own, with one possible exception . . . a sagging belt, caused by a log log duplex decitrig drooping down toward wobbly knees."[68]

Undergraduate men returned fire, responding with some asperity to the audacity of a female student daring to criticize Tech masculine culture. Several men defended their choice of jeans as comfortable, inexpensive, and low maintenance. Male students suggested they might consider donning sports jackets if women began wearing formal gowns to class. They called baggy T-shirts "traditional" at Tech and suggested that if Lillard disliked it, she should transfer to Tulane, "the coat and tie school." Linking sloppiness to Tech culture, student George Greenacre wrote, "If they [female students] think they can change our way of dress which is older than they are, I think they're going just a little too far." Sheldon Till said that Lillard's comments drew his "tears of pity for the author . . . realiz[ing] how terrible it must be to be a coed in a school full of . . . the masculine gender. The coeds will probably never obtain the enjoyment and esteem which goes with wearing faded levis, sloppy T-shirts and country loafers." The most direct shot came from John Wise, who asked, "What's wrong with dungarees and levis?" and continued:

I don't know why Miss Lillard is going to Tech, but I *do* know why I am. I am not here to impress the coeds, the Georgia Tech secretaries, the professors, or my fellow students. I am here to get an education so that some day I may be able to support a houseful of kids and some nice young woman who, I hope, is *NOT* an engineer.[69]

Meanwhile, administrators wrestled with complications stemming from the regents' decision to limit women's options for enrolling at Georgia Tech to a specific list of majors. Within weeks, at least one potential female student expressed interest in entering other programs. In September 1952, Helen Thompson petitioned to be allowed to come to Tech to study industrial management. Van Leer wanted to stretch the rules to raise female enrollment, especially since Tech had managed to attract only two women that first quarter. Describing Thompson as "a mature woman who is endeavoring to take work which will increase her earning capacity," Van Leer emphasized that "we at Georgia Tech would be glad to permit her to take these courses." Just as firmly, the regents refused permission, reiterating that Tech could admit female students only to engineering and architecture programs.[70] In another case during that first coed year, Elizabeth Newbury asked to pursue an electrical engineering degree on a part-time basis. The demands of her job as a technical assistant at Tech's Engineering Experiment Station made it impossible to handle a standard course load. Van Leer supported Newbury, but the regents rejected her petition, insisting that women could enter only as full-time students. The Atlanta Women's Chamber of Commerce protested such discrimination, complaining that the Board's decision for coeducation "was not an all-inclusive action as we had presumed it to be."[71]

At roughly the same time, the Women's Chamber of Commerce underlined its ongoing support for coeducation by establishing the first (and at that time, the only) scholarship for female students at Georgia Tech. Van Leer publicly welcomed creation of this $300 award. He repeated his conviction that "in view of the fact that our country is at present some 30,000 engineers and scientists short . . . the only source for making up this deficiency seems to be by encouraging young women to enter the engineering profession."[72]

In the minds of many in the Tech community, "engineer" still equated to male; although female students had been admitted, they were so few as to be ignorable. The school did not particularly try to attract the interest of women as potential attendees—just the opposite. In a special issue for high school students, the 1953 alumni magazine assumed that Tech's pool of future engineering students would be boys and described engineering as by men, for men, offering the satisfaction of macho conquest and leadership:

In this issue we have attempted to give you . . . information . . . of interest to the young man entering college. . . . The engineer is the pioneer of this era. Through his work, our way of life has been greatly enriched. From his efforts to make use of the great scientific advances of our age have come our superhighways, plastics, wonder drugs, electronic devices, miracle fabrics and many others. He is the man who is developing atomic energy for peacetime use, opening a brand new era of progress.[73]

Although neither Lillard nor Sargent chose to continue at Georgia Tech after their first quarter, Herndon and Michel were joined in fall 1953 by "six new coeds invading the Georgia Institute of Technology campus" (as the *Atlanta Constitution* put it). Three freshmen—"green-eyed" Lola Sonia Friedman, plus "vivacious" and "pretty . . . brunette" Virginia Ann Brown (daughter of a Tech alumnus), and Helen Kimbrough from Georgia—were all single, the *Technique* reported. "Statuesque" chemical engineering major Caroline Seale was the first winner of the Tech women's scholarship created by the Atlanta Women's Chamber of Commerce. "Because people told me I couldn't" study engineering at Tech, Seale switched her original plans to study medicine at Tulane and aimed to prove everyone wrong for scoffing. Elizabeth Newbury, the Tech lab assistant who previously was denied permission to enroll part-time, entered electrical engineering, as did Shirley Clements, a transfer from a teachers' college after she became intrigued by opportunities for female engineers.[74]

As in the previous year, these Tech women were swamped with attention from both male classmates and the media, which acknowledged females' intellectual ability but preferred to dwell on their physical attributes, feminine interests, and romantic potential. The *Technique's* feature on Seale noted that she had graduated third in a class of 117 and intended to seek a doctorate but soon proceeded to her "vital statistics: . . . what everyone has been waiting for. Caroline is single, not going steady and seventeen years old. She is five feet eleven inches tall and weighs 140 pounds. As an interesting sidelight, Caroline is quoted as saying that she would appreciate any help she can get in all her subjects." The accompanying photograph showed Seale wearing an off-the-shoulder top, with the caption, "Miss Caroline Seale, Tech's fifth coed, strikes an alluring pose."[75]

In practice, the "belle of an engineer" or "Nell of an engineer" (as these women were soon nicknamed) still did not have an easy time. The *Atlanta Constitution* reported seeing "good-natured hazing," with a crowd of upperclassmen badgering Clements for dates. At first, Seale reported, the boys' stares made her feel uncomfortable, as though she was an outcast. But by midyear, she described her class as "one big happy family" where male students were "very polite—always opening doors for me." Kimbrough noticed that officially women remained invisible: "at meetings, where the entire fresh-man class has to appear, we're usually addressed as 'gentlemen' or 'men' or 'boys.'" Seale described faculty as "all . . . helpful and polite" but added she had been "frequently embarrassed by subtle remarks." Clements later remembered facing one professor "who wouldn't tolerate girls in his classes and got away with it. . . . I didn't want to be in his classes anyway, so that was fine with me. It really didn't bother me. . . . The minor acts of discrimination were far outweighed by the people who were sympathetic." Clements remembered with gratitude the support that she received from the electrical engineering chair and from women in the Tech community, particularly Ella Van Leer and librarian Dorothy Crosland. She appreciated that female students' special status gave her a chance to become acquainted with president Van Leer, "so that made up for the professor who didn't want to have me in his classroom."[76]

Yet by the second year of coeducation, one aspect of the situation was already dif-ferent. Female students who entered in 1953 could consult sophomores Herndon and Michel, gaining the benefit of their experiences. During orientation, Michel gave new women essential advice about the location of ladies' rooms. Michel also helped indoctri-nate her successors into the feminized versions of Tech tradition that had been cobbled together in her first year. New women felt "self-conscious" about donning freshman "rat caps," but Michel hinted that the hats were "good for keeping your hair in place."[77]

Numbers made a difference; at orientation, the women started discussing the pos-sibility of creating a Tech sorority. Fraternities had existed at Tech since its founding, and by spring 1954, female students organized the first sorority. The president's wife was an alumna of Alpha Xi Delta. "Miss Ella" (as the young women called her) helped make arrangements with the national organization and acted as the institutional spon-sor to win approval from Tech's governing body for student activities. The Van Leers hosted an April initiation at their house where the sorority's national president offici-ated at the ceremony welcoming new pledges. Their chapter held the special status as the first of any sorority to be located at an engineering school. Clements explained, "We were supposed to have a minimum of 25 girls to form a chapter, . . . but the committee decided we might never have a sorority if we waited. We have so few girl students." Members immediately threw themselves into campus tradition, entering a jalopy "wreck" in the homecoming parade. The *Atlanta Journal* wrote, "The girls claim they've come in for a lot of kidding from the male students while working on their wreck, 'but we can take it.'"[78]

In 1954, seven new female students arrived at Tech for the third year of coeducation, raising the total number of women attending to double-digits (fifteen). "Any of you guys need a date? Well, the answer to your problems may be solved," the *Technique* commented: "It is time once again for the Tech male population to appraise the facts and figures of the current crop of female rats." The paper described the incoming women as "all A" students with real "feminine charm."[79] Other press coverage continued to emphasize the physical attributes, romantic availability, and domesticity of Tech women. An *Atlanta Journal* article updating Elizabeth Herndon's progress emphasized that she had devoted her summer to pre-preparing family dinners for the semester. The accompanying photo showed her in the kitchen rather than the lab, cooking rather than studying, with the caption, "Georgia Tech's first coed . . . stocks freezer for winter meals."[80]

With their new numbers, Tech's female students began to speak up, gradually pushing the institution to address at least some of their needs. In 1954, two women expressed interest in joining the U.S. Air Force ROTC unit. Jackie Easton, an aeronautical engineering major, and Teresa Thomas, a chemical engineer who was planning to specialize in rockets, emphasized their love of aviation. Although cadet training was officially male-only and there was no chance that the air force would offer women commissions, the colonel who headed Tech's program agreed to let the two sign up for training as an academic elective. Unable to wear the free ROTC uniforms that were given to Tech men, the women bought themselves expensive uniforms from the Women's Air Force auxiliary to "fit in better." Male cadets "resented" the female presence and made their life awkward. When Thomas was assigned to lead drill, her male squad mates laughed, but "afterwards, they told me I did all right. In fact the one who laughed loudest said it was the best anybody had drilled." Easton's officer teased her about needing a haircut, but "at the final evaluation, he asked me, just like the boys, if I thought I were officer material and if I were a gentleman."[81]

In spring 1956, Shirley Clements and Barbara Michel became the first two women to complete Georgia Tech regular undergraduate degrees. President Blake Van Leer, whose support had played an instrumental role in facilitating coeducation, had passed away the preceding winter and so could not celebrate the landmark with them. Michel graduated with an industrial engineering major and Clements in electrical engineering. Clements noted that her diploma was printed with a "Mr." before her name. Looking back, Michel remembered that "at first, it was plenty tough . . . we definitely weren't accepted like we are now. And the great publicity fuss made over the two of us sure didn't help. The fear that somehow we might change Tech infected the boys. And although they were never nasty to us, they didn't go out of their way to ease our situation." Although critics had feared that women's presence would ruin Tech's traditional character, both Michel and Clements sought to reassure observers that the school's culture remained unaltered and indeed had converted its female graduates into the equivalent of true Tech men. They joked about celebrating graduation by brazenly flouting campus traffic rules, a "statement [that] might easily have been attributed to any Tech senior," the alumni magazine commented, showing "how completely the girls have been indoctrinated with the mores of Georgia Tech."[82] The 1956 yearbook depicted Clements not as an odd invader or a feminine curiosity but as just another engineering student preparing for the rigor of exams. A photo showed Clements in bare feet and bathrobe, still bent over her books at 2:30 a.m. The caption read, "The coeds have to use the slide rule and spend many long hours studying too. Shirley Clements crams late."[83]

Even as Clements and Michel claimed their degrees, the two still felt obligated to reassure those who held lingering doubts about their suitability as Tech graduates. Michel said, "How about straightening one thing out for us? Tell the alumni not to worry. We are proud to be alumnae of Tech. We intend to be career engineers, and we have no intentions of trying to bring about any changes in the profession." The two had achieved respectable grade-point averages and, in a period of strong business expansion, did not anticipate difficulty finding employment. Michel was hired by Shell Oil. Clements found work in IBM's research library, a position that applied her engineering knowledge in a nonthreatening manner; many female scientists of the era were similarly employed in corporate technical libraries.[84]

Although both Clements and Michel had demonstrated their ability to handle Tech engineering coursework, after graduation the two still saw a need to address suspicions about their motives for attending a mostly male institution. Michel said, "One thing I'd like to get straight right from the beginning, . . . I didn't come to Tech to find a husband." Clements echoed, "Neither did I, and any girl who does is getting one the hard way. . . . This is such a tough school, and the girls who come here for a lark don't last long enough to get married."[85] After a year at IBM, Clements returned to Atlanta to marry Duke Mewborn, a fellow electrical engineering 1956 Tech graduate. When he learned about Clements's engagement, the alumni secretary joked, "I guess with our record system the only thing we can do when a coed graduate gets married is to mark her deceased."[86]

Other Tech females felt a similar burden as observers speculated about their reasons for choosing a traditionally male school. Textile engineering major Paula Stevenson vented her frustration at "people always sounding off in the press about you," especially the common implication that girls came to Tech to gain easy access to potential

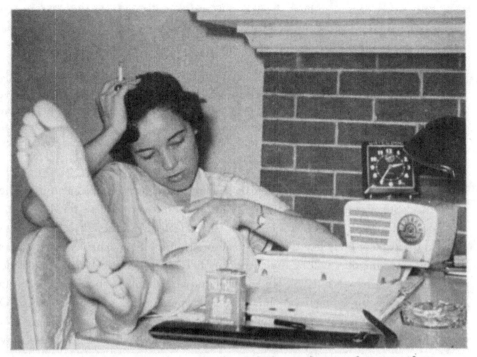

The co-eds have to use the slide rule and spend many long hours studying too. Shirley Clements crams late.

FIGURE 4.4
The first generation of female undergraduates at Georgia Tech felt obliged to prove to men that their unusual presence would not compromise school traditions. This photo from the 1956 Tech yearbook showed the striking novelty of a woman as she prepared to take engineering exams, but the sight of Shirley Clements as she pulled an all-nighter reassured doubters that Tech's standards remained intact. *Georgia Tech Blueprint*, 1956, p. 242. Used with permission.

boyfriends. Stevenson scornfully replied that any woman who primarily was seeking to land a husband was likely to flunk out first. She stated that she had come to Tech knowing that "there's a definite need for women in the engineering fields today" and wanting "the best school to prepare yourself for a career."[87]

As Stevenson indicated, women at Tech in the late 1950s still felt on trial: "Acceptance of the first coed didn't help me or any of the present group of girls a bit. The way the Tech men look at it, each of us must prove herself." The institution had made "a number of outstanding advances . . . in the acceptance of coeds," Stevenson wrote, but female students continued to feel "the resentment that the Tech men have for you because you have dared to strive toward entering a basically male profession":

They seem to look upon all of us as walking "Univacs" or mental cases. "A woman's place is in the home," they reason. "And she should stay there." . . . In fact, in the past few years, women engineers have proven very successful working in industry. I guess the Tech men sometimes still prefer those "sweet, helpless little girls" that flatter them to death and then trap them at the same time. Coeds at Tech aren't given much incentive to make good grades. If a girl is making poor grades, the boys laugh about it. And if she makes good ones, the boys resent it.[88]

With the low numbers of female students, each one felt conspicuous by her difference. Stevenson compared it to "living in the proverbial glass bowl." Using a more pointed metaphor, the 1962 *Freshman Girls' Handbook* warned that "coeds still stick out in the student body like the proverbial sore thumb. Your achievements, your successes and your mistakes will be much more noticeable." Individual men could fade in with the crowd, but each female student was known and talked about. Stevenson found the men's constant staring to be "annoying" but opted to look for a silver lining: "Once you get over your initial embarrassment it becomes very valuable. It forces a girl to eliminate any self-consciousness in her makeup and . . . helps a Tech coed to get along with most any group of people."[89]

Stevenson agreed with Clements and Michel that coeducation had not altered "the rough, tough, masculine atmosphere that is Georgia Tech":

The 5,600 male students . . . are as proud as the alumni of the School's masculine reputation. And, believe me, they don't intend to see it changed one bit. . . . Men are still wandering around wearing the standard campus costume of beat-up "T" shirts, blue jeans and that slide rule on their belts. . . . The main reason that the female-type student will never change Tech can be found in the toughness of the curriculum. The whole academic program is geared to the men, and its pace is such that a girl has to work so hard to keep up with her class that she has little time to think of changing things around the campus.[90]

Stevenson did not withdraw from student life in disillusion; just the opposite. She became the first band majorette, the first female member of the Ramblin' Reck Club (the student organization devoted to fostering school spirit and tradition), and the first woman to serve on the Student Council, where (the alumni magazine noted) she was "always outnumbered in voting."[91]

Although she denied having any intent to change the nature of Tech, Stevenson expressed frustration with school routines that made a woman's life difficult. For yearbook photographs, all students were ordered to report to the football stadium men's locker room. Stevenson later recalled, "I had my freshman picture taken looking at the urinals. . . . I had never been in a men's room in my life. . . . It was really awful, a terrible experience. It was my senior year before women were permitted to go to the photography studio to have their pictures made." Similarly, on registration day, Stevenson was sent through the men's dressing room (in use at the time), only to find a high divider blocking her entrance to the main gymnasium. Unable to climb this wearing a skirt, Stevenson got a boost from two men: "One guy took one arm, and one guy took the other and they dragged me up that six foot thing and I went on in and registered. That was the type of thing that went on with regularity."[92]

School life still revolved around assumptions that a Tech student was by definition male. The institution had few established procedures to address needs that might be more specific to female students and the peculiar, often unclear policies governing their presence at Tech. Stevenson complained that "the administration . . . was definitely not set up for girls. Whenever you need to find out anything specifically concerning the girls, you generally make the rounds of every dean before you get your answer." Rules governing the women's dormitory had been completely revised four times in a single year, she noted. But in the end, Stevenson indicated, the superior engineering education available at Tech was "enough reason for any girl bent on pursuing this career field to ignore all of the school's disadvantages."[93]

In the early years of coeducation, female students from Atlanta lived at home, and Tech found apartments for the few out-of-towners. In 1955, the school designated a converted house as a Women's Residence Hall, but students found its location and design unsatisfactory. Residents complained that the single bathroom became ridiculously congested on Friday and Saturday nights. Women paid higher living fees than men, even though their hall had no kitchen; they had to walk six blocks to eat meals in one of the men's dormitories. Stevenson complained about this injustice and added that the women's weekday 10:30 p.m. curfew put them "at a distinct disadvantage with the men," forcing them to leave the library before its 11:30 p.m. closing time and making it harder for them to collaborate with male classmates on lab reports.[94]

Georgia Tech required all nonlocal freshman females to live in the Women's Residence Hall, which had a capacity of just eleven students. Demand soon outgrew available space. For the 1958–59 academic year, the school rejected seven otherwise-qualified women simply due to lack of housing, Tech's registrar announced. Administrators hesitated to commit to building a larger women's dorm, worried about the cost of construction and paying housemothers. But women pointed out that Tech would never raise its numbers of female students unless it addressed the housing crisis. First-year student Emily Hill said, "I know Tech could get more coeds if we had better facilities. It's really a shame that more coeds can't enter; I believe . . . coeds really benefit the school by adding dignity and color to the campus." The *Technique* called the space shortage "unfortunate," since

> coeds added much to our campus life. Through participation in such activities as Student Council, the Ramblin' Reck Club, and the Tech Band, they have demonstrated their potentiality. . . . Coeds deserve to be congratulated on the manner in which they have adapted themselves to Georgia Tech's way of life. . . . If the school advocates these students, it certainly should have adequate housing facilities for them.[95]

Some Tech men agreed that the unfair situation should be corrected so that female "morale would be improved and the girls would feel wanted." One industrial engineering senior said, "The school should realize that the co-eds here have shown they can do the work, and housing should be provided for them." A mechanical engineering junior declared, "Coeds accepted into school should be given equal opportunities, including housing. If the school doesn't provide housing, it will force the co-eds into the disadvantage of commuting." It seemed particularly galling that Tech was finalizing plans to add four new men's dormitories, complete with modern conveniences such as an intercom system. The *Technique* editorialized, "We cannot help but wonder of the impact of the news of these new men's dormitories on our coeds. What will they say when they see the new dorms and then turn to gaze upon the crowded house in which many of them have lived? How would you, Mr. Student, feel if you were in their position?"[96]

## WOMEN'S CAMPUS CULTURE: EVOLUTION AND DIFFICULTY

Despite the artificial cap on female numbers due to the housing shortage, seventeen enrolled in September 1958, almost double the number of women who were in previous entering classes. By fall 1959, the total number of undergraduate women at Tech reached forty-three, which was enough of a critical mass to begin defining a Tech coed culture. Since 1952, female students entering Tech were automatically signed up as members of the campus Young Men's Christian Association, an influential center of student religious and social life. In 1957, the chapter created a women's group called Gamma Psi, a name coined to mean "coeds of the Y." Gamma Psi gave women elected representation in the YMCA cabinet and sponsored an annual Coed Freshman Camp to provide orientation for incoming female students. Also in 1957, Tech established a

Women Students' Association to promote the general well-being of female students, coordinate their activities, and maintain "the finest standards of character and conduct." As Tech's only sorority, Alpha Xi Delta held membership on the Inter-Fraternity Council and joined campus projects alongside the twenty-six men's Greek societies. Meanwhile, the Society of Women Engineers, which had established an Atlanta chapter in 1956, also became a major presence in the lives of Tech women, thanks largely to the efforts of Maryly Van Leer Peck, the daughter of Tech's former president. As SWE's national chair of student affairs, Van Leer Peck hosted a dinner to welcome 1958's incoming female students, which reinforced their interest in establishing a campus SWE chapter.[97]

Since the start of coeducation, Tech's campus paper had assigned male reporters to evaluate the appearance and personalities of incoming females for curious male readers, but in 1958, women wrote articles themselves for the first time. "Coed Chatter" columns reinforced the sense of female students as a group with inside jokes. The column gave women a public voice to speak about their everyday lives, including the "friendly razzing" that they received from male classmates. More than that, women reporters documented the ways that they and their classmates tried to solidify their place in the overwhelmingly male campus climate. Four of the 1958 freshmen created a majorette corps that performed with the band at football halftime shows. Female students adapted Tech tradition to fit coed life. At homecoming, the "Ratesses" (an unlovely nickname given to freshman women, which fortunately did not stick) held a hundred-yard dash known as the "cupcake race," corresponding to the traditional half-mile "cake race" for freshman males. The female winner earned a kiss from the man who was president of the student body.[98]

Observers remained curious about whether women could handle engineering studies that usually were associated with men, and grade-point averages offered a natural mark for comparison. In 1958, Tech's female students scored an average of 2.4006, beating the men's 2.1874 average. With a headline "Coeds Surpass Males," the *Technique* editorialized, "we are quite discouraged to learn that the stronger sex has been forced to take a back seat in the scholastic category." But one year later, the undergraduate female average declined to 2.1056, falling below the men at 2.1313. Under the headline, "Coed's Average Hits New Low," *Technique* columnists offered mock-helpful advice, telling women to "forget that the ratio of male to females here is 117.33 to one. Remember that you all came here to get a BS, not a MRS," and avoid hanging out at bars drinking beer: "If a Tech man should happen to have a date with a coed, he should be sure to get her in by 9:30 p.m. This would give her time to study a bit before she gets to bed."[99]

Significantly, a high–profile argument that the United States should welcome more women in nontraditional fields came from Georgia Tech's successor to Van Leer as president, Edwin Harrison. In 1958, Harrison stated that the post-*Sputnik* era placed unprecedented pressures on the American educational system to keep up with the Soviets, who "are ahead of us . . . turning out graduates . . . two to one." Harrison continued, "[I]t's urgent that we . . . interest as many young people as possible in gaining technological education. . . . EVERY CHILD educated today, male or female, should be given an opportunity to take courses in science and mathematics from competent teachers in order that a wise choice of career can be made."[100]

Meanwhile, Georgia's Board of Regents was fielding a growing number of requests to reconsider the original restrictions that limited Tech women to studying only engineering and architecture. Some fathers appealed on behalf of their daughters. In 1959, Marie Kleine commuted twenty miles into Atlanta to take Tech evening classes in chemistry, arriving home as late as midnight. Her father told Tech's president that although her parents felt "quite uneasy for her to drive home alone at such a late hour, we agreed to it because we believed that this would be only temporary." He was surprised to realize that Tech would not allow her to transfer into regular daytime undergraduate chemistry classes, and the school's president, Edwin Harrison, confirmed this with "regret."[101]

Under pressure, Georgia's regents agreed to open one new major for women at Tech. Atlanta father Walter Skelly declared that his daughter Lois, an A student in accelerated high school math, "has had her heart set on" attending Tech to get a degree in applied mathematics, "with a view to becoming an Electronic Computer Programmer." Skelly argued that it was impossible for his daughter to receive proper training anywhere else in the state and carefully presented his argument: "No other college in Georgia offers the opportunity of taking such a comprehensive mathematics curriculum, including Georgia Tech's Math 425–Principles of Digital Computers, and Math 426–Computer Programming and Coding, combined with the advantages of various electives in electronics." Skelly suggested that opening such areas to women was a matter of not just personal, but national significance, "in order to advance the output of those who can make direct contributions to our national defense. Computer programming is, of course, an integral part of missile research and production. Since we don't have enough young men going into the field of higher mathematics, why not give young women of Georgia the opportunity to begin plugging this gap in our national defense against *Sputnik*?" President Harrison agreed that Tech's major should be open to women, since "there is no other institution in the State of Georgia capable of teaching this type of work, and . . . shortages in the scientific, technological, and mathematical fields can be in part alleviated by making use of properly qualified women graduates." The claim persuaded the regents to allow women to enroll at Tech as applied math majors, of which Lois Skelly became the first.[102]

Some important regional leaders supported the cause of making Tech fully open to women. In 1961, Fulton E. Callaway Jr., chair of the board of one of Georgia's largest textile corporations, wrote to Tech's registrar about Julia McKenzie Hodges, one of his female workers who was following her husband to Tech and wanted to take chemistry courses herself. The Callaway family was among the state's most influential businessmen, employers, community builders, public leaders, and philanthropists. Fulton E. Callaway Jr. had earned a 1926 Tech degree in textile engineering and helped establish the industrial-development foundation that grew into the Georgia Tech Research Corporation. Tech president Harrison used the Hodges case to recruit Callaway's support for the "difficult problem" of persuading the regents to ease limits on coeducation, "without antagonizing some of our good friends with whom we happen to disagree." Harrison told Callaway that the primary opposition came from regents chair Robert Arnold, "who just doesn't think that women ought to attend Tech." Harrison urged Callaway "to discuss the problem casually" with Arnold and "press him a little toward relaxing the restriction on women" because he "might be more receptive to the suggestion" from you than to another appeal "from among us 'academic' people." Within one week, Callaway sent Arnold a letter suggesting that women be allowed to enter all majors at Tech, as "an advantage to the State [to] . . . relieve the shortage of scientific and technical people."[103]

In replying, Arnold voiced his continued doubts about the wisdom of having any Tech coeducation at all:

When Mrs.Van Leer began her drive to have Georgia Tech made co-ed, I opposed it with all the effort I had. I have always felt the worst thing that ever happened to the University of Georgia was making it co-ed. There are too many play boys and play girls . . . on campus in Athens. I didn't want to see this happen to Tech—a campus full of "debutants" [sic] looking for a husband. I lost my fight, but I did get inserted into the regulations the rule that no girl could get into Tech except for an Engineering Degree or Architecture. I felt that other subjects were offered at Emory and Agnes Scott. . . . I still feel very strongly that too many girls at Tech will all but destroy the seriousness of purpose in the lives of many young men. Perhaps I am an old fogey on this subject but I am sincere.[104]

Callaway replied in equally frank terms that when the idea of making his alma mater coed first arose, "I felt very much the same way you did . . . [that] it would be a bad thing. . . . I am still not sure whether it was a good thing to 'let down the bars' or not." But given that women had been admitted to study engineering and architecture, "the damage of female diversion has been done." The regents should now lift all limits on female enrollment, Callaway suggested, and limit the "damage" by increasing academic rigor so that "we won't have many girls at Tech who are just looking for a husband but instead the girls will be those who are seriously looking for a real education. . . . Tighten up the curriculum to the point that neither girls who are merely looking for a husband nor boys who are only looking for a good time can stay in school."[105]

But the rules did not change, forcing each woman who wished to major in anything outside of engineering or architecture to plead her case separately for judgment by the entire Board of Regents. President Harrison supported the majority of appeals. He was committed in principle to complete coeducation, with his only "major reluctance" centered on the problem that Tech still did not have much on-campus housing space for women. Other administrators helped game the system to attract and assist good women. Non-coed departments arranged to "farm out" women students to other majors, where they would load class schedules with electives in their desired subject. Some faculty remained impatient, pressuring Harrison to work harder to persuade the regents to make Tech completely coed. In 1962, the dean of Tech's graduate division, Mario Goglia, urged Harrison to ask for permission to admit women to physics and applied biology on the grounds that "one of the fastest moving fields of endeavor is that of the 'Medical engineer'—we are singularly adept to meet this and other challenges, especially where young ladies are involved." Harrison refused to approach the regents at that particular time, protesting, "Frankly, there is some strong resistance on the part of a few members of the Board, and I am convinced that to bring up the subject again this soon after being rebuffed would be a mistake."[106]

The greatest resistance continued to come from regents chair Arnold, who remained skeptical about women's commitment or ability to be good engineers or scientists. In 1962, Tech sophomore Dorothy Jean McDowell requested permission to convert her major from chemical engineering to chemistry, arguing that no other in-state school offered facilities that were appropriate to her interest in inorganic chemistry. Georgia Tech had four professors in that specialty, but the University of Georgia had none, McDowell wrote in her appeal to the regents. Tech planned to construct a neutron diffraction spectrometer, she noted, and its faculty had access to two nuclear magnetic resonance spectrometers. Arnold responded by asking McDowell, "Did you formulate this letter, yourself? It surely sounds mature for a sophomore so maybe you are an exceptional case. I am not implying that I understand the technical phrases used . . . , but, even so, I am impressed. . . . I will appreciate very much if you will satisfy my curiosity as to whether or not you formulated the letter of April 23rd."[107]

Between 1962 and 1964, the regents granted at least twenty-one women who were in "unusual circumstances" permission to major in Tech fields that otherwise were forbidden or to pursue studies part-time, emphasizing that each "special exception" did not indicate any broader change in policy. Arnold in particular required female students to prove that they were dedicated scholars rather than flirts. Applied math major Julia Bouchelle appealed to transfer her major to physics, explaining that Georgia Tech's program uniquely fit her desire for applied rather than theoretical work and also that she could not afford to move out of her grandmother's home in Atlanta since she had lost her father in the Korean War. Arnold decided that Bouchelle's "request sounds reasonable" and told Harrison, "I am entirely sympathetic in this case and in all cases which indicate a serious desire for learning. Tech is a fine school . . . and I don't want it ever to degenerate into a co-ed nine months 'house-party.' If this is a deserving case, let's see what we can do?" The regents also permitted two female applied math majors to transfer to applied psychology, acknowledging that no other public school in Georgia offered an equivalent program. One of these women, Varner Jo Blackshaw, told the board, "I realize that there are many difficulties involved when a woman student wishes to enter a field heretofore reserved almost exclusively for men. However, if I am permitted to come into the Psychology department, I will work hard to justify all the trouble I may have caused."[108]

In December 1962, Harrison decided it was safe to ask the board a second time to open all Tech programs to women on the same terms as men: "I feel that this action is justified on the basis of the merit of the special exceptions which have been granted in the past, and because of the role which women can play in helping to relieve the shortage of technical and scientific manpower requirements."[109] But the regents again turned down Harrison's request, annoying faculty who favored full coeducation. After Georgia's chancellor asked Tech's physics chair why relatively few physicists chose to pursue work as teachers, Vernon Crawford answered that to get more physics majors into teaching, the regents should open all Tech enrollment to women:

Salaries for women in industry are lower than salaries for men, hence the public schools are in a better competitive position for the services of women physicists. The teaching profession has always attracted more women than men at the elementary and secondary levels, and probably always will. Many girls have tried to enroll as physics majors at Tech. . . . To liberalize the admissions policy seems to represent a very small sacrifice (of what, I'm not quite sure) in view of the possible rewards (of which enlarging the source of potential teachers is only one.) I hope you will consider this suggestion seriously. Apparently considerable effort is required to persuade the Regents of its wisdom.[110]

A 1963 Georgia Tech Institutional Self-Study expressed similar frustration at limits on women's enrollment, which the committee criticized as

out of keeping with the best thought in educational circles and . . . unrealistic at a time when the need to expand our supply of technically trained personnel is urgent. . . . . The present hodgepodge of rules regarding the admission of women to various curricula and degree programs serves no useful purpose and damages the institution in its relations with the public. Why an undergraduate girl may major in electrical engineering or mathematics but must have special permission from the regents to major in chemistry or physics is difficult to explain.[111]

Policy change finally appeared possible in 1965, when Tech president Harrison again proposed revisiting the question, and new regents chair James Dunlap responded favorably. With Dunlap's encouragement, Harrison submitted a formal request, stressing that as the state's only technological university, Tech was likely to remain male-dominated:

The number of women electing programs like ours will never be large, but it is difficult to justify denying those so inclined an opportunity. If fear exists that Tech would become overrun with women students, it can be allayed. The present enrollment of women at Tech is approximately one per cent of the total. Removal of all restrictions would have little effect on this figure. I feel confident that the proportion of women students at Tech would never exceed ten per cent of the total, and I believe it would take many years to reach this figure.[112]

At its March 1965 meeting, Georgia's Board of Regents approved a measure to admit qualified female applicants to all Tech majors, with one remaining exception. Women wishing to concentrate in industrial management had to continue appealing to the regents for individual permission until 1968, when Tech officials convinced the board to lift that final restriction.[113]

Housing issues remained the major practical obstacle to expanding female enrollment in the mid-1960s, given school rules that required first-year women to live either on campus or with Atlanta-based relatives. Given the limited capacity of the Women's Residence Hall, Tech's controller warned admissions personnel to remember that they could enroll no more than ten or eleven women from outside the city each year. In 1962, the Women's Student group began lobbying for construction of a new dormitory with at least one hundred beds. The leading advocate, civil engineering senior Beverly Ann Cover, noted that when she moved into the existing hall as student counselor to incoming women, a shortage of rooms meant that she "had to sleep on the porch until nearly Christmas." Another woman, who arrived on campus to find the women's residence hall already full, inquired about sleeping on the porch, only to have an administrator reply "that for all he'd cared, I could go home."[114]

Beyond the housing difficulties, other issues throughout the 1960s led a number of Tech's female students to regard their treatment as second-class. Women's restrooms remained difficult to find in many parts of campus. The old gymnasium pool had no women's changing room, which meant that women walked to their residence hall in swimsuits. In academic terms, Tech's freshman honor society and a number of other scholarly honor organizations still did not officially admit women as full members. Most noticeably, Tech women complained about a "somewhat antagonistic and indifferent attitude of the administration toward the girls." Female students sarcastically noted the "warm" welcome offered right from the beginning, including an acceptance letter that read in part as follows:

I wanted to let you know some special factors about life as a coed at Tech which may or may not affect your desire for attendance here. We have a student body of about 7,300, only 100 of whom are girls. This may at first sight seem attractive to you, but many . . . have found it to be quite disconcerting. Very few concessions are made for coed students here at Tech.[115]

The default Tech student was masculine, and Lee Diane Gostin laughed about receiving a list of men's dorms from which to choose her housing preference.[116]

By 1968, Georgia Tech had begun construction on its first women's dormitory, with plans that included a study lounge, kitchen, laundry facilities, and air conditioning. Earlier Tech housing had been designed with men in mind, so after the female student representative on the President's Advisory Cabinet, Mikki Hodor, looked over the blueprints, she recommended replacing the space set aside for urinals with extra shower stalls. In contrast to the isolation imposed on women in their old residence hall, the new dorm promised to make it easier for female students to join school activities. Georgia Tech located the new dorm near the campus commons, which served as a convenient spot for women to meet men for joint study sessions. The initial building phase accommodated sixty female students, but construction director Odell Williamson said, "If we get more, we can move the men. . . . The old alumni are completely amazed. It's hard to get adjusted to the idea of ladies over here." Tech immediately began preparing plans to build extra women's housing, including about another 140 beds for female students inside a large new dormitory complex. That physical commitment to a larger female presence reflected the clear intent of Tech's 1968 institutional self-study, which declared, "Women students have found their way into the membership of almost every campus organization and activity, and their academic achievements continue to be equal to or above their male counterparts. Georgia Tech anticipates an increased coed enrollment in the near future." One female student that year put it more bluntly, declaring that with additional dorms, "this campus will become like a [real] coed campus. . . . They are heading in that direction so fast. Of course they are not going to predict that. It would kill all the alums."[117]

Women's presence indeed showed a clear numerical upswing in the 1960s. In 1962, Georgia Tech had fifty female undergraduates enrolled with about eighteen engineering majors and, by mid-decade, around one hundred undergrads. Following the completion of the first women's dorm in 1969, enrollment of female students jumped from about 132 to 212. The class entering Tech in fall 1969 contained eighty-five women, the largest group to date.[118]

Even as women's presence passed two hundred, they remained outnumbered by men roughly forty to one. That lopsided ratio meant that Tech life was still effectively defined by and for men. Alumni in particular maintained that rather than altering institutional culture, coeducation simply inserted "women in a man's world." The alumni magazine of 1970 ran a cover photo of a woman carrying both a purse and drawing board, which was meant to represent both her ability "to adjust easily to the male tempo of the place and her own loneliness as she walks the campus. . . . [D]espite the fact that it now graduates more female engineers than any in America, [Tech] still remains in the minds of the outsider (and many alumni) as the most masculine institution of higher learning in the country."[119]

In the early 1970s, women were not actually the smallest minority at Tech. During the same years that women had been struggling to gain access to the school, the issue of racial integration also came to a head. Throughout the pre–World War II decades, opportunities for young African American men and women to enter higher education had remained sharply limited in both the North and South outside the network of historically black colleges. Patriotic World War II rhetoric and posters promoted an idealized vision of black and white Americans working together toward victory, but discrimination remained pervasive in both military and civilian life. Many black college students, along with members of the National Association for the Advancement of Colored People (NAACP) and other groups, objected to the ironies and unfairness of wartime segregation in a nation that was fighting to defend democratic freedom. In 1948, President Harry Truman issued an executive order to integrate the U.S. armed forces. In 1954, the U.S. Supreme Court unanimously ruled in *Brown v. Board of Education of Topeka, Kansas* that public school segregation violated constitutional protections of citizenship because separate educational facilities for blacks and whites were "inherently unequal." Implementation and enforcement of that landmark decision remained slow and problematic, giving opponents time to mobilize, protest, stall, and seek to block change at schools that had long been white-only by law and tradition. Across many parts of the South, Confederate flags symbolized a passionate resistance to desegregation that often turned hostile or violent.[120]

In Georgia, authorities strove to retain the color line in higher education. State law prohibited school integration and threatened to cut off funding for any institution that adopted desegregation. In 1959, the University of Georgia made excuses to reject applications from teenagers Charlayne Hunter and Hamilton Holmes, honors graduates of an elite all-black Atlanta high school. On January 6, 1961, federal judge William Augustus Bootle ruled that Hunter and Holmes deserved immediate admission to the University of Georgia, an order that was upheld by the U.S. Supreme Court. Sentiment at the school split, and many students and faculty welcomed or at least accepted the decision. But as they arrived for classes, Hunter and Holmes were confronted with crowds chanting anti-integration slogans. A few evenings later, angry students joined Ku Klux Klan sympathizers in rioting outside Hunter's dorm. They shouted racist slurs and threw bricks through the window of what *Time* magazine described as the "pretty Negro Coed." Hunter and Holmes remained at the University of Georgia amid an atmosphere of isolation, finding corners of support, especially from faculty. The two graduated in 1963, having defied repeated episodes of harassment and threats of physical harm.[121]

The January 1961 storm at the University of Georgia prompted officials at Georgia Tech to consider how their school might handle the process of desegregation more smoothly, before any federal court forced the issue. In May 1961, president Edwin Harrison declared that Georgia Tech had inspected applications from thirteen African American candidates and decided to accept three. Atlanta black high school graduates Ford Greene, Ralph Long Jr., and Lawrence Williams arrived in fall 1961, joining a first-year class of about 1,200 white students. The three registered and started attending class without confronting any antagonistic crowds such as those gathered nine months earlier in Athens, Georgia. President Harrison threatened to expel any students who created disturbances and barred reporters from campus. A small KKK protest did not escalate, and one paper summed up the situation with the headline "Georgia Tech Is Desegregated with No Fuss."[122]

The 1961 racial transition at Georgia Tech remained relatively low-key and peaceful for a number of interconnected reasons. Tech administrators were eager to dodge the negative national publicity that the University of Georgia encountered due to the rioting on its campus. Also, urban Atlanta offered a climate of everyday diversity and civil rights dialog that was significantly different from the conditions in Athens, Georgia. Although Atlanta had experienced a notorious race riot in 1906, was home to a large postwar KKK contingent, and continued to display racial distrust and dislike, the city also was a historic center of educational, cultural, and business opportunities for African Americans, the birthplace of Martin Luther King Jr., and the headquarters for his Southern Christian Leadership Conference. Politically progressive and moderate white authorities informally negotiated with black advisers and aimed to differentiate Atlanta from other southern cities through its racial tolerance. Pragmatic business boosters sought to avoid race-related disturbances that risked scaring away investors. The members of this commercial-civic partnership hoped that nonviolent, gradual desegregation would maintain social peace, thus promoting both prosperity and national admiration for Atlanta. Mayor William Hartsfield described Atlanta as "a city too busy to hate," a phrase subsequently embraced in pro-business public relations. In a 1959 *Newsweek* interview, Hartsfield said that he sought to "undo the damage" that "loud-mouthed clowns" had done, so that "no business, no industry, no educational or social organization [need be] ashamed of the dateline, 'Atlanta.'" In the late 1950s, Atlanta hired a few black police officers, desegregated a public golf course (following a landmark lawsuit), and removed the labels "White" and "Colored" from airport bathrooms. In August 1961, just a few weeks before Georgia Tech's integration, nine black students entered previously white high schools without open conflict. President John F. Kennedy and the U.S. attorney general, Robert F. Kennedy, praised Atlanta as a model for desegregation elsewhere, a positive theme that was echoed in the *New York Times*, *Life*, *Good Housekeeping,* and other high-visibility press coverage. The months surrounding Tech's enrollment of black students also brought desegregation of many Atlanta restaurants, public parks, and city pools, under pressure from activists' sit-ins and from court orders. White opponents to desegregation sought to delay change and often made concessions only grudgingly. Later years would witness plenty of racial tension and trouble in Atlanta, white flight to the suburbs, and black frustration with tokenism and a lack of economic and political justice. Nevertheless, at the time of Tech desegregation, many city leaders felt proud that they had kept their city calm.[123]

The admission of black students at Tech may well have triggered less antagonism than it did at the University of Georgia because Tech's students, faculty, leaders, alumni, and the broader community had recently exhausted themselves over the coeducation debate. The arrival of women had opened discussion about the nature of institutions in changing times. In the end, Tech's desegregation posed a striking contrast to episodes on other campuses, most notably the deadly racial rioting that erupted the following year, 1962, when African American student James Meredith arrived at the all-white University of Mississippi.

Georgia Tech administrators and professors offered the newcomers vital support, and dean of students Jim Dull extended his personal assistance to them. The obstacles to integration at Georgia Tech were less blatant than the violence in Mississippi or at the University of Georgia, but nonetheless very real. Ralph Long Jr. noted that as a black student, he could not share his white classmates' access to fraternity test files or other insider information. Atlanta-born Ronald Yancey, who arrived at Tech as a transfer from Morehouse College, had difficulty finding lab partners. After getting some menacing telephone calls, Yancey avoided studying at night in Tech's library. Instead, he and others felt more comfortable spending their spare time around nearby historically black Atlanta University. Tech did not allow Long and Greene to join the football team, and the two ultimately transferred elsewhere to pursue degrees; the military drafted Williams. Yancey became the first African American student to graduate from Tech, finishing in 1965 with an electrical engineering major.[124]

There were no African American women at Tech in the early years, but the school's few African American male students and the few white female students shared some experiences. Each black man and white woman visibly stood out on campus due to their rarity. Each faced the repeated reality of being the only one of their kind in the room and often were literally set apart. Yancey recalled, "What really bothered me was the ring of empty seats around me in class"; white men also sometimes avoided sitting next to the first female students. White women and black men alike were isolated, intellectually and professionally, from the school's distinct world of engineering culture. But ostracism by race was ultimately also different from that by sex. Black men's isolation extended to much of the wider campus world. They were initially barred from fraternities, football, and casual social mixing, where white classmates ignored them as if invisible. Tech's second African American graduate, Frederick Espy, remembered, "Nobody ever spoke to me." Women's isolation was relieved because white male students would talk with their female counterparts, but women sometimes perceived that male classmates approached them with skepticism, as a type of test, or with ulterior motives, thinking ahead to date night. White women did not confront anything resembling black men's fears that integration might compromise their safety. Southern male chivalry generally protected female students from public threats or routine physical hazing. But although white women were normalized by race, their gender was repeatedly portrayed as inherently incompatible with engineering. Women were trapped in a double bind of two false assumptions—that they must not be serious about their studies and had come to Tech only to land a husband and that any female who genuinely was interested in technology could not be appropriately feminine and attractive.[125]

It took almost a decade after black male students initially arrived at Tech for the first African American women to enroll, starting around 1970. Attrition rates among the first black men and women to matriculate at Tech remained high due in part to the multiple stresses on that population. The first black women to complete degrees at Georgia Tech were Grace Hammonds and Clemmie Whatley, who both earned mathematics master's degrees in 1973. In that year, Tech still only had about fifteen African American female students altogether, for whom the double-minority status multiplied the difficulties of belonging. Freshman biology major Brenda Gates reported, "I really get some odd stares," along with questions about why she chose to come to Tech since "there's nobody for you to date."[126]

Tech officials seized every opportunity to boast about the school's production of female engineers and black engineers. In 1966, dean of students James Dull declared, "Georgia Tech continues to have the greatest number of women engineering students of any school in the U.S.A.; an almost unbelievable first but don't forget it. (And in only 15 years too!)" Administrators pointed to evidence from both the Society of Women Engineers and the *Journal of Engineering Education* that suggested that Tech far outranked any other American institution in training coed engineers. According to *Southern Engineering*, Georgia Tech graduated fifteen female engineers in 1966, while their closest competitors, MIT, Colorado, and Brooklyn Polytech, each produced only five.[127]

Georgia Tech bragged about not only the numbers, but the abilities of female undergraduates. Women joined men in graduating with honors; in 1963, four won the right to be inducted into Phi Kappa Phi, the school's top academic society. Nevertheless, some observers remained concerned that given the limited number of young women in American high schools with cultivated ability in math and science, a rapid increase in the quantity of Tech female enrollment might endanger quality. Tech officials and faculty insisted that the institution maintained high academic expectations and did not pretend that every woman who enrolled would automatically manage to survive. There were also dropouts among Tech's male students, of course, but the unique pressures on women drew special attention to their attrition rate. Placement-center director Neil DeRosa said in 1961, "We have noticed . . . if girls don't fail as freshmen, they're exceptionally good upper-class students. We have found men will stay here with mediocre grades. Girls realize, since they are girls, they need to have above-average scholastic standings." Twenty-one women entered Tech in 1968's class, but their ranks rapidly shrank to ten.

Former valedictorian Kathy Coldren went from earning As in high school math and science to Ds in college, with a first-quarter grade-point average of 1.8. Coldren blamed inadequate preparation at her small high school, which had not even offered calculus, and she changed her major from chemical engineering to industrial management. Others spoke about needing remedial assistance, especially in calculus, trigonometry, and physics, to catch up with classmates. One told the alumni magazine, "I think Tech lowered its standards this year to get girls to fill the dorms but I don't think that will be a problem in the future."[128]

Some female students hoped that a rise in numbers would naturally equate to female strength, opening larger campus roles for women, specifically in student government. Two women won seats on Tech's student council in 1969, and Susan Clemmons, the first female candidate to run for vice president, drew more than a third of the vote. Soon, one Tech woman commented, "I think it will become a tradition for the student council vice-president (or secretary) to be a girl." Aiming for the top leadership spot (at least publicly) did not yet seem practical.[129]

The "Techettes" (one 1960s nickname) maintained separate governance structures and activities that both paralleled and interacted with the male-dominated main organizations and events. All entering females automatically joined the Women's Student Association. Tech's female engineering majors also had the option of joining the campus chapter of the Society of Women Engineers. The student group occasionally met with the Atlanta section, which connected current students to engineering graduates in the area, including a growing roster of Tech alumnae. One such veteran, Shirley Clements Mewborn, explained her willingness to help counsel female students by saying, "I know I had problems at Tech and I think we can help the girls by putting our past experience to use."[130]

Gamma Psi also continued to grow. Its activities included producing "faculty frolics," talent shows that raised money for charity and raised the profile of female students in the Tech community. These organizations continued building what came to be a Tech women's culture, complete with its own rules (both written and unwritten). Gamma Psi ran summer-camp sessions each year to welcome incoming women. Such programs aimed to ease the transition for new women by pairing them with a "big sister" who could familiarize them with the norms of campus life (in its different reality for women). The orientation advised women that they would be "needing help," since "in your first year at Tech you are going to meet special problems . . . unlike those of the rest of the students."[131]

Even as such programs sought to support new women, they also confronted female students with difficulties lying just ahead. Rhetoric underlined the pressure created by the many levels of expectation imposed on women by school tradition, by officials and faculty, by male alumni and students, and not least, by other Tech women. In Gamma Psi's 1966 message to women, dean of students James Dull wrote, "Membership in a decided minorities group has, does, and always will demand added responsibility and greater attention to the welfare of the group. Strive to make each individual contribution a meaningful one . . . that reflects a challenge met successfully and in the best interests of the advancement of women students at Georgia Tech."[132]

Gamma Psi published an annual handbook of rules, which they emphasized had been created by and for female students. For enforcement, women were "honor bound to report any infraction" to the Women's Student Association because "we govern ourselves." Some rules were couched in gender-neutral language, declaring that all Tech students should display "respect for order, morality, and the rights of others . . . honor, integrity, and dignity." But other principles made clear that females had to hold themselves (and were held) to particularly high standards as a way of validating their presence: "Women students should behave at all times in such a manner as not to bring justified criticism to them or to the school. Unacceptable conduct not only discredits the individual, but also reflects unfavorably on all women students and upon the institution." Alcohol consumption on campus was forbidden, the rules noted, adding, "Drinking in excess is unladylike and improper." Clothing prescriptions defined decency in detail and banned tight or low-cut blouses, high-cut shorts, and strapless dresses without a jacket. Femininity remained a prime concern; women were not allowed to wear jeans or slacks on campus on weekdays or on Saturday and Sunday mornings. Regulations did allow concessions for the Georgia heat. In spring and summer, women could wear Bermuda shorts in campus dining halls and even to lab, but "only if the instructor has been asked beforehand and has consented. A skirt or non-transparent coat must be worn over the attire when not in lab."[133]

In earlier years, when the number of female students on campus could still be counted just in double digits, Georgia Tech assumed that any special needs that women might have for advising could be handled informally by mentors on campus (such as President and Mrs. Van Leer) or in the community (SWE members Maryly Van Leer Peck and Shirley Clements Mewborn). But as women's ranks grew to three digits, complaints about neglect or even disdain from school administration demanded attention. In 1969, Tech created a new position for an assistant dean of students for women, which the school called "the latest and most timely recognition of the status of women students at Georgia Tech . . . who can help, advise, direct and assist in this involvement toward full participation in what is still a man's world."[134] By the early 1970s, Tech also had the state's only female campus minister, Caroline Leach, who chose to work at Tech because she empathized with the stress facing its few female students. As one of five women when she studied at the Columbia Theological Seminary, Leach recalled a "downright hostile" reception from male classmates who refused to "acknowledge me or [would] walk up quoting Scriptures trying to say I had no business studying to be a leader in the church. Many . . . asked if I was there to catch a husband." Speaking of her pain at being "ostracized," Leach declared, "It's very frustrating for a female to compete with men. A woman is socially isolated . . . very lonely. . . . So, I thought the women at Tech would want someone there to talk to."[135]

Regarding treatment inside the classroom, a number of women in the 1960s and early 1970s reported that most faculty members treated them fairly but that they still encountered a few "throw-backs to another age who think a woman's place is anywhere but at an engineering school." Susan Clemmons, who graduated in the upper third of her 1968 class, remembered, "Sooner or later, you got a professor who said he never passed a girl yet and didn't intend to start. You know you just have to work 10 times harder." Karen Borman said in 1973, "There are a few profs who will melt to a coed. Others don't think you belong here." But other female students believed that Tech had outgrown any problems of discrimination. Kathy Coldren said, "I'd like to think that there isn't [any more prejudice], so maybe I'm not seeing it if it is here. I know a couple of girls who said they had profs that didn't like girls, but I don't know if that is true."[136]

Paula Stevenson, who remained on campus as a researcher after receiving her 1958 degree, believed she saw substantial alteration in male undergraduates' attitudes toward female counterparts over the course of just a few years. Stevenson commented in 1962 that "men students frequently showed resentment toward us when I was a freshman. . . . they looked down on us. They would go out of their way to show us we weren't wanted. But all that is changed today. I'd say resentment against girls on the campus has almost disappeared." Observers credited Stevenson herself with helping erase bias through her commitment to both academic success (she was the first Tech coed selected to *Who's Who in American Universities & Colleges*) and school spirit (she was involved with the band, student council, and Ramblin' Reck Club).[137]

Other observations, however, suggested that some Tech men remained skeptical about women's presence. In 1962, the student-body president, Moe McCutchen, said, "Some [of us] like them [Tech girls] and some don't. . . . They don't bother me. As a matter of fact, their representative on the student council has some good ideas. They're good students. They're doing a fine job. They help us—I think." Some women blamed the campus paper for continuing to disparage them. A 1965 article evaluating the desirability of "our infamous coeds" concluded:

These girls . . . are well worth meeting and definitely cannot be categorized as being below the standards of any other school. In fact, the only thing which can be held against them as a group is that they are Tech coeds. Whether or not this is a detrimental remark is a matter of individual . . . opinion. It is simply a matter of whether or not you feel coeds have a place on our campus.[138]

Some female students believed that attitudes would inevitably evolve over time and that each year's turnover of male students reduced the number who doubted whether "coeds have a place on our campus." One 1968 female student said she saw resentment dropping each year as the men who had been exposed to heated debates over coeducation left campus: "I have noticed that the older boys are much more against coeds. . . . These boys say 'I don't think there should be any girls here at all, this should be a man's school,' and they still think that."[139]

Daily relations between the sexes on campus still sometimes remained uncomfortable. Michelle Murphy admitted in 1970 that the "overabundance of stares" felt "frightening sometimes," especially during her first months: "I used to feel that either I was not completely dressed, my slip showed, or the zipper to my dress was down. I then realized that Tech guys just enjoy staring. . . . A girl feels like crawling in a hole or wearing a maxi to cover her legs." Psychology graduate student DiAnne Bradford said in 1973 that the staring made her so self-conscious that she began wearing longer skirts and, for the first time in years, a bra. Remembering how male students had hurled water balloons at her during freshman year, 1970 graduate Patty Durant interpreted such harassment as men's stress test of women: "If you stick around for a while, people start to take you seriously as a student."[140]

Beyond the staring, women perceived a Tech culture that fostered gender separation and a lack of comfortable, casual interactions between men and women. Female students noted that men often avoided sitting beside them in lecture hall, leaving a lone woman "surrounded by a circle of empty seats which act . . . like a buffer zone isolating them from the other classmates." Other women regretted the absence of "a closeness which exists between boys and girls at other campuses. . . . Here you don't see a crowd of boys and girls standing and talking and laughing between or after classes."[141]

## WOMEN ENGINEERS: STEREOTYPES AND STRESS

Tech's female students frequently maintained that they did not intend to undermine their engineering classmates' masculinity. Women noted with pleasure the polite manners of Tech's southern gentlemen, who held open doors for them and picked up their dropped books. Linda Craine "certainly [didn't] object to the common courtesies afforded a lady.

Being one of the guys just isn't very soul-stirring! If anything, Georgia Tech girls are more . . . aware of the necessity of preserving their femininity in a man's world." Craine's contemporary Michelle Murphy, however, claimed that "trying to keep your feminine identity among 8,000 guys is similar to surviving in an African jungle."[142]

Although several Tech women publicly praised Tech men as being more mature than those at Emory or other nearby schools, male classmates did not always return the compliments. Opinion reportedly denigrated Tech women as being "short, fat, and ugly." A 1965 article investigating "the world of Tech 'femininity'" (with "femininity" in quotation marks) feigned surprise at finding "real, honest-to-goodness girls," rather than duds or walking computers: "They walked like girls, talked like girls, and, most assuredly, looked like girls. There certainly was not a football player in the lot." Disparaging nicknames for them included "co-odd" and "Ko-tech" (a pun on "Kotex"). One sophomore man commented that the number of women without Saturday-night dates showed they must be "either ugly or screwed up." A junior added, "If she's alone in her room on Saturday night, she's definitely not a whore—or else she's really ugly."[143]

Reflecting an assumption that "normal" femininity did not involve spending hours poring over engineering textbooks, a 1965 *Technique* article reported that Tech's "girls while away their many free hours playing bridge rather than the famous Tech pastime of studying," a comment that both reflected and fanned lingering skepticism about whether women could succeed at Tech. But female students generally emphasized their academic commitment. "I love my major," industrial-design major Linda Linn said, and "I work in my lab constantly. I don't have much social life." Others mentioned that rather than going out on weekend dates, they valued the extra time to study and make up on lost sleep.[144]

Female students themselves picked up on the bias that engineering women must be by definition unfeminine and therefore unappealing. Expressing distaste for a commonly heard pickup line, Jo Anne Freeman said, "We get pretty tired of hearing 'you're different—you don't look like a lady engineer.' If we were smart enough to get in here, we can certainly figure out how to make ourselves look attractive." Tech women sought to establish that they combined attractiveness and brains. In 1969, the Women's Student Association sponsored several entries in the Homecoming Queen competition, and the striking six-foot-tall Kathy Coldren won. Pictures showed her both as a "tomboy" chatting with shirtless men after a touch-football game at freshman camp and as a "beauty" in short skirts holding a bouquet.[145]

Women often complained that many Tech men were not open to dating female classmates. Some women said that men never bothered to ask them out and assumed that their social calendars were already full. Other female students worried that they intimidated the men. Linda Linn said, "Most of the guys are afraid that girls will shun them." Campus minister Caroline Leach agreed that she saw many Tech men reluctant to approach their female counterparts: "Tech guys seem to only date girls from Agnes Scott College and Massey (Junior College) because they think they're less threatening."[146] But some Tech women and men did date, and by the 1960s, coed life included engagements and marriages (one to a Tech professor). Two out of ten women graduating in 1966 had already wed Tech men. Coldren, who was going steady with a Tech engineering classmate, said in 1970, "I'd like to finish Tech before I get married, . . . but I can't guarantee that at this stage of the game." Publicity about Tech's female students who married and had children (even twins) underlined that they were "normal" feminine women, not freaks of nature who had been contaminated by attending a traditionally male school.[147]

Tech women who married during or immediately after college helped to validate the idea of female engineers as feminine and even desirable, but also underlined continued suspicions that women students lacked seriousness. Throughout the 1960s, many observers still suggested that women came to Tech primarily to take advantage of the skewed gender ratio. Female students repeatedly countered that any party girl would soon flunk out. Linda Craine wrote, "I want to let the male student body of Georgia Tech in on a little secret. There are better reasons for females coming to Tech than the ones you people generally accept. It doesn't make sense that a girl would come here seeking a mate, now does it? After all, we suffer through six-hour labs and tread halfway across the campus to wash our hands."[148]

Occasionally Tech women themselves denigrated classmates as husband hunters. Coldren told the 1970 alumni magazine that "A lot of the girls might have come here for a husband and you can begin to pick them out right about now. They're the ones who are letting their studies slide, and we have our share of them just as any other college has." In 1973, several female students suggested to the *Atlanta Journal* that some classmates had chosen Tech hoping for a rich social life but "are gone by now. They were so dumb."[149]

Although Tech women wanted to be seen as attractively feminine rather than ugly eggheads, female students also commonly separated themselves from other women by pointing to their intellectual achievements and willingness to venture into nontraditional fields. Some explicitly scorned what they considered to be the shortcomings of stereotypical femininity. Craine said, "There are some women in this world who are simply differently motivated than others. Not all women find intellectual satisfaction in a soap opera written on the level of a six-year-old. . . . Some women would prefer to exercise their minds, and to seek relevance in the world. We don't all enjoy giggling or gossiping (as is rumored) our lives away." Female students repeatedly emphasized that like men, they attended Tech to pursue a love of science or technology. Given a social stigma in certain quarters against women displaying their intelligence (especially if they outshone some males), female students appreciated being able to embrace an academic challenge openly. One said, "It's refreshing to be around a school where you aren't afraid to be smart." Some also spoke of a desire to test themselves against the challenge of surviving as a female student at Tech. Coldren declared, "I never felt that anything was worth doing unless it was hard." Her father had opposed Coldren's plans to attend Georgia Tech, "and that's one of the reasons I chose it, because if somebody tells you not to do something that's usually what you want to do."[150]

Like their male counterparts, many women who chose Tech had alumni relatives and absorbed school loyalty from them. After the school began accepting women, some male graduates began grooming their daughters to follow in their footsteps. The father of 1968 freshman Lee Gostin took pride in her ambition to attend his school, having sung "Ramblin' Wreck" to her since childhood and "brainwash[ed]" her with Tech lore and reminiscences. Other female Tech students picked up a fascination with engineering from male relatives. Civil engineer Anita Reed credited her career choice to the influence of her civil engineering father and uncle: "That's about all I've known all my life."[151]

Tech women pursuing nontraditional studies such as engineering still faced a common popular-culture message that the "natural" wish for a woman was marriage, quickly followed by child bearing. Some Tech undergraduates criticized traditional patterns of motherhood as anti-intellectual and personally suffocating. Jo Anne Freeman said, "We [Tech women] could never be content to stay home and let our minds stagnate while we concentrated on diapers and dusting. . . . When we took the place of a boy who might have gotten this education, we accepted a responsibility to use that knowledge." Freeman and fellow students of the 1960s did not renounce a desire for marriage and children, nor did they agonize (at least in public) about the trauma of trying to juggle career and family. Rather, most declared that they intended to work for a few years, take time off for babies, and then reenter employment. Few overtly anticipated any problems with this strategy.[152] Many stopped short of declaring an intent to pursue lifetime careers, especially if that might correspond to an undesirable singlehood. While noting that a college degree could provide economic security for divorced or widowed women, Michelle Murphy quickly added, "Every girl has fleeting thoughts of walking toward the altar, since coeds are females and we would like the role of wife and mother someday. Nevertheless, we have intelligence and drive; otherwise, schooling would have stopped after receiving the high school diploma. So, if the benefits of Tech happen to include a husband, who are we to argue?"[153]

The message that female Tech graduates could and should alternate family and profession suited the school's hope that the women it trained could contribute to national science and technology needs while retaining feminine desires and instincts. In 1970, the alumni magazine profiled Mary Anne Jackson Wright, who had pursued her love of engineering, despite being discouraged by a high school guidance counselor who did not believe that she had the intellectual capacity to handle Georgia Tech. With a 3.7 grade-point average, Jackson ranked in the highest 2 percent of her 1964 Tech graduating class and atop the aerospace engineering department's roster of fifty-eight students. Far from burying herself in books, Jackson also pursued an active social life, capped by marrying fellow aerospace engineering major Terry Wright. After graduation, the pair remained at Tech to earn their master's and doctoral degrees. The alumni magazine emphasized both Mary Anne's scholarship and her domesticity, describing how she sewed clothes for herself and her husband, knitted and crocheted, and after returning from campus each night, put together dinner "to feed a hungry husband quickly." Mary Anne planned to work for a few years and save money, withdraw for an indefinite time to have two babies, and rejoin the workforce as her children grew older. Mary Anne expressed her horror of becoming "the absolutely trapped housewife" and believed that engineering gave her flexibility: "If I should decide that I need more time with the children, . . . I could always find a job teaching part-time."[154]

Despite the low number of women in America's engineering profession in the 1960s, Tech's female students publicly expressed optimism that good employers and male coworkers would welcome them. Through persistence and professional ability, the women felt sure that they could overcome any occasional episodes of prejudice. Showing faith in a meritocracy, chemical engineering major Audry Brickhouse said, "A woman's sex will not be a drawback in engineering if she is dedicated. Promotion for a woman will come slowly, but if you stick with it I'm confident it will come." Jacqueline Hill added, 'I have heard that if a woman graduates from Tech, she will be classified on the same level as a man. In business it's whether you're an engineer, not whether you're a female." Mine Enginun, a Tech student from Turkey, questioned the notion that engineering must be inevitably characterized as male, saying, "In my country, more than half the chemists and chemical engineers are women. They are said to be more patient with detail work." Mary Anne Jackson Wright said, "Women have traditionally held desk jobs . . . and the work of an engineer is today for the most part a desk job. It requires the ability to think, but I'm not aware of any natural law which separates the sexes on this count. But I can see some areas in engineering—like hardhat construction work—where I don't think a woman would fit in well."[155]

During years of an expanding job market in business and government, the demand for well-educated specialists did foster a certain openness to hiring women engineers. Tech officials boasted about their most accomplished female students, such as 1962 engineer Beverly Ann Cover, who joined the U.S. Bureau of Roads. Neil DeRosa, director of Tech's placement office, bragged that ceramic engineer Dorothy Vidosic, who graduated in 1960 with the school's third-highest grade-point average, received multiple job offers.[156] In 1962, Tech civil engineering graduate Anita Reed became the first female engineer to apply to Georgia's Highway Department and the first hired. Reed's supervisor worried that curmudgeonly contractors might balk at taking directions from a female engineer and planned to adjust standard fieldwork training to avoid exposing Reed to anything inappropriate: "We wouldn't want to send her into the Okefenokee Swamp in charge of a survey party and have her wearing hip boots and dodging snakes." But the state needed engineers so badly to accelerate its road-building drive, her new boss said, that "we don't care whether they're men or women as long as they're qualified." Reed appreciated that she would earn the same salary as her male counterparts. She noted that some of Tech's other female engineers were not given fair treatment on wages, "but if they hold out for equal pay, they usually get it."[157]

In 1968, Tech females remained positive about their professional futures, declaring, "Engineering is opening up for women so fast. It has just increased fantastically in the last ten years." But the economic downturn of 1970, combined with reductions in government spending and a dim outlook for the aerospace industry, suddenly made employment prospects seem grimmer for all Tech graduates. About one thousand alumni signed up for assistance from the school's placement service, which normally handled half that number. Several women who finished degrees that year experienced disheartening difficulty in securing work. Industrial management graduate Mikki Hodor complained that "The big companies sometimes bring in two or three interviewers to explain in great detail they aren't discriminating against you by not hiring you," while other firms simply announced their refusal to employ women.[158]

Within a few years, however, experts predicted a dramatic upswing in the job picture. Tech engineering dean Charles Vail said in 1973 that he foresaw a "drastic shortage" because the country would see "a need for about 400,000 more engineers in the next decade." Tech observers particularly anticipated a boom in hiring for minority engineers (counting women as a minority) due to the civil rights movement, feminism, and the passage of equal employment opportunity legislation. The placement-center director, Bill Pickel, said that an African American engineer could "get almost anything" and that the school's female engineers were also starting to benefit from preferential treatment: "Women are coming on just as strong as the blacks," especially since "it's easier to find a girl" in engineering.[159] The suggestion of affirmative action made some Tech women recoil. Wanting to be judged strictly on technical merit, these female engineers insisted that they could succeed by competing with men on an even basis. A 1970 graduate, Ann Patterson, said that she "wouldn't want any job because they need a token woman and it would be even better if I were bright. I want to be accepted as a good aerospace engineer."[160]

In the early 1970s, school officials proclaimed Tech to be "one of the Nation's fastest growing women's colleges." Numbers bore out that assertion as female students' fall enrollment grew from 212 in 1969 to 276 in 1970, 365 in 1971, 452 in 1972, and 600 in 1973. Not coincidentally, that expansion came at a time when Tech began devoting more attention to the challenge of both increasing the number of female students and improving their position. By 1973, Tech had hired its first female engineering professor. Thomas Stelson, assistant engineering dean, commented, "We are recruiting for more women engineer professors, but they are extremely hard to find."[161]

In 1972, Tech hired Esther Lee Burks, a master's student in industrial and systems engineering, to serve as a special assistant to the dean of engineering in charge of developing programs to support and advise women. To encourage more high school women to consider studying engineering at Tech, Burks helped produce a recruiting pamphlet to show how "our facilities and atmosphere become continually more congenial to the intelligent young woman of today."[162] The brochure opened by reassuring potential students that they "could make it as a woman" in engineering and earn higher salaries than any other major: "[M]any companies are anxious to hire Tech students. Currently, women . . . engineers have a particular advantage." To emphasize that Tech welcomed women in nontraditional fields, the pamphlet stressed that 20.5 percent of current female students were concentrating in engineering, 19.7 percent in industrial management, 16.9 percent in architecture, 37.5 percent in math and science, and 5.4 percent in computer science. Both the campus chapter of the American Society of Mechanical Engineers and the engineering honor society Tau Beta Pi had female student presidents: "Tech is a good place for women to study. You can be yourself—feminine and intelligent. There is no need to pretend to be a silly girl; here you are expected—and respected—to be bright and to be able to use your brains."[163]

A later version of this brochure, revised with input from current female engineering students, sought to dispel any lingering belief that the profession was not suitable for women:

These women are not Amazons, but are intelligent persons looking for better ways of doing things [and] . . . to make a difference in the world . . . from outer space to city traffic. . . . In the past, one main hindrance to women entering engineering fields has been the image of the engineer as a rugged outdoor type with a transit or a greasy he-man clambering about a huge machine with wrench in hand. It's true that such jobs still exist, and some women enjoy working in these jobs. However, today's engineering is more likely to be carried out in an office or lab where problem-solving, both scientific and financial, is the essence of the work. Being a woman in engineering still has aspects of pioneering, but the doors to the profession are now completely open.[164]

Women could succeed in engineering without having to adopt a masculine personality, the leaflet emphasized, and they could maintain and even relish gender differences. It quoted one female Tech student:

I find an extra zing in life from dealing with men in a professional role. The issue . . . is how knowledgeable, proficient and competent you are . . . not how muscular your frame, how colorless your emotions, or how fluent your four-letter word vocabulary. The fact that men and women approach and react to situations in sometimes typically masculine and feminine ways is, to me, part of the joy of living and working, and I'm not at all in favor of changing it.[165]

Tech's brochure suggested that many women approached engineering studies differently than men did and that therefore the institution gave students wide scope in electives to combine courses from separate fields, such as industrial engineering and psychology: "This flexibility can be especially important to women students in engineering, who seem to want to explore a wider variety of interest areas than do men." The leaflet pointed out that engineering training could prepare women for multiple pathways, including graduate study, medical school, law school, teaching, sales work, freelance business consulting, and industry management. The booklet took pains to note that engineering education also offered advantages to future homemakers: "The wife and mother who is also an engineer can be a citizen able to cope with and understand more of our complex society than one who is not so widely educated. Studying engineering gives you great freedom for growth all your life."[166]

CONCLUSION

Changes both at Georgia Tech and in the broader educational and social context of the United States continued to increase women's presence at the institution. In the 1981–82 school year, Tech enrolled 1,883 female undergraduates and awarded 372 bachelor's degrees to women. Women continued to break barriers for academic recognition. Helen Gould, who received her industrial engineering degree in 1982, became the first female president of the school's top senior honor society, ANAK. Outside the classroom, Tech's undergraduate women enjoyed opportunities for involvement in a widening list of campus activities, including women's varsity sports, club intercollegiate teams, and five sororities. The prediction that larger numbers of women would result in a more visible presence in student government seemed to come true. In 1982, three female engineering majors swept the top three posts of president, vice president, and secretary; three other women were chosen as presidents of the sophomore, junior, and senior classes.[167]

Coeducation at Georgia Tech owed much to the activism of advocates such as Blake and Ella Van Leer, Dorothy Crosland, Anne Bonds, and the Atlanta women's clubs. They embraced the language of feminism and equality and demanded that Georgia provide in-state engineering training for women. Yet even those who spoke most passionately about fairness did not expect that Tech would find enough women who were qualified and interested in engineering to match its male students any time in the near future. Indeed, they used low projections of women's enrollment numbers to reassure doubters that Tech could admit female students while retaining its historically male spirit and traditions. The women who arrived at the school felt obligated to acknowledge its male-dominated character, even when it spilled over into derision of the prospect of female engineers. Tech did not readily allow much deviation from its traditional masculine character. Without the threat of lawsuits, the prospects for public engineering coeducation in Georgia might well have stagnated for years.

Even as Georgia Tech began to enroll a small but growing number of women in the early 1950s, leaders at the California Institute of Technology assumed that their under-graduate program would and should remain all-male. The first female undergraduates did not arrive at Caltech until more than a decade and a half later. Moreover, debates over coeducation played out very differently at the two schools, reflecting contrasting institutional cultures and the changes in cultural context between the early 1950s and late 1960s. As a private school in California, where other places offered women op-portunities to pursue in-state engineering degrees, Caltech did not face legal or external pressures for coeducation. Instead, the drive to open Caltech completely to women came from inside, reflecting the rebellious cultural context of the late 1960s. Whereas Georgia Tech's students and alumni often resisted or questioned change, at Caltech it was undergraduates who pushed for coeducation on social grounds as part of a broader campaign calling for reforms to make their school's curriculum and climate more hu-mane and more relevant to the real world.[1]

## CALTECH'S MASCULINE IDENTITY

Amos Throop founded Throop University in 1891 as a coeducational institution for manual training and basic education from the equivalent of fifth grade through college. In the early 1900s, George Ellery Hale led a campaign to reinvent the undistinguished institution with a new name, a new location, and a new mission focused on higher education. Throop College of Technology sought to establish intellectual excellence virtually overnight and, not coincidentally, committed itself to an all-male campus. Hale concentrated first on building up the electrical engineering department to train men to master a growing state's power needs. Over succeeding decades, as Throop evolved into Caltech, its dual commitments to science and engineering education and to an all-male identity became intertwined. Caltech's cultural notion of masculinity centered around academic intensity, imposing stressful calibrations of success on students and faculty alike.

Caltech's masculine engineering code reflected decades-old gender patterns in which both formal rules and informal social prescriptions barred most women from the intellectual and cultural centers of technology and science. Stereotypes painted women as illogical and overemotional, whereas leading educators defined ideal engineers as the diametrical opposite, rational and tough.[2] Popular culture reinforced the image of engineers as heroic and proud figures who used daring and knowledge to help modern civilization conquer hostile wilderness.[3] Early twentieth-century California celebrated engineers, innovators, inventors, politicians, and entrepreneurs as creators of massive urban infrastructure and expanded farming. The development of ports, roads, bridges, agribusiness, oil facilities, and the Los Angeles aqueduct all took an environmental toll but promoted confidence in seemingly endless technological progress. Even during the Depression, admiring observers marveled at the scale of epic engineering that was involved in building the Central Valley water system, Los Angeles parkways, Caldecott Tunnel, Bay Bridge, and Golden Gate Bridge.[4]

This context of California ascendant underlay Caltech's evolving character as educator of a selective elite, defined by intellectual accomplishments and an ambition to seize a national lead in science and technology. Pasadena drew a roster of faculty and research luminaries, including astronomer Edwin Hubble, chemist Arthur Noyes, and future Nobel laureates Linus Pauling (chemistry) and Robert A. Millikan (physics). World War II reinforced Californians' sense of the possibilities that were being opened by population growth, industrial might, and geographic advantage. Along with the state's massive programs in airplane and ship construction,[5] Caltech lent its scientific and technical expertise to weapons development and other defense efforts. It joined almost all good-size colleges across the country in offering short-term technical classes for women through the federally funded Engineering, Science, and Management War Training (ESMWT) program. These females did not enroll as regular students and did not gain credit toward Caltech degrees, but nevertheless used Institute facilities; in classes such as "Engineering Fundamentals for Women," Caltech staff taught drafting, surveying, and technical mathematics.[6]

After World War II ended, California continued to thrive as the cold war brought unprecedented government support to the state's "military-industrial-academic complex." In an age that closely tied scientific and engineering strength to the democratic world's struggle against Communism, California assumed leadership in jet-age aviation and aerospace, electronics, missile development, atomic power, and computers.[7] None of these new technical fields proved particularly welcoming to female participation.

Caltech, which took a leading role in rocket research through its management of the new Jet Propulsion Laboratory, remained a masculine club.[8] As in Victorian-era fraternal organizations, traditions and rituals dictated that male novices endure a passage of indoctrination, which linked initiates into a brotherhood founded on shared endurance.[9] Caltech students also displayed a notorious fondness for "pranks" that celebrated technical ingenuity and fostered male bonding, while confining troublemaking within acceptable boundaries.

The school occasionally hired a female researcher or staffer, but both undergraduate and graduate degrees remained reserved for men. In 1947, biology faculty challenged those limits, unanimously voting to place their division on record as supporting the admission of female graduate students. They pushed fellow professors, Caltech's executive committee, and the Board of Trustees to approve what they acknowledged would be a drastic policy change. Although "in general women should be discouraged from going into science because of the difficulties they will encounter in becoming professionally established," Caltech biologists declared, "there are individual women . . . so exceptional in ability, drive and other characteristics that they should be given every encouragement to follow . . . careers in science." Biology faculty members saw "no good reason" that Caltech should be "one of the few remaining institutions of higher learning in this country that denies women equality with men at the graduate level." The biology division announced that it had two female candidates whom it "might well recommend" for graduate training but quickly reassured skeptics not to worry about women taking over: "[W]e would in fact probably admit only an occasional woman student in Biology [and i]nitially, at least, . . . recommend that women not be given graduate assistantships involving teaching duties."[10]

Caltech's president, Lee DuBridge, made clear that "[e]ven now there is no suggestion that we admit women undergraduates," but he worried that opening the door to female graduate students might invite pressure for full coeducation. DuBridge regarded MIT, coeducational since the 1870s, as a useful model for comparison because it was a parallel high-powered institution oriented toward science and technology. DuBridge queried MIT vice president James Killian about whether MIT would choose to admit women "if it could reconsider the problem from scratch" and whether the school had any difficulty keeping the matter "under control."[11]

Killian reassured DuBridge, "I know of nothing in our experience here which would lead us to feel that a mistake had been made in admitting women students without restrictions." He declared that MIT felt "proud of our women students, and many . . . have made outstanding records in professional practice [even though m]any . . . get married either during their work here or shortly thereafter." Killian noted that "the number [of women] who apply continues to be small, but there is some tendency for it to grow" and that MIT even started letting them attend Freshman Initiation Camp after "[s]pecial housing was found for them, and they apparently were cordially and properly received" by male classmates.[12]

In truth, realities of life for MIT's few female students were less rosy than Killian indicated, but in any case, in 1948, the idea of Caltech female graduate admission went nowhere. Doubters cited expensive roadblocks such as the scarcity of women's restrooms on campus. But the issue resurfaced in 1953, when (according to some accounts) distinguished chemist John Roberts, then a professor at MIT, refused to move to Caltech unless permitted to bring his graduate research assistants, including Dorothy Semenow.[13] Caltech's Division of Chemistry voted by a seventeen to three margin in favor of admitting Semenow, and chair Linus Pauling pushed the matter forward through various special meetings. Later that year, Caltech's faculty as a group voted to admit female graduate students, provided they matched men in academic qualifications, and the Board of Trustees approved. The welcome remained half-hearted. Until 1967, Caltech catalogs announced that "women students are admitted only in exceptional cases" for graduate standing.[14]

The notion of female undergraduates remained essentially unthinkable, especially to conservative board members. In 1951, Charles Newton, assistant to Caltech's president, DuBridge, inserted into admissions booklets a few pictures of "nice-looking girls walking with boys through the arcades" to "convey the impression that there was a social life on the campus."[15] Trustee chair Albert Ruddock soon complained about the "very attractive photo of two handsome fellows and a good looking gal descending the steps of a student house. Does this suggest that both sexes occupy the student houses—and simultaneously??? . . . What . . . was the underlying purpose in publishing this, otherwise, attractive view? (Aside from the rather obvious but hardly admittable printed purpose of attracting well-sexed as well as well-brained students to the Institute)."[16] Chastened, Newton promised that future versions of *Facts about Caltech* would not feature any images of women. As commentary, he or some other administrator composed a poem:

Lines written to A. Ruddock . . . :
Sir, yours are most surprising taunts.
Why shouldn't nymphs be seen with fauns?
Follows it then as dusks do dawns

They live (in sin) in common haunts?
Sir: *honi soit qui mal y pense.*
And yet why *not* share common haunts
And view the same delicious dawns
Through the same windows? Nymphs and fauns
Differ so little! *Et puis, je pense*
*Viva la petite difference!*
Alas! Such moral nonchalance
Is as fatal for Tech as maiden aunts.
I'll see that the next edition flaunts
No cute but suspect debutantes.
*A bas la joie. Vive la science!*[17]

After 1953, Caltech's number of female graduate students slowly grew, reaching a "new record" of seventeen enrolled in 1963 (five in biology, four in chemistry, three each in physics and aeronautics, and one each in math and astronomy). The campus newspaper editorialized, "The girls are ambitious; . . . twelve are trying for the Ph.D. . . . One thing remains unknown, though: why would any girl come to a school with 1300 men?"[18] Because these women were generally over twenty-one years old, the Institute left them to locate off-campus apartments rather than face the complications of arranging official female housing.[19] By 1968, the thirty-seven matriculated women made up exactly 5 percent of Caltech's graduate population, and Caltech had awarded women nineteen doctoral degrees. Although few in numbers, female graduate students felt that their intellectual bent fit well with Caltech's single-minded science drive. Chemist Nancy Rathjen said, "As undergraduates, we [women studying science] really seemed much odder because there were conventional means of comparison, whereas at Caltech all the women around are more or less like you.'"[20]

## UNDERGRADUATE UNREST: CALLS FOR COEDUCATION, LIBERALIZATION, AND SOCIAL RELEVANCE

Serious calls for undergraduate coeducation first appeared in a 1963 survey of Caltech alumni in which ninety-one respondents declared that the "absence of co-eds" had "limited or interfered" with the quality of their education. Alumni criticized fellow graduates as socially unaware, a defect that they blamed on the Institute's "overemphasis on technical subjects, a need for more liberal arts, and for more attention to . . . values." Admitting women would promote the "humanization of students," many believed.[21]

Current undergraduates were more interested in the issue of "sex on campus," as the student newspaper put it. Campaigning for student-government social chair, Steve Card declared that Caltech's main social problems were "participation by the men and obtaining women with whom the men can 'participate.'" Rival candidate Stewart Hopkins added, "Caltech has acquired a reputation as the home of the mad scientist and the troll. Many bitching broads still believe this. . . . I promise at least three open exchanges . . . [and] will prepare posters and fliers for distribution to all girls in nearby schools to encourage attendance at a more personal level. With these vigorous methods, Stu will have a girl for you. . . . Vote for Stu; it's best for you!!"[22]

This focus on women as social and sexual partners appeared in the 1965 student paper's joke issue with a fake story that the administration's "Ad hoc Committee for More Sex on Campus" intended to convert the faculty club into an official campus brothel: "[C]oeds who live there will not . . . attend regular classes, . . . [but] will perform a needed service to Caltech." Senior men would be allotted the most visits to this "undergraduate play pen" because "after four years at this place, the tensions must be relieved often and quickly." The satire portrayed access to prostitutes as a "natural continuation of present policy," the real-life steps Caltech had taken to ease student stress—abolishing first-year grading and subsidizing tickets to cultural events. An accompanying pinup illustration contained a poem: "I may be small / But I'm clean. / I'm not a coed / I'm a lovin' machine. / Burma Shave."[23]

Although such articles suggested that male students were not focused on women as intellectually promising classmates, Caltech within a few years began a broader self-examination of undergraduate life. Administrators saw troubling signs of apathy. Participation in club activities remained low, and students turned down free tickets to Los Angeles concerts. In response to a 1966 questionnaire, undergraduates depicted their own campus as an "oppressive . . . anti-intellectual environment." These students blamed the dormitory atmosphere for fostering "childish goofing-off" instead of cultivating interest in ideas and culture: "Talking seriously about science is a rarity (unless it is haggling over a test) and anyone who tries to do so is a 'troll,' 'snake,' or 'eagerbeaver' . . . an untouchable." With students "noted not only for their brains" but also for poor "personal appearance . . . table manners [and] . . . attitude," faculty members felt reluctant to eat or mingle in "dull and distasteful" student houses. The survey bemoaned this waste of top minds:

Scientists and engineers, as superspecialists in an age of specialists, are becoming more and more important in modern society. . . . [I]t is Caltech's responsibility to produce men who can see the danger in atomic testing, as well as design atomic weapons; men who can make automobiles and airplanes safer, as well as faster; men who can see dangers of chemicals in our food, as well as find new ones to use. . . . Caltech is not meeting its responsibility to produce such scientists and engineers.[24]

Editorializing on this corrupted psychological climate, the student newspaper directly blamed single-sex education for creating "Eunuchs of Science": "The brutal facts are that the basic, 'skewed' nature of Caltech's population . . . foment[s] boorishness and anti-intellectualism and will continue to do so unless the 'skew-factor' is removed, e.g., building a neighboring women's college and sharing dorms with them. . . . [A]cademic policies and the all-male makeup of this Institution are to blame for what's wrong with Teckers."[25]

Officially, Caltech's administration was not ready to consider admitting female undergraduates. In 1967, when MIT's Association of Women Students inquired about prospects for Caltech full coeducation, DuBridge replied that the question "presents some difficulties. Even if we received applicants from women in the same percentage as does MIT, we would be unlikely to have more than five or ten women in each class, and it would no doubt take a number of years to reach even that level."[26] Yet behind the scenes, administrators paid attention to rising student dissatisfaction. Caltech's master of undergraduate houses sent seven students on observational visits to Harvard, Yale, Amherst, Wesleyan, Williams, MIT, Swarthmore, Bowdoin, and Rice. Their reports uniformly painted Caltech as a comparative "social wasteland." Richard Wright said that the only other college he saw with equally low morale was Wesleyan, where "most of the students aren't scientists or engineers, and therefore cannot be told by their administration that they are supposed to be socially inept or inactive by nature. . . . It's sad to realize that one is always at least subconsciously, ashamed of attending CIT. Perhaps even scientists want to be human. . . . A real education is needed . . . in all parts of life, not [only] acquisition of sterile formulaic knowledge." Wright warned that if Caltech failed to change, it would pay the price later when bitter alumni withheld donations.[27]

These undergraduates faulted a lack of normal everyday interaction with women for fostering "unhealthy attitude[s] towards the opposite sex" and for making men either overly shy or scarily aggressive. Scientists and engineers had trouble communicating with liberal arts girls; one futilely tried to "snow" (impress) his blind date by showing her the Synchrotron. Some felt too awkward or intimidated even to invite women to dance; Caltech once contemplated offering free Arthur Murray lessons. Other men, out to "get as much as one can while it is around," treated women as targets for conquest.[28] A student cartoon satirizing such exploitation showed a large-bosomed, short-skirted woman being pursued by a leering caveman. It was captioned, "[T]ypical cultural activities pursued by Teckers during visit to Scripps."[29]

## Caltech-Scripps Exchange

**Artist's conception of typical cultural activities pursued by Techers during visit to Scripps.**

FIGURE 5.1

Although some Caltech administrators and faculty prized their school's lack of feminine distractions as the ideal for serious academia, a number of students complained that the unnatural monastic environment prevented them from learning how to form healthy relationships with women. For those critics, coeducation offered the logical answer to problems such as those depicted in this campus cartoon. *California Tech*, January 30, 1969, 6. Artist: G. Used with permission.

These seven student reports presented coeducation as a simple panacea to cure the social and intellectual stagnation of "Millikan's monastery." One simply said, "Full blooded American young men cannot be expected . . . [to live] four years apart from girls." Terry Burns wrote:

[The] Techer . . . would feel much more at ease and willing to work if he were not frustrated by the lack of a full, rich social life. The presence of women on campus would . . . be a great civilizing factor. Presently, the Caltech male loafs around in sloppy clothes, unshaved, . . . ignoring . . . social graces. Hopefully, this would change for the better if an occasional female walked by. . . . [H]aving the girls eat in the student houses . . . , [the] effect on the dining room decorum would be remarkable. So let's go coed.[30]

Wright promised that coeducation would "end the frantic concentration on girls as weekend sex goals" and rejuvenate men's spirits, creating a renaissance in drama, art, and debate: "Students would not have to turn to marijuana because there would be many activities to engage in."[31]

The image of bored, stressed Caltech geeks turning to drugs when deprived of female company fundamentally called into question the school's equation of scientific and technical strength with masculine power. Early twentieth-century heroic engineering had positioned technical experts alongside rugged cowboys in taming the frontier West, where their work defined action, accomplishment, and the bending of nature by determination. Yet as the twentieth century went on, engineering became increasingly theoretical and corporate. Management desks and offices replaced field camps and shop floors as engineers' default location.

Post-*Sputnik* competitiveness led the federal government to pour funds into scientific and technical education, urging America's young people to keep pace with Soviet youth. Yet the cold war also triggered a certain anti-intellectualism, as when Senator Joseph McCarthy linked "eggheads" to softness, Communism, and homosexuality. Since Victorian times, critics had decried the flaws of "sissy" boys and effeminate weaklings.[32] Advocates of wholesome all-American culture praised "authentic" teenagers who excelled at sports, fit in easily with social peers, and exuded a healthy appeal to the opposite sex. Caltech's reputation for producing socially incompetent, emotionally crippled nerds, bound to the library and laboratory, seemed to be out of step with a masculinity that prioritized personality and sexuality.[33] Pasadena's campus positioned students only a short drive away from southern California's hedonistic beach culture and Hollywood's mass-marketed macho role models, but it was worlds apart from them psychologically. In a society that set geeks apart, Caltech men hoped that coeducation could offer them quick validation of "normality."

Students recognized that the issue of making Caltech fully coeducational connected to wider questions about women's place in the world of American science and technology. Undergraduate Dennis Schneringer argued that if Caltech simply opened up admissions, it would naturally attract enough female students. The "problem of where to find the girls to go to school here is not nearly as great as it is played up," he declared; given that MIT managed to draw over one hundred women mainly from the eastern

United States, Caltech would probably have equal luck attracting women from the western states. But James Woodhead echoed many faculty and administrators in worrying that the "number of women who would come to a science-and-engineering-oriented school would be small" always. Rather than struggling to attract women to science culture, William Hocker suggested, Caltech should expand its humanities programs as the draw for a newly created women's college that would be separate but coordinated with the men's, along the lines of Harvard College and Radcliffe College. Woodhead similarly advocated widening the curriculum as a social corrective for Caltech men. At present, he wrote, science and engineering majors felt "afraid to match their opinion on James Joyce, Rama Krishna . . . or Pieter Brueghel" with potential girlfriends who were specializing in history, literature, or art. Men majoring in humanities could "'bridge the gap'. . . between the science-oriented mind of the Caltech male, and the liberal-arts-oriented mind of the girls. A Tech student who has . . . discussions with a roommate majoring in sociology would not feel nearly as much out of his own territory when talking about the same . . . things with his date."[34]

Although student expressions of malaise were impressionistic, evidence that confirmed Caltech's low morale came from a 1967 survey that the American Council on Education administered nationwide to students who were exiting freshman year. One statement—"The student body is apathetic and has little school spirit"—got 80 percent affirmative votes at Caltech versus 43 percent at other universities. Only 1 percent of Caltech respondents agreed that their "college atmosphere is social" versus 44 percent among undergraduates overall.[35]

Picking up on such negative signs, Caltech's Faculty Board undertook a special initiative to examine students' experiences during their first two years on campus. After extensive discussion, interviews, and study, this Ad Hoc Committee on the Freshman and Sophomore Years warned that Caltech's insistence on technical precision harmed students by stifling creativity and social consciousness. The narrow orientation spurred some to transfer out and left others feeling alienated from science. Visiting professor of philosophy Abraham Kaplan sharply contrasted the campus climate to that at his University of Michigan home. He described Caltech men as "the most intellectually mature undergraduates I've ever known" but who suffered "emotional impoverishment" that left their lives "gray" and "de-humanized."[36]

The Ad Hoc Committee considered establishing a separate but associated girls' school oriented toward humanities, for which "the primary advantage . . . seemed to be . . . it would greatly facilitate the ease with which CIT boys could acquire dates." Professors perceived undergraduates as demanding "that the faculty should consider serving up girls as at least as important as serving up more culture to improve the students' overall educational experience." But in an academic setting that prized intellectual immersion, the idea that an active social life benefited student development proved controversial: "Some faculty voiced the opinion that too much savoring of social intercourse may be antithetical to success as a research scientist or engineer. Most of the CIT faculty . . . spend relatively little time with their wives. Conspicuous lack of agreement was apparent on this issue."[37]

The Ad Hoc Committee ultimately recommended to the full Faculty Board that Caltech admit female undergraduates, taking the the first bureaucratic step toward official policy reversal: "In this day and age, it is immoral to discriminate against equally able, promising, creative, and potentially contributing applicants on the basis of their sex, as it is on . . . creed or color. Caltech has something unique to offer candidates for careers in science and engineering, and it is no longer justifiable to restrict that offering to men. . . . [W]omen in turn may have much to offer . . . [in] varieties of viewpoints and interests, . . . social and behavioral sciences, and in the humanities and arts."[38] The subcommittee acknowledged that Caltech might face "significant difficulties" in finding sizable numbers of well-qualified women who were interested in intense science and engineering work: "To avoid an awkward and uncertain initial period during which women students constitute too small a minority," the group recommended bolstering their numbers through an "exchange arrangement" that would permit women from nearby Scripps and Pitzer colleges to take Caltech classes.[39]

Significant skepticism remained about the wisdom of admitting undergraduate women. Faculty Board secretary H. C. Martel estimated that "Probably a third of the faculty would flatly say no while another third would say it is a wonderful idea." Faculty Board chair Norman Davidson cautioned, "I don't believe Caltech will become fully coed because there are more men interested in this type of education than women." Trustee chair Arnold Beckman similarly warned that by virtue of its intensity and specialized mission, the Institute might never interest enough female students: "Caltech can't be all things to all people . . . we are not mass education. We have an elite type of education."[40]

Amid faculty uncertainty about coeducation, lobbyists from the student government body, Associated Students of Caltech (ASCIT), immediately mobilized to pressure professors who were undecided or opposed to full coeducation. ASCIT's Ad Hoc Committee on the Admission of Women issued a statement declaring that "systematic discrimination against all females in the admission policy at Caltech is morally unjustifiable" and possibly also legally unsound: "[I]n the Middle Ages, it would . . . simply be . . . our unshakable conviction that women are second-rate human beings, merely a derivative of a man's rib. But to carry the trappings of monasticism into the 20th century and to impose them upon an institute which purports to be in the vanguard of scientific knowledge is surely an intolerable anachronism."[41]

While condemning Caltech's exclusion of half the population on principle, male students did not elaborate on the ideal of helping female peers maximize their intellectual gifts. Few spoke of women establishing careers in science and engineering. Instead, Caltech undergraduates argued that because women were by nature more emotional and attuned to the liberal arts, coeducation would improve men's perspective on the humanities: "[I]n discussing love, life, tradition, etc. . . . , a woman's viewpoint would be a positive educational benefit." Student government leaders promoted full coeducation primarily to allay the "chronically depressing" effects of single-sex life on men: "Many people seem to feel there is something fundamentally wrong with life at Caltech; but, like the weather or an act of God, no one seems to do anything about it. . . . [T]he attitude that it does not matter whether Caltech students mature or grow socially, as long as they can solve partial differential equations, is as myopic as it is commonplace."[42]

Undergraduates warned that Caltech's all-male status increasingly harmed the institution, driving away promising high school boys who wanted to attend institutions that offered both a strong social life and extraordinary academics.[43] In an October 1967 poll by Caltech's Educational Policies Committee, 79 percent of current undergraduates favored coeducation.[44] After hearing students' pro-coeducation arguments and after several weeks of consideration and debate, the Faculty Board voted by a fifty to ten majority in November 1967 to recommend to the administration and trustees that Caltech "proceed with all deliberate speed toward the admission of women undergraduates."[45] Davidson described this agreement as "decisive," even though "different members of the faculty favor the admission of women undergraduates for different reasons." In particular, many "feel that a more coeducational atmosphere would be an improvement in the psychological environment for our students. They recognize that the small number of women undergraduates likely to come to Caltech would not produce a major change."[46] Davidson himself saw no chance that Caltech would ever attract many women; he believed that a goal of making classes even 10 percent female would be tricky to achieve.[47]

President DuBridge and other administrators immediately endorsed full coeducation, in principle. The dean of admissions, L. W. Jones, said, "We want to get the brightest people we can get, and certainly some of the brightest people are women." DuBridge sought to tamp down unrealistic expectations that coeducation would magically reform campus culture. Responding to alumni complaints about students' unkempt appearance, he wrote, "Maybe when women are admitted, the boys will dress up a bit more, but . . .

informal atmosphere and the long periods in the laboratory encourage informal dress."[48] Significantly, DuBridge used the word *when* rather than *if*. Administrators remained worried about two particular problems—managing to bring in a critical mass of female students and finding proper housing for them. On the first question, Jones echoed Davidson's aim of a 10 percent female undergraduate population and broached the idea of expanding the size of entering classes so that adding women did not subtract openings for men. Provost Robert Bacher remained dubious about numbers. He noted that MIT had only recently managed to raise its proportion of female undergraduates above 5 percent after building women's residence facilities.[49]

Caltech's admissions office had already started investigating the national pool of potential female applicants. Troubling data from the U.S. Office of Education suggested that high school women on college-preparatory tracks usually took biology but mostly avoided chemistry and physics, which were Caltech prerequisites. Caltech entrance decisions placed much weight on the required College Entrance Examination Board's math level 2 test, but during the most recent test cycle, only 1,891 young women had taken that exam, which was only one sixth the number of boys who took it. Among Caltech's recent male applicants, half scored over 700 on science tests, but nationwide, just 161 out of 4,178 women taking chemistry college-board achievement tests scored above 700, as did only 25 out of 552 women taking physics exams. Admissions officers wrote, "[I]f we are to have the same latitude in selection of girls as we have for boys, the above figures lead to the fantastic conclusion that we would have to induce all of the girls in the United States who have the requisite preparation in high school mathematics and science to apply for admission here." Caltech observers blamed "environmental influences" and "social . . . handicaps" for influencing girls to outscore boys on verbal exams while underperforming in math. They did not judge women's minds to be inferior, declaring, "It is generally believed that girls are equally as capable as boys of high-level accomplishment in scientific fields *if* given adequate preparation and motivation."[50]

Looking beyond test scores, Caltech admissions officers found hopeful indications in the high school records of potential students. Although "boys with low high school GPAs are three times more likely than girls to enter college, . . . boys and girls with A averages are equally likely to enter. . . . It can therefore be concluded that there is a reservoir of female students of the caliber from which Caltech selects equal to, if not exceeding, the number of potential male students. Most of these girls, however, are not interested and do not prepare themselves for work in the fields that are emphasized at Caltech . . . [but] could be encouraged to enter science if the barriers were less formidable" and if high school counselors were more supportive.[51]

Complementing this numerical assessment of possible female recruits, researchers collected questionnaires from twenty-eight of Caltech's female doctoral students. Results supported admissions officers' inclination to blame high schools for failing to nurture women's scientific curiosity. Separate studies of male science majors showed that most "become interested in science quite early in life and accordingly pursue an appropriate high school curriculum," but thirteen of Caltech's twenty-eight women decided to concentrate in science only after entering college. Caltech also found that a high proportion of its women graduate students came from towns under 25,000 population: "It is possible that in small high schools science courses and laboratory space are not preempted by boy students, that teachers are better able to encourage bright girl students as well as boys in smaller classes, or that social pressures to conform to a feminine vocational stereotype are not as strong as in an urban environment." The survey reinforced a sense that female candidates might present a typical profile that was different from men's, with lower SAT math scores but higher grades. The graduate women ranked roughly forty SAT points lower than the mean for Caltech freshmen. But 85 percent had earned straight As in high school, versus just 67 percent of Caltech's recent freshmen.[52]

Beyond crunching the numbers, admission officers asked current female graduate students whether they thought they might have encountered "any special problems . . . if you had been able to do your undergraduate work at Caltech?" Several thought that women would face only the same challenges as men, mostly the difficulty of coping with extreme academic pressure. But others emphatically warned that as long as the Institute remained so closely identified with the "masculine world" of science and engineering, this would make it "*very* difficult for a girl to be really a part of Caltech." If Caltech failed to draw a decent number of undergraduate females, the skewed ratio would make women's lives unpleasant: "To expect a girl of seventeen to enter an environment in which she will be an outcast (however politely disguised) is unrealistic." Nevertheless, twelve women graduate students favored undergraduate coeducation, several on the principle that it was "ridiculous to deny qualified women the opportunities available at Caltech. . . . They certainly would not lower the academic standards. Any woman contemplating a scientific career is well aware that she has to be twice as good to go half as far."[53]

Graduate women expressed dismay that Caltech men dreamed that coeducation could "immediately solve all their social problems—it won't. Any coeds admitted are not coming to keep the undergrads happy. They come for the same reason as the men—for the CIT education, and will tend to be the same type—dedicated . . . [to] studying. I would expect very few Rose Queens to come." Like faculty and administrators, female graduate students strongly cautioned that undergraduate women would be miserable at Caltech unless the school could admit a critical mass of at least 10 percent women per class and house them together:

If you only admit a few on an experimental basis and have them live off-campus, they will all feel like freaks and be terribly lonely. . . . It is very important . . . to accept them as people—not turning them into pseudo-men scientists or into some super-woman who is not supposed to have the longings and desires and instincts inherent to her sex. She must be made to feel at home with herself and not constantly be pulled to be a "male" scientist and at the same time a "socially acceptable" woman. . . . Hope the girls prove able to put up with the strain until Caltech adjusts! . . . If they survive here, . . . they will contribute greatly to . . . erase the commonly held view that women are somehow inherently incapable of serious scientific pursuits.[54]

Caltech leaders also regarded housing as the thorniest challenge, again looking at MIT's experience. Despite talk that science and engineering were not appealing to females, MIT's admissions office had artificially capped the number of women that the school accepted and evaluated them more selectively than men. In a typical year, MIT rejected at least four qualified women solely because it had nowhere to house them. The problem persisted until 1960, when alumna Katherine McCormick pledged $1.5 million to build MIT's first on-campus women's housing, and the accompanying publicity splash sparked a 50 percent jump in women's applications.

In the late 1960s, Caltech was in the middle of a development drive, with plans to construct four new graduate houses. Administrators asked the architect to assess possible modifications for redesigning one as a residence for both undergraduate and graduate women. Steel magnate Earle Jorgensen, a Caltech trustee and long-time donor, offered $625,000 to underwrite such a women's residence. But after examining housing arrangements at other schools, some Building and Grounds Committee members worried that single-sex dorms isolated women. "Harvey Mudd has . . . lost a number of girls who were unhappy about living in the Scripps dormitories where they felt excluded from the main stream of life at Harvey Mudd." The committee favored creating coed dorms that would mix undergraduate women and men while providing special accommodations for feminine needs: a "den mother" living onsite, women's laundry rooms, hair dryers, and a "gathering place for . . . girlish gossip without the intrusion of men." The separate section for women's bedrooms would need window grills and doors that locked at night for "[p]rotection from intrusion by outsiders." But despite locks and bars, "[e]very effort must be made to make the girls feel as much a part of undergraduate life as possible, hence coeducational dormitories and dining facilities. This is especially true of girls majoring in science or engineering—fields . . . still not considered as the kind that women ordinarily enter. Just being girls at Caltech will make them feel out of the ordinary and they need assurance of belonging."[55] Architect Robert Alexander similarly favored creating mixed-gender housing rather than perpetuating the "anachronism" of segregation: "[I]n the past five years there has been a rapid change in attitudes and practice throughout the country. . . . Rules . . . and concepts of institutional responsibility have been undergoing profound changes, not only demanded by the students, but generally understood by many parents."[56]

Discussions of potential housing for undergraduate women in early 1968 revealed ongoing uncertainty regarding how many might enroll. Administrators spoke about adding admissions staff who could undertake "an intensive information campaign" to recruit high school women and female transfer students.[57] Caltech's Committee on Admission of Women Undergraduates concluded, "All the advice we have received, including actual experience at other colleges, indicates that serious adjustment problems result in a coeducational situation when the number of women is very small, and we should, therefore, bend every effort to start with as many as possible who can meet our admission standards."[58]

Caltech had not previously conducted aggressive recruiting, so these plans fed growing debate among students, faculty, and administrators about broadly revising the direction of Caltech education. Alarm about student "homogeneity" sparked calls to raise the number of entering freshmen to make room for humanities majors and others "with more diverse interests . . . to cut down on the large proportion who . . . want to major in physics or math."[59] Opponents feared that class growth would nullify Caltech's small-size advantages, such as close contact between students and faculty, and also divert precious resources from engineering and science.

Activists simultaneously pressed for other steps to correct what they condemned as the elitism of white male science students who were out of touch with minorities and even "ordinary citizens." In 1968, Caltech had only five African American students out of seven hundred undergraduates. The Subcommittee on Admission of the Culturally Deprived proposed a six-year goal of raising the presence of "disadvantaged" students to at least 10 percent. But as they had with women, skeptics worried that "even the most intensive [recruiting] program may not show great results" because otherwise-qualified minority students tended to display an "apparent lack of interest in scientific and engineering careers."[60]

Student-body president Joseph Rhodes took a high-profile stand in favor of expanding admissions to include more women and minorities in keeping with the goal of broader campus liberalization. Rhodes argued that although Caltech was "not a scene of student violence [where g]reat throngs of angry students . . . imprison administrators," the Institute faced the same "urgency for re-examination" that other colleges were embroiled in. Rhodes complained that although Caltech's latest fundraising drive carried the slogan of "Science for Mankind," faculty and administrators set poor examples of social awareness. He criticized school leaders for allegedly neglecting community involvement and shunning discussions about the human ramifications of scientific and technical change. Striking at the heart of Caltech's self-justification, Rhodes sought to debunk the "myth" that CIT provided the world's most perfect technical training. He asserted from personal observation that Stanford and Berkeley offered equally "good, if not better taught science courses" and research opportunities, plus "added advantages of fully developed non-science departments," junior-year abroad programs, "social benefits aplenty," and a resulting "enriched . . . intellectual fervor." By comparison, Caltech "fails miserably" in "responsibility for the growth of the entire student," Rhodes concluded, and thus "perhaps no one should come to Caltech."[61]

Even as coeducation remained on hold (pending the trustees' approval), Rhodes spearheaded an initiative to invite young women to campus while fostering student-driven social consciousness. Caltech's student government created and sponsored a Smog Research Project that was intended to fight Los Angeles pollution. The choice of this project was telling. Smog had posed increasingly visible problems in California since World War II, marring the state's lure of environmental beauty. By the 1960s, California had the country's highest death rates from respiratory illness, sparking citizen protests. California's self-celebration as a golden land of limitless opportunity was running into the less pleasant realities of overcrowding, economic uncertainty, and environmental degradation. National social turmoil over inner-city decay, racial inequalities, and class tensions manifested in California in the 1965 Watts riot and in the rise of Oakland's radical Black Panther Party. In San Francisco, the Haight-Ashbury neighborhood's 1967 summer of love symbolized drug counterculture and hippie rebellion. Berkeley became a radical catalyst for free-speech protests, sit-ins, and strikes, culminating in the violent 1969 People's Park confrontation with law enforcement.[62]

As students elsewhere seized university buildings or set them on fire, many Caltech undergraduates voiced their own disaffection. Like activists in Students for a Democratic Society, Caltech men condemned their ivory-tower elders for wearing blinders. Caltech men's unrest focused not on the Vietnam draft but on discontent with a sterile curriculum that university leaders seemingly had designed to turn out corporate automatons rather than address urgent needs of society. At an institution that equated engineering with progress, undergraduates saw their campus as being out of step with a public that was increasingly pessimistic and cynical about technology. The smog project let Caltech students exercise their self-proclaimed expertise in science and technology even as they sought to prove that engineering, if approached in the right way, could solve problems rather than worsen them. The student-driven undertaking let Caltech men join their nonengineering counterparts in questioning the established order, embracing the new "normality" of youth-culture activism while adapting it to the Institute's technical focus.

ASCIT invited a dozen undergraduate women from Wellesley, UC Irvine, and other schools to join its smog project in the summer of 1968. The structure did not model ideals of gender equality. Women commonly filled clerical and public-outreach roles rather than contributing to research or sharing in technical work. Several edited reports or worked on fundraising, while another served as project secretary, where "typing and other grunge keep her well occupied."[63] Nevertheless, six stayed at Caltech through the fall while on leave from their home colleges. The women joined incoming freshmen at orientation camp, where observers commented, "Only six of them, but their influence was tenfold. . . . Though . . . trustees [have] yet to give formal consideration to . . . women, it appears that the undergraduate student body has already admitted them." Official Caltech still made it clear that the women were merely visitors; one camp photograph was captioned, "Girls move into the background as faculty and upperclassmen dispense traditional advice to freshmen around the fire circle."[64] Once the semester began, these "ARPettes" (the nickname derived from "ASCIT Research Project") took graded Caltech courses in math, biology, anthropology, languages, literature, and history; two also audited basic physics with faculty permission. The women did not feel entirely accepted, complaining about "a minority of males who definitely feel that females are inferior and fit certain stereotyped categories." Student newspaper coverage did not help; a front-page photo showing two women in short skirts was captioned, "Part of ARP's project to improve the scenery."[65]

Moreover, even some observers who approved of the ARPettes' presence remained unconvinced that full coeducation could succeed. While praising the smog project's "bright, lively, hard-working and on-the-ball girls," geography professor Ned Munger suggested that Caltech continue coeducation on a visiting rather than permanent basis by having "interested girls come here for a 'junior year abroad.'" He reasoned, "Caltech is a rather foreign place . . . , especially to non-science majors. By doing this, we would not be bound down by some entrance requirements in science, nor by the heavy, though desirable, science emphasis in the first two years. [W]e would not be giving a Caltech B.S. to people who obviously had not and usually would not want to have had the requisite science training."[66]

But top administrators were increasingly committed to undergraduate coeducation and cited the existence of female Nobel Prize winners as proof that "exceptional scientific and engineering talents are to be found among . . . many women [who] would profit by the . . . undergraduate educational experience . . . at Caltech . . . not available at other institutions." Given the nationwide trend toward coeducation, Caltech leaders also feared losing ground in competing for top applicants. In the preceding year, eighteen formerly all-male colleges and thirty-five all-female colleges had become coed. The brightest students had come to disdain single-sex limitations, DuBridge believed, citing a study showing that "more than 80 per cent of the high school men who rank in the upper two-fifths scholastically feel that coeducation increases the attractiveness of a college."[67] Caltech leaders particularly noted Princeton's recent self-study conclusion that "clear[ly] Princeton would be a better university if women were admitted to the undergraduate college. . . . [T]o remain an all-male institution in the face of today's evolving social system would be out of keeping with her past willingness to change with the times." Caltech administrators admitted that the situations were not entirely comparable because Princeton might aim for a perfectly gender-balanced student body within its broad range of majors, while science-oriented Caltech was more likely to parallel MIT.[68]

DuBridge saw continued resistance to undergraduate coeducation among several trustees. One or two "brought up the age-old question of whether we are wasting our efforts by training women who will then not make professional contributions."[69] To avert a "grave misunderstanding," DuBridge warned trustees that any veto of coeducation might have "grave consequences indeed in relations between the Board, administration, faculty, and students." He stressed that MIT's long record of coeducation and Caltech's own "excellent" experience with visiting women on the ASCIT smog project had led both faculty and administration to "overwhelming" support for change: "Women no longer face the barriers in pursuing professional careers in science and engineering that were once present. . . . [W]e have a distinguished woman on our faculty, and a considerable number of women . . . research fellows, and . . . the presence of women in graduate classes has proven to be a perfectly happy and fruitful experience."[70]

DuBridge's assistant, Charles Newton, favored coeducation but opposed broader calls to "humanize undergraduate life," which he dismissed as a "follow-the-leader" campus fad comparable to goldfish swallowing. Newton saw no evidence behind the accusations that Caltech's depressing social climate drove good students to drop out and hence no justification for "an elaborate overhaul" of admissions: "[U]nrest is rife . . . across the nation, both in and out of college (hippies, ghetto youth);   the whole world is . . .

operating at elevated temperature. There is nothing in this so specific as a mandate to change the . . . undergraduate population. What if we should do that and then find . . . discontent . . . still present—or even worse?"[71]

Fundamentally, Newton argued, scientists and engineers were by character different from nontechnical counterparts. Questionnaires showed that compared to some students elsewhere, Caltech students tended to be more introverted, more shy around women, and less likely to have many close friendships. Thirty-seven percent of Caltech students (versus 14 percent of Amherst's nonscience majors) said that they had little or nothing to do with girls during high school. Fifty-seven percent at Caltech (versus 27 percent at Amherst) said that they were no good in sports, and 65 percent at Caltech (versus 33 percent at Amherst) spent more time in individual than group activities. Newton wrote, "One's first thought, on reading these self-descriptions is, 'all the more reason to humanize these poor fellows!' . . . I'm not so sure we can really change people's natures. . . . In seeking social adroitness, social involvement, we may [force] our students . . . [to miss] the main point of their lives." In a world that prized human connections, Newton admitted, any "suggestion that there might be some people who are less 'involved' than others is just about as welcome as a fart in church." But he praised Einstein's "lone traveler" temperament and rejected the call for college to develop "well-rounded" men: "Well-roundedness never won a battle, decided an issue, or made a discovery; and as for its having anything to do with happiness, that is a fiction perpetrated by a two-thousand-year-old conspiracy of morose . . . philosophers . . . from Plato to Hutchins. The happiest people are the specialists." Caltech's typical student was not a pathetic "basket case," Newton insisted, but "already a uniquely successful human being. To try to change that nature . . . might well destroy the efficacy of a Caltech education." Broadening Caltech's mission and admissions would necessarily reduce student quality, given "the fact that high intelligence [is less common in the social] sciences than it is in Physical Science." Enlarging humanities departments would add expenses for small tutorials, and students "will realize that no matter how much we expand, we'll never match a good university. A large part [of non-science-oriented students] . . . will transfer out. And they won't be very cheerful about their subordinate status while they're here."[72]

At least some students agreed with Newton that Caltech should not force personality makeovers on men who felt passionate about pursuing science. In a 1968 survey, one wrote that

I . . . came here so sure I would have to work continuously that I had no trouble starting out right. . . . I have not had any personal social life. Plenty of time for that after Graduate School and the military. I would not *consider* marriage until finished . . . about thirty years old. I *was* a bit surprised to find at this institution that not all students think like that. As a school with high academic standards . . . , my institution is unsurpassed, . . . exactly what I want.[73]

Some students, alumni, and professors prized academic toughness and high expectations. Richard Feynman added notoriously hard treatments of quantum mechanics to raise the rigor of Caltech's mandatory physics class.[74] But other undergraduates decried the extreme pressure, calling Caltech "a perfectly functioning machine . . . [that] turns out . . . very bright fellows with a lot of precision, but *no soul.*" Another student described the school as "a factory for turning out sterile (socially and intellectually) but competent . . . scientists."[75]

At the November 1968 Board of Trustees meeting, DuBridge requested official approval of resolutions to admit women "both as entering freshmen and as upperclass transfers, on the basis of the same academic qualifications as . . . men; the [starting] date . . . contingent upon the availability of suitable housing and appropriate administrative arrangements . . . not earlier than September 1971."[76] Faculty chair Davidson admitted that while "the vast majority of the Faculty" favored coeducation as almost essential for improving Caltech, "there were then and are now stoutly held opinions to the contrary." Davidson repeated his goal for classes to reach 10 percent female, judging it "unlikely that we will be able to attract enough qualified women applicants to exceed this."[77] DuBridge set the bar lower, warning that "even the most energetic recruiting is not likely to bring us more than about five percent . . . female," since there were "still fairly few women interested in pursuing . . . science and engineering."[78]

The trustees' "vigorous" discussion reflected no overwhelming reservations about coeducation in theory but instead centered on problems of implementation. Trustees unanimously approved the admission of undergraduate women, pending later consideration of administrators' proposals for housing, costs, and numbers.[79] Most faculty and staff responded approvingly; the campus YMCA director, Wes Hershey, called the step "the greatest thing that [ever] happened to Caltech." Institute psychologist Kenneth Eells predicted "a very considerable improvement in student life."[80] Undergraduates focused on the arrival of potential dates rather than classroom peers. Under the large bold headline "Girls Coming to Tech!," the student paper ran a photo of a boy carrying a laughing girl in his arms, captioned, "Coeds are sure to get a very warm welcome at Caltech. But will there be enough to go around? Don't fight, boys!"[81]

Administrators proposed to start full coeducation in 1971, but young men who already were on campus resented the delay. A vocal set, including the student-body president, Rhodes, wanted female classmates within a year. DuBridge insisted that Caltech needed more time to design and build a separate female residence. Student critics denounced that housing plan as a mistake that would only "make Caltech more monastic . . . [by] adding a distant nunnery." DuBridge fought student demands to house the sexes together, insisting that it would be costly and difficult to retrofit additions for women-only lounges, kitchens, and a new house-mother's apartment. Lack of private bathrooms, he said, would compel women "to go down the hall, day or night . . . possibly only lightly clad" in sight of men: "I would not want my daughter or granddaughter to be thrown into such a situation, and I think other parents would feel the same." Moreover, he added, any dorm with females as half its occupants "clearly . . . would be out of competition in all . . . interhouse affairs. It could not field an athletic team, a singing choral group, and would not be competing on an equal basis for athletic, or scholarship, or other trophies. It would not even have enough manpower to build the facilities . . . required for the Interhouse Party."[82]

Male students intensified their lobbying, citing the existence of apparently successful coed dorms at Stanford, UCLA, Antioch, and the University of Oregon. In fact, they pointed out, Caltech had already integrated women into male living space. Dabney House had installed a married couple as resident associates in an "experiment" to justify moving ASCIT's female research corps into one hallway of the otherwise-male dorm.[83] The student newspaper promoted the great services these "Darbesses" offered men, asking, "When was the last time an on-campus house member sewed on a button for you? In Dabney House it's not an uncommon occurrence anymore." An accompanying photo showed the "improved scenery"—four young women in miniskirts.[84]

In this age of liberation, undergraduates asserted, there was a "normalcy" to having women and men live together "in a total environmental experience." They dismissed concerns that shared housing caused promiscuity, while arguing that it would help "relieve sexual hangups and inhibitions." Male students shrugged aside fears that parents would refuse to let daughters move into men's dorms, arguing that modern young women controlled such decisions themselves: "Coeducational housing is an experiment . . . which would uphold the progressiveness of scientific inquiry at Caltech. . . . After-dinner discussions and midnight bull sessions will attain greater relevancy with . . . participation of female minds."[85]

Amid nationwide trends of colleges ceding more control over student behavior, the administration (under the new president, Harold Brown) conceded by fall 1969 and agreed to set aside areas for women inside existing dorms. Newton recalled, "[W]e gulped, and . . . an interior decorator was employed to feminize a few alleys . . . that had doors and could be cut off from the rest of the dorms."[86] (The first women to arrive appreciated nice paint and comfortable chairs but immediately noticed that transoms over their doors were nailed and glued shut although the men's opened for ventilation.) With housing matters finally settled, Caltech's trustees agreed to advance the coeducation schedule by one year, admitting up to twenty-five women to start in September 1970. Student-relations director Lyman Bonner praised the recognition "that it was unfair to deny a good brain a top quality education solely because that brain is housed in the body of a woman."[87] But in other quarters, jokes about Caltech becoming husband-hunting ground suggested lingering doubts about female brains and scholarly commitment. A student cartoon showed male administrators gathered around a conference table while one announced, "Now, gentlemen, those are the fifty female applications! Twelve plan to major in physics and mathematics—and thirty-eight in physicists and mathematicians!"[88]

## THE REALITIES OF COEDUCATION: WOMEN'S EXPERIENCES WITH CAMPUS LIFE

Thirty women entered Caltech in fall 1970 as first-year students (of these, twenty-seven ultimately completed bachelor's degrees), and two female sophomore transfer students joined them.[89] Two "Co-Techs" (an early nickname) later recalled overhearing men discussing their presence like "strange creatures from outer space." The women reported encountering boys who "swoop down on you like vultures . . . all trying to introduce themselves at once. . . . I have never felt so self-conscious in my whole life."[90]

The Caltech class entering in 1970 had thirty women out of 220 total students; in 1971, twenty-nine women out of 217; in 1972, twenty-six women out of 230; and in 1973, twenty-nine women out of 214 freshmen, exceeding the goal of 10 percent. The school carefully examined these first female undergrads, collecting data on everything from their feelings about premarital sex to their sensitivity to criticism. Reinforcing expectations that were based on prior studies about gender differences, these entering women had average verbal SAT scores that were as much as forty-seven points higher than the class overall but average math SAT scores ten to twenty-eight points lower. Again following predictions, these first sets of Caltech female students were more likely than their male classmates to major in biology (63 percent of women entering in 1971 versus 16 percent of their class overall) and less likely to choose physics (zero women versus 23 percent overall) or engineering (13 percent of women versus 20 percent overall in the 1971 entering class; zero women versus 17 percent of the 1972 entering class). Female freshmen were more likely than males to come from homes closer to Pasadena, more likely to believe that their career choice would change, but also more likely to believe that they would find employment in the field for which they trained. Female students rated their social self-confidence higher than men did but rated their intellectual self-confidence and leadership ability lower.[91] Despite that academic insecurity, undergraduate women achieved a higher "survival rate": 73 percent of women entering in 1970 completed a degree in four years versus 63 percent of their overall class. "[Y]ou can inform tormented fathers that girls are making as good if not better . . . grade point averages as their classes as a whole," administrators concluded. Twenty-nine percent of Caltech's first-year women did not plan to marry (versus 8.5 percent of the men), and over half did not expect to have children (versus 28 percent of the men). Administrators stated, "[O]nly 15 percent of the [women] are married by the time they receive their baccalaureate degrees—but maybe that's not as reassuring as it might be."[92]

Although optimists had hoped that the number of women who qualified for and chose to enter Caltech would rise steadily, trends proved more frustrating. Women's representation dropped from 13.5 percent of the total students entering in 1970 to just 7.8 percent in 1975 (twenty women out of 257 new students). For the class that entered in 1976, 117 women applied, fifty-five were admitted, but just eighteen actually came. The year 1977 brought only another twenty, still well under the goal of 10 percent.[93] Of the women admitted in 1977 who ultimately opted to attend MIT, Stanford, Harvard, or other schools, the majority who cited specific factors pointed to Caltech's lack of intellectual diversity and the inflexible science-heavy curriculum.[94]

But the entering class of 1978 jumped to thirty-three women, perhaps owing to intensified recruiting. Currently enrolled women wrote personal letters to the female applicants who were offered admission,[95] and the school distributed a brochure containing the lyrics of a song ("What's a nice girl like you doing in a place like Caltech?") that had been composed by faculty member J. Kent Clark:

What's a nice girl like you doing in a place like this?
A nice girl like you should be doing something better;
A nice girl like you ought to be as free as air,
Waving at a star now, cruising in a car now,
Strumming a guitar now, laughing in a bar now,
Something light and easy, something bright and breezy,
Pleasing a lovely miss, Never in a place like this.
What's a nice girl like me doing in a place like this?
A nice girl like me happens to be fond of physics.
A nice girl like me wants to be an engineer.
Fond of mathematics, fond of hydrostatics,
Any new dimension, any new invention. . . .

What's a nice girl like you doing in a place like this?
A nice girl like you should be playing golf and tennis; . . .
Dreaming through a tango, doing a fandango . . .
Waltzing toward a kiss, Never in a place like this. . . .
A nice girl like me loves to mess around in labs.
Measuring attraction, getting a reaction, . . .
Checking out a meter, any kind of heater.
Bunsens to me are bliss, Being in a place like this![96]

The brochure talked directly to high school women who had scientific or technical interests:

Possibly you have been kidded . . . about being a brain. Or you may have seen people react with surprise—even amusement—when you talked about having a career in science or engineering. [B]y now, both you and the people who know you take it for granted that a woman with keen intelligence and a strong interest in math and science will naturally be looking for the best possible education. . . . [O]n this once all-male campus, female undergraduates are no longer as novel—or suspect—as they once were. . . . Elsewhere on campus you may be identified as a Caltech female, but in class or in the lab, you are a Caltech student. The intense academic pressures and the high expectations for performance apply in the same degree to men and women. . . . We'd be delighted to have more qualified women apply.[97]

Illustrated with photographs of women working in laboratories, the text put the best possible spin on a severely skewed gender ratio. The brochure quoted one current student, "Sometimes it's awfully hard being a girl in a male world, and you wish there were more women around. But . . . usually it's great just being with people of your own ability who share similar interests." In fact, Caltech argued, the lopsided distribution did female students a favor by toughening them up:

[T]eams for interhouse competition are made up of men and women players. Some women . . . seriously interested in sports feel that teamwork with—and competition against—men considerably sharpens their skills. A number of . . . alumnae view their experience with the male-female ratio as an important factor in their careers—enabling them to better relate to men in the outside world, and preparing them for the women-to-men ratios existing today in the fields of science and engineering.[98]

Outside of public-relations propaganda, Caltech women had more mixed, even painful feelings about campus life. One of the first women to receive a Caltech undergraduate degree, Sharon Long, wrote in the alumni magazine about the "special challenges" of her experience. On the academic side, Long felt "afraid" to ask for help the way that male students casually did, "because I didn't want people saying, 'Dumb girl!'" Long objected to the lack of female professors as role models and the problem of "male professors here who definitely let their prejudices show." Women faced challenges even in matters of everyday appearance, she noted, citing the case of a fellow coed who wore a fancy outfit to class one day and was embarrassed when the professor treated her "extra nice . . . leap[ing] down to her chair to light her cigarette."[99]

Driven by her own freshman and sophomore unhappiness, another woman distributed questionnaires to all fellow female undergraduates. Of sixty-four responses, eighteen women said they had not witnessed sexism at Caltech and that the "professional" atmosphere of lectures "mainly" kept out sexism. One wrote, "I have never come across anyone at Tech who has expressed, verbally or otherwise, that I . . . had less right or was less able to become a scientist because I am a woman." But seven reported sexism in classes and research labs, encountering "nasty references in lectures" and professors who either ignored female students or picked on them more often. Nine women saw sexism in recitation sections, with teaching assistants who lavished extra attention on women, a different burden. Female undergraduates commented on discrimination against Caltech's female staff, noticed how advisers "dominated" female graduate students more than males, and called it "scandalous that there are no tenured women professors." Thirty-nine women complained about sexism in dorm life and "derogatory" comments from male classmates who became "upset if you score higher than they do" and told women, "Why are you worried about how well you do in school? . . . you will just get married anyway."[100]

Although some female undergraduates admitted taking advantage of favoritism from eager teaching assistants, more worried that Caltech's eagerness to recruit women amounted to a reverse discrimination that fostered unhealthy resentment among men that "girls have it easy here." Every April, the Institute hosted expenses-paid weekend visits for all the women who had been accepted, hoping that the welcome would induce them to choose Caltech. Some current female students favored any measures to attract more women, but others saw this "catering to women" as an insult. Some also grew "annoyed at the guys who followed the visiting pre-frosh women around, drooling."[101]

Some women traced their unhappiness to the fierce academic pressure that boosters claimed made Caltech education so successful—the "taking-a-drink-from-a-fire-hose" pace that made life "brutal . . . for [both] guys and girls." One junior wrote, "The stressing workload . . . makes me forget that I am intelligent, and breaks down my self-esteem." Women mentioned that Caltech's value system prioritized research and graduate students above undergraduates, which led them to feel lost and unsure about where to find help with problems such as choosing majors. One said, "The Deans have rose-colored glasses and . . . my adviser wouldn't know what to do with me." Other women hesitated to blame Caltech for their discomfort, admitting that they might have experienced loneliness and self-doubt at any college on their first time away from home.[102]

Academics aside, women overwhelmingly complained that Caltech's skewed sex ratio made daily life miserable. A student-newspaper cartoon showed three men chasing two slide-rule-clutching women, one of whom complained, "My God! They said we'd spend four years 'pursuing academic excellence'—not the other way around!"[103] Women felt that male students were competing to get a "monopoly on me." Initially, some first-year females found this rush an ego-boosting novelty, but many later felt trapped, defined as trophies: "If a woman does not choose to date, she is considered strange and/or a prude. If she dates several boys, she is automatically a slut. If she breaks up with someone . . . , she is a bitch. If she sticks with *one guy only* for her entire four years, she is treated fairly. . . . [S]he must be clearly someone's property before she can be related to as a person." Many female students reported feeling pressure to pair off quickly, which was the only strategy for fending off "puppy dogs"—men "who hang around and will not leave [a girl] alone." Several echoed earlier complaints from alumni, faculty, and male students themselves that Caltech's distinctive science and technology orientation attracted young men whose intellectual agility outpaced personal skills. One woman recommended that Caltech find "a more socially mature freshman class, and reject those nurds who will spend the rest of their lives being the same eminent bores." But another female student accepted assumptions that brilliant minds were inherently antisocial: "Tech could not conceivably change their admission procedure to favor more emotionally stable people because we then would not be a technical school of as high a caliber." Some resigned themselves to the strange climate, saying that unbalanced gender "ratios are to be expected in my chosen field; best to learn to handle it."[104]

Ironically, even as individual women reported being pestered to distraction by romantic and sexual demands, many men complained that Caltech females as a group were unattractive and unsatisfactory. Women heard male classmates "refer to 'real women' vs. 'Tech women'" and knew that interest in science and engineering made them seem abnormal and less than ideally feminine in many eyes: "[T]here is not so much pressure—on the upperclass girls at least—to be 'pretty and popular.' Rather, it is vaguely accepted that Tech women are not and will not be."[105]

Women found it hard to become platonic friends with male classmates, whose claims extended beyond dating. One wrote, "Too often in the student houses, we are regarded as special tokens whose purpose it is to please the desires of the men at any time."

**FIGURE 5.2**
With only thirty women entering Caltech in 1970 along with almost two hundred first-year men, the skewed ratio made life awkward, even unbearable, for some female students. Many felt pressured by the men who competed for their favors, especially when that attention spilled over into harassment. *California Tech*, October 15, 1970, 2. Artist: TJO. Used with permission.

Women complained, "It's hard to be so many things to so many people." Men "seem to think that the girls *owe* them . . . attention" and expected women to be perpetually pleasant. One senior liked becoming a confidante and "mother figure," but another objected to the "large emotional burden attached with being a girl here; sharing people's problems." Dorm leaders pressured women to join activities and "fill a quota of female bodies for parties," which required them to sacrifice valuable study time. Especially after Caltech stopped clustering women's bedrooms in separate hallways, some women objected to the resulting lack of privacy. Seven complained of actual harassment, while others resented "rude jokes" and "prejudiced attitudes of what a woman's behavior ought to be." The climate of specific dormitories varied. Female students credited a few dorm leaders for integrating women into student leadership roles, but one "jock house" was condemned as particularly hostile. Not surprisingly, proportionately more women than men moved off campus. One junior recommended her personal "therapy" of escaping campus as often as possible, by going to the beach or bicycling eighty miles alone. A senior said that since leaving dorm life, she had not thought about suicide once.[106]

Citing such stresses, a third of female freshmen and sophomores and half of juniors and seniors said that if they could choose again, they would not or might not come to Caltech: "There are many aspects of life here that no one can tell you about . . . subtleties which can only be experienced by day-to-day living at Tech . . . that wear people down and rob them of their youth." When asked if they would help Caltech try to attract more women, several agreed that they would do so, citing the value of cultivating female science and engineering talent. One senior said that she already had spent time promoting Caltech to women, but doing so made her "feel guilty because I don't really like it here." Others worried that Institute efforts at courting potential female students deepened male suspicions of a double standard and reverse discrimination. One wrote, "I was once told by a guy . . . that [being female] was the only reason I had been accepted, even though my SATs, GPA, were well in excess of his. That . . . attitude engendered . . . insecurity which lasted for a long time."[107]

In the early 1980s, Caltech's female undergraduate enrollment slowly rose, only to plunge again by the middle of the decade. Female applications dived to a new low of 173 in 1985 (seventy-one admitted, thirty enrolled). But applications soon rebounded, as did the yield. The number of women who chose to come to Caltech suddenly jumped from thirty-seven in 1988 (16.4 percent of the total class) to sixty-five in 1989 (27.4 percent). Women surpassed 30 percent of the first-year class (eighty out of 259 new students). Over the next several years, the number of women's applications remained above 400, but the number admitted did not rise accordingly, and the number who enrolled actually dropped. Meanwhile, women's representation in Caltech graduate school rose from eighty-three out of 993 graduate students in 1980 (8.4 percent) to 240 out of 1,065 in 1994 (22.5 percent). In 1970, Caltech's faculty was 0.4 percent female; by 1995, women accounted for 8.6 percent of the professors, concentrated in the humanities, social sciences, and biology. Out of seventy-six total faculty members in engineering departments, Caltech had four women in 1995, while women comprised just 3 percent of physics and geology faculty.[108]

Paralleling earlier trends at MIT and Georgia Tech, Caltech's female faculty, students, and staff consciously organized, hoping that solidarity would increase women's institutional influence and improve their personal well-being. The Organization for Women at Caltech was created in 1980, and the Caltech Women's Center was founded in 1993. Those new groups focused on career development, child care, and fair pay. Undergraduates developed their own Women's Forum, and female graduate students formed an Affirmative Action Committee. Given Caltech's nature, special momentum grew around local chapters of the Society of Women Engineers (SWE), the Association of Women Geoscientists, and Women in Science and Engineering (WISE), which all devoted attention to mentoring, networking, and other professional and social outreach activities to provide support in male-dominated fields. Many Caltech women, especially the politically apathetic or socially conservative (not uncommon among female engineers), ignored these programs or resisted joining them. Others considered it futile to fight campus gender problems. According to one observer, most women considered "[p]etty sexual harassment [and] discrimination . . . 'background noise' of the Caltech environment."[109]

Even those who resisted using the term *feminism*, however, often demanded respect for women's intellectual equality and took pride in the "amazing" power of Caltech's female students as a counterargument to society's misogynist stereotypes. One chemistry junior declared that prior to attending Caltech, she had been "hung up on [the idea] . . . that women are inferior to men. Since I have been here, I have met some extremely talented women . . . [who] run neck-and-neck with men academically . . . active in athletics, . . . music, drama, journalism, politics. . . . After three years, I am convinced that women can be electrifying, energetic people. I have even come to believe it about myself."[110]

## CONCLUSION

At Caltech, as at RPI, Georgia Tech, and elsewhere, unwritten rules and long-standing traditions tied technical expertise to masculinity. Caltech's battle over admitting women took its own unique form, shaped by the broader national social context of the late 1960s. For decades, Caltech had thrived as a masculine retreat where some faculty took pride in devoting evenings to laboratory conferences and admitted that they spent little time with their wives. Professors deliberately made classes difficult to "weed out" the unfit and forge the survivors into true scientists and engineers, toughened by stress. The nickname "Millikan's monastery" emphasized the institution's expectations of single-minded dedication within this homosocial environment. Like medieval monks, Caltech's promoters relished their exclusive culture that was set apart from outsiders and feared that women students would pose harmful distractions from serious work.

By the 1960s, even as their teachers equated masculine identity with an elite dedication to scientific and technical excellence, undergraduates began to question what this one-dimensional value system cost them in social measures of masculine and human identity. Amid a youth culture that was pushing the limits of the so-called sexual revolution, Caltech students embraced heteronormativity and sought to correct their own stereotypical existence as asexual eggheads and apathetic duds. Defenders of traditional Caltech valued its intellectual obsession as setting the institution above others, but students sought to bring their lifestyle closer to what they and popular culture saw as normal campus life, characterized by a healthy pursuit of the opposite sex. Caltech undergraduates joined counterparts at Columbia and Berkeley in challenging institutional authority but without doing anything so extreme as to endanger their professional futures. Like fellow Vietnam-era students elsewhere, Techers sought to redefine the campus world on their own terms. Most of these future scientists and engineers were not politically radical, but many hoped for the opportunity to become sexually liberal. Advocates of change linked the coeducation issue to deeper angst about the narrowness of scientific and technical education, its relevance to the humanities, and its responsibility to serve public well-being. Discussions pigeonholed women as emotional beings who could translate the unfamiliar world of standard social behavior for Caltech men and help them establish a healthier balance of laboratory and life.

Although male students, faculty, administrators, and trustees made all the right noises about the justice of opening opportunities to both sexes, a profound philosophical commitment to feminism was not what brought coeducation to Caltech. Observers never expected that many women would want to pursue intensive scientific or technical excellence. Caltech went coed to make men more human, more humane scientists and engineers. They deigned to share masculine academic privileges with women, hoping to gain social and sexual benefits in exchange.

Prospects for suddenly incorporating women into a physical and institutional environment built for men would have been difficult under any circumstances, but obstacles were undoubtedly worsened by Caltech's insistence on defining women as foreign creatures who were strangers to the technical world. The institute's masculine academic culture encouraged male students to assess female classmates as potential girlfriends, amusements, or sex objects rather than as intellectual equals and future professional colleagues. Mixed male-female dormitory arrangements facilitated men's hopes for heterosexual encounters but forced some women to vacate campus to escape harassment. The school's smallness and insularity exacerbated complications of the heavily skewed gender ratio for the first female undergraduates, both inside and outside engineering classrooms.

Unlike the previous two case studies of Georgia Tech and Caltech, the Massachusetts Institute of Technology had become coeducational (at least nominally) within a decade of its establishment in 1861. Founder and first president William Barton Rogers steered the new school in the direction of useful knowledge, combining scientific principles and technical applications under the motto *Mens et manus* (mind and hand). Early degrees included civil engineering, mechanical engineering, mining engineering, architecture, and industrial chemistry. Initially, MIT declined to accept women in regular degree courses, despite pressure from female supporters. Interest did exist, and the school received inquiries from potential students, including women who were already taking night classes through the nearby Lowell Institute.[1]

As noted in chapter 1, MIT's self-described 1871 "experiment" allowed Ellen Swallow to pursue chemistry studies, which paved the way for her creation of the MIT Women's Laboratory, which in turn evolved into permission for women to seek full degrees alongside men. But in the decades before World War II, the number of female students at MIT remained very small (or, in many years, nonexistent), allowing observers either to dismiss them as a curiosity or to overlook their presence altogether.

## POSTWAR MIT: WOMEN AS "THE FORGOTTEN MEN"

In the mid-twentieth century, MIT made it clear to incoming students right from their first week that the institution equated campus life with brotherhood. It welcomed first-year students through events designed to bond incoming men together through sports competition and smokers. No one expected that such activities would feel comfortable or congenial for any incoming female students who might appear. Aware that female students' presence at such a macho immersion would discomfit both women and men, MIT's Association of Women Students organized a separate weekend orientation, where established "big sisters" offered hints on how to survive a hostile environment. "You will probably be on your own among a relatively large group of men" at extracurricular events such as smokers, the AWS handbook told newcomers. "Don't let this frighten you. . . . there really *isn't any problem* unless you make one. On the whole, a group of men is easier to get along with than a group for women." At the same time, the AWS warned each female student that any missteps would not only reflect badly on herself but also endanger the prospects for all other MIT women: "If you do get on the boys' nerves, they are likely not only to dislike you but also to decide that they don't want *any* coed" involved in their activities; therefore, "you must maintain a good impression for others as well as yourself. A helpless female is a nuisance, and her counterpart, 'one of the boys,' is resented." Given the tentative nature of their welcome at MIT, female students felt obliged to make their presence as acceptable or at least as inoffensive as possible.[2]

World War II absorbed MIT professors and leadership in a heavy commitment to top-secret research and military training, which left little opportunity for them to cultivate female engineering students. But even without formal recruitment, the proportion of female undergraduates choosing engineering jumped to twelve out of thirty-eight women. Five enrolled in aeronautical engineering; others studied mechanical, electrical, and chemical engineering; and one entered the relatively rare field of naval architecture and marine engineering. In a 1944 report, Compton anticipated that postwar conditions might bring more female students to the Institute. "For reasons, some logical and some traditional, technology has been predominantly of interest to the male of the species," he commented. Nevertheless, "the female of the species continues to display both interest and effectiveness in technological pursuits . . . slowly but definitely increasing."[3]

Aside from those gendered associations of technology, Compton suggested, practical deterrents also cut into MIT's potential pool of female students. The president noted that under existing conditions, parents felt nervous about letting their daughters attend MIT, given the "serious" problem that the Institute had never helped women find housing in Cambridge. One mother worried that without the "good influence and balance" of supervised living, her daughter might "develop into a queer sort of person interested only in her work." To counter that problem and make the school more hospitable, Compton recommended that MIT rent or buy an old house to fix up as a women's dorm. As women's adviser, Florence Stiles seconded that idea as a step toward helping female students succeed in a hostile environment. She noted that although women entered MIT with high school records at least equal to their male counterparts, only one in twenty finished a degree. Despite that discouraging figure, Stiles maintained that female students had a place at MIT as long as they could adjust to the same terms as males. She wrote, "The women are treated academically exactly as are the men and like it. They learn to work *with* men—not in competition." In asserting that MIT's intellectual climate did not and should not discriminate between men and women, Stiles blamed women's difficulties on factors outside the classroom. The few female students felt isolated and frustrated; Stiles hoped that by creating an "esprit de corps" among women, a centralized residence could prevent many from dropping out.[4]

Although Compton, Stiles, and others recognized that MIT was a hostile climate for young women, they could not wave a magic wand and transform the school overnight. In 1945, as a "small scale experiment," MIT opened a women's house at 120 Bay State Road in Boston. Female students living there faced sizable disadvantages compared to male dormitory residents. The inconvenient location forced women to travel a half hour by subway train and trolley to reach campus. More unfortunately, since Bay State bedrooms could accommodate only fourteen new women each year, MIT capped female enrollment at that limit (plus an unspecified number of married and commuter students). Like many schools nationwide after the war, MIT was soon swamped by men who were leaving the armed forces. In 1946, three thousand out of MIT's total five thousand students (the highest enrollment to date) were veterans. Such trends crowded out women. In the postwar period, MIT admissions staff actively discouraged many high school women from applying and ultimately evaluated women more selectively than men. In a typical year, MIT rejected at least four qualified women due solely to lack of housing space.[5]

Those constrained circumstances shaped postwar gender attitudes at MIT. As long as the Institute could fit in a few women without much trouble, it would do so and then ignore their existence as a minor anomaly. Female undergraduates were always defined first and foremost by gender rather than academic potential. A 1945 campus article, headlined "Freshman Coeds Reveal Their Interests Are in Men as Well as in Chemistry," noted in passing that incoming females were "very serious about their work here." But the piece emphasized that "most" entering women were "looking around for a man to marry" and interested in meeting anyone matching their described ideal dates. By default, the Institute was still gendered male. In 1947, the dean of students defined an MIT education as intended "to prepare men for particular fields of engineering; . . . to educate . . . men for self-reliant, responsible, cooperative citizenship." As one observer later summarized it, "Before 1960 women entered MIT at their own risk. If they succeeded, fine; if they failed—well, no one had expected them to succeed."[6]

As female adviser, even Stiles did not question women's peripheral place at an institution that was defined by male leaders and professors for male students and that revolved around engineering. In 1946, she wrote, "Women in general do not make acceptable engineers, although they have the intellectual ability to be proficient academically. However, they are acceptable in the so-called 'white apron' jobs in foods and hospitals," plus architecture and city planning. Stiles's assumption that women did not belong in engineering ignored the immediate past record of American educators' wartime success in raising female enrollment.[7]

Before World War II, having a few women at MIT seemed an odd but harmless phenomenon. But in the postwar climate, when cold war concerns made American engineering and science triumphs a deadly matter of worldwide import, many MIT leaders felt compelled to reexamine the issue of coeducation. If few female students could meet the Institute's high expectations, then MIT was doing itself, the women, the male students, professors, and the entire country a disservice by admitting them, wasting time and energy that would be better invested in men. MIT's medical director commented, "At this time, when there is such a shortage of engineers, one wonders if we are justified in taking positions away from male students for female." In 1952, the dean of students suggested that the Institute had reached a crossroads. He called for making the decision either to "eliminate women students, at least undergraduates; or decide we really want women students, plan an adequate set-up, and then deliberately go out and get more good girls."[8]

In the mid-1950s, when MIT's provost appointed a special committee to revisit the question of undergraduate coeducation, a number of MIT figures argued for the first option—making MIT male-only. The committee chair, chemistry professor Leicester Hamilton, noted in November 1956 that six out of twenty-three first-year women were headed for academic trouble after just three months, failing physics in particular. Hamilton declared that in his personal opinion, MIT should leave female undergraduate education to "specialists in that field," such as Wellesley and Mount Holyoke, and accept only "able mature young women" for graduate training.[9]

Hamilton's alarm over the dismaying prospects of female freshmen in fall 1956 reflected an unusual circumstance that had temporarily made existing inadequacies of MIT female housing even worse. An unexpectedly high number of women, twenty-two, had accepted admissions offers for that year's class, and MIT could fit only about half in the Bay State Road house. Sheila Widnall, one of those women who entered MIT in fall 1956 (and went on to become one of the school's most influential female professors), later recalled that MIT squeezed the remaining women into a brownstone that it rented at the last minute. Upset that the women "weren't keeping their rooms clean," the landlord then "literally" threw them "out on the street." MIT moved the displaced women into Boston University housing that proved "just so noisy, people running up and down the halls . . . they couldn't study as much as MIT students have to study . . . so all of the women . . . involved in that housing problem flunked out." At the time, Widnall was intensely aware that MIT female undergraduates were still in many ways on trial after eighty-five years of coeducation. They knew that "the class before us had had twelve women students in it, and six had flunked out their freshman year. So needless to say, all of us were absolutely terrified."[10]

For some MIT observers, such catastrophic housing complications could not excuse women's poor academic performance. Margaret Alvort was one of the staff members who had the most daily contact with female students. As supervisor of the Bay State Road residence, Alvort wrote that her "doubt as to whether . . . [women] belong in the undergraduate school has grown into certainty that they do not." If MIT aimed to serve the United States by producing as many superb scientists and engineers as possible, she declared, then "there is little in the records of the girls who have lived in the dormitory in the past ten years to justify their continuance."[11]

Skeptics doubted whether MIT would ever find it "possible to provide a small group of women undergraduates with a sound environment for study in an institution primarily designed for men." MIT engineering programs appeared to present particular obstacles to the inclusion of female students. To master surveying and other field techniques, civil engineering majors spent weeks living at a rough camp that lacked any accommodations judged appropriate for women. Mechanical engineering exercises required round-the-clock observations of engine performance. Generations of male students had turned the "twenty-four-hour boiler tests" into beer parties. The prospect of women hanging out with men in the lab overnight seemed disreputable.[12]

Critics also worried that MIT's inherent character as a masculinized technical environment inevitably twisted the proper femininity of the few women who chose to attend. They warned about the "emotional conflicts . . . raised" for female students, especially those who insisted on competing with men rather than seeking to "contribute as women." MIT's medical director praised the "healthy" aggressive drive of MIT's "well-adjusted men" but associated competitiveness in women with a rejection of femininity. His Freudian interpretation pigeonholed those who chose MIT as the type of "young women who reject identification with their mothers and the feminine life . . . [with] deeper seated conflicts as to their own status in society." Although women brought "pleasure and ornamentation" to campus, he commented, women almost inevitably failed to hold their own against "high-grade intellects" of male classmates and wound up "suffer[ing] a severe blow to their pride." Contradicting his earlier assertion that MIT women shunned femininity, he then accused some of resorting to female wiles "to compensate for their failure at intellectual competition." In short, he pronounced, "except for the rare individual woman, [MIT] is an unsuitable place."[13]

Defending coeducation in this 1950s debate, other voices within the Institute community insisted that it was genuinely worthwhile for MIT to continue welcoming at least a few women. After conducting a survey of alumnae, the MIT Women's Association concluded that significant numbers had applied their degrees to solid careers in research and teaching, medicine, law, business, and government service. Over half of the graduates belonged to at least one professional society, and 30 percent had contributed to the literature in their fields. To counter any impression that the women who chose and managed to survive MIT had no proper sense of femininity, the alumnae and the MIT president's office stressed that in addition to putting "their technical training to excellent use," many female graduates had also "found time for . . . successful raising of families." The Women's Association survey calculated that 70 percent of respondents had married (and 40 percent of those to MIT men), with an average of 1.83 children each.[14]

In the absence of a universal welcome or an intensive support network on campus, MIT's women improvised, reaching out to the Boston chapter of the Society of Women Engineers for encouragement. Female students held afternoon teas that featured inspirational speakers, including an MIT alumna who had won SWE's national award for most outstanding contribution to raising women's status in engineering. In the mid-1950s, female students teamed with SWE to organize several outreach gatherings for local high school women who were interested in science, engineering, and math, where MIT undergraduates gave presentations about educational and employment opportunities.

In the late 1950s, the few female undergraduates at MIT felt afraid to rock the boat by demanding too much. Christina Jansen, who received a bachelor's degree in electrical engineering in 1963 (and later returned to MIT for her master's and doctorate), recalled, "I was really very conscious of having to represent women in each class. If I did anything wrong, if I said anything stupid, it would be ammunition for all the men who didn't want us to be there in the first place." Only three out of twenty women in the previous class had survived freshman year, leading many at MIT to insist that female students had no place in high-powered science and engineering. Jansen poured extra effort into studying and tutoring any first-year women in trouble so that the female GPA would not be a disgrace. "Discriminatory events were so common that it didn't occur to us to object," she remembered. Besides, "other engineering schools weren't accepting women at all, so even though MIT was only accepting twenty a year, . . . I still felt that MIT was doing us an enormous favor to have us there at all."[15]

Despite the lingering skepticism in some quarters within the Institute, MIT's leaders issued a 1957 "Statement of Policy on Women Students" that endorsed continued undergraduate coeducation. Chancellor (and soon MIT president) Julius Stratton wrote that "to close our doors to the admission of women at any level after these many years would appear . . . directly contrary to the whole main stream of contemporary social development." Stratton acknowledged that throughout the foreseeable future, female students would "continue to be grossly outnumbered by men in classroom and laboratory." But he maintained that even if their numbers remained small, "for the relatively few girls seriously interested in certain specialized areas of study, MIT offers possibilities that are still difficult to duplicate." Despite housing difficulties, admissions discrimination, and internal doubts, the number of female students had actually crept up over the years. In academic year 1940–41, MIT had a total of fifty-eight undergraduate and graduate women on campus, and in 1958–59, it had 125. However, because overall enrollment jumped to more than five thousand by the late 1950s, women's proportion of the total student body actually fell to about 2 percent. The number opting into engineering rose and fell over the years but always remained small. In 1952, thirteen out of fifty-seven undergraduate women majored in some form of engineering, but in 1961, only five out of seventy-eight undergraduate women enrolled in engineering.[16]

Ruth Bean, assistant dean of students in charge of women, grew frustrated with the lack of more rapid progress. For ninety years, she complained in 1959, MIT had "merely tolerated the very few women who have been able to convince a reluctant faculty and administration of their scholastic achievement." The Institute generally kept the presence of coeds "secret"; "only if a direct question is asked, do we admit that we have a few." To critics who scorned females' academic performance as lower than men's, Bean pointed out that the differences were statistically insignificant and based on a very small sample size. Furthermore, she noted, MIT's women students surpassed the national average in terms of their percentage of degree completion. Failure to acknowledge that success not only discouraged women and harmed the school, Bean told fellow staff members, but also undermined national well-being. Given the race between the United States and the Soviet Union for supremacy in technology and science, Bean considered it a disaster that American "women are not yet fully accepted in these fields and society is still confused as to its expectations for women." The United States was "losing large numbers of women who, if properly guided and made aware of the opportunities available, could be persuaded to go on in science and engineering."[17]

Bean's perspective resonated with the two key MIT presidents of the late 1950s, first James Killian and then Julius Stratton. Cold war competition against the Soviets demanded that the United States develop all professional talent, Killian emphasized in 1956, and he fully agreed that women could become outstanding scientists or engineers. Killian squelched any talk about returning MIT to its mid-nineteenth-century single-sex undergraduate status: "I do not see how the Institute, having admitted women for so long, can now change its policy," nor should it, considering "the growing feeling that women should have access to our great universities." In his annual president's reports, Killian repeatedly highlighted concerns that MIT's current environment did not help female students "realize their full potential," especially given "makeshift arrangements" in housing. One 1959 memo concluded, "Women are the 'forgotten men' at MIT in terms of the kind and character of facilities at their disposal."[18]

It was high time for MIT to embrace a new path for educating women, Killian declared, "to think more boldly than we have [so far] about recognizing [females'] presence and providing for them adequately." The new leader broached the idea of setting up a women's college inside MIT along the lines of Oxford's system or the Harvard-Radcliffe arrangement. Women would share classes with men but live in a separate dormitory with self-contained eating and recreation facilities. Killian hoped that such an undertaking would create room for about two hundred female students, a novel project that could attract outside financial support and for the first time "really justify" admitting women to MIT. Moving to bring at least part of Killian's vision to fruition, philanthropist Katharine McCormick (MIT class of 1904) pledged $1.5 million in 1960 to build MIT's first on-campus women's dorm.[19]

## WOMEN'S FUTURE AT MIT: A DEBATED VISION, 1960 TO 1969

MIT's 1963 dedication of the new McCormick Hall, with its capacity to house 116 women, attracted national publicity. "Hardly anyone imagines girls attending mighty MIT," *Time* reported, but the school had "dedicated its first women's dormitory to go with its first women's dean, an attractive blonde lured from nearby Radcliffe." *Time* rhapsodized that "Tech girls have brains" plus "looks," yielding "striking equations—long legs, wind-blown hair, fresh faces—attached to creatures who turn out to be working on doctorates in fluid dynamics." The article paid more attention to Tech women's preferences in boyfriends than to their preferences in classes. Female students "have made a virtue of their small numbers," it noted; being the sole woman in a class was guaranteed to attract a professor's attention as well as dates. Further coverage of MIT's new dorm in *Seventeen* magazine reached a target audience of high school readers ready to consider college possibilities. The teen magazine showed a photograph of a woman working alongside two men at a lab bench and complimented MIT's "luxurious new women's dorm overlooking the Charles River."[20]

Such publicity surrounding the opening of McCormick Hall alerted many members of the public for the first time to the fact that MIT accepted women. Specifically mentioning up-front that MIT was coed, the school's 1963–64 catalog emphatically stated that "opportunities for women in science [and] engineering . . . are clearly increasing." In 1964, the number of female undergraduate applicants jumped to four hundred, up about 50 percent from the 1959 to 1963 level of about 275. The opening of McCormick allowed MIT to double the number of undergraduate women admitted each year from approximately twenty per class to forty. Advocates of coeducation welcomed such trends as a "vote of confidence," "immutable testimony to the decision that women are to remain a permanent part of MIT's student body." The new associate dean of student affairs for women, Jacquelyn Mattfeld (the former Radcliffe administrator), wrote, "It is safe to assume that within the next ten to fifteen years the number and academic qualifications of women students enrolled at Tech will rise steadily."[21]

In addition to taking responsibility for all academic and campus functions affecting women, Mattfeld felt "called upon to raise a clarion voice in . . . speaking for the 'needs of women' in a predominantly male technological institution." Confident that the school finally had a physical place to house female students, Mattfeld took MIT to task for its ongoing failure to integrate coeds intellectually and socially. Mattfeld savaged MIT for clinging to "late 19th century attitudes so loudly voiced in those first years when higher education was made available to women. The idea that to educate women displaces . . . men can only be held if one assumes both university and universe to be the exclusive and rightful 'property' of men." Drawing an even more pointed analogy, she added, "Over the years a convenient myopia has succeeded in blurring our vision to the anomalies in our attitudes and policies in much the same way that token integration and intellectual tolerance in the North has shielded us from observing our discriminatory racial practices." Although some faculty and male classmates were "staunch supporters and admirers" of female undergraduates, "the loudest voices" scorned MIT women as "incompetent, unnatural, and intruders." Those male students "let it be known that *they* prefer the 'Wellesley type'" and suggested that "MIT coeds, like Johnson's lady preachers, simply are not to be taken seriously."[22]

Drawing attention to the lopsided gender balance in the front of classrooms, Mattfeld noted that only ten women held regular faculty appointments at MIT in 1964—nine in the humanities or economics and one in food technology. She worried that the absence of role models undermined coeds' self-confidence, deterred them from considering academic careers, and conveyed "incontestable evidence that as an institution of higher learning we have little faith in their ability." Without having to work with female professors, male students "can take from Tech, unchallenged, whatever prejudices they entered with"—assumptions that women could not and should not handle high-level science or engineering. "No woman should be given an appointment *because* she is a woman," Mattfeld wrote, but she felt sure that if MIT expressed interest in hiring women, it could locate good candidates. Society must "constantly affirm that whatever reforms or improves the educational opportunities for one portion of the population cannot help but be beneficial to the whole." The Soviet Union, Israel, and Scandinavia had forged ahead with educating girls in science and engineering, Mattfeld declared, and the United States had to catch up.[23]

Mattfeld called on MIT to lead the way and become a model academic community that encouraged bright women to fulfill their potential in traditionally male fields. Mattfeld recommended that MIT raise female undergraduate enrollment to four or even eight hundred and set up a large women's "honors college" as a special subset of the regular institution: "The national impact of the creation of such a comprehensive program at this moment in our country's history would be tremendous." Mattfeld believed that by including more photographs and discussions of women in its brochures and by targeting special recruitment efforts directly to high school women, MIT could succeed in drawing many good female students. She also recommended calling attention to MIT's commitment to the humanities and liberal arts, which she (and many others) believed that female students, more than men, wished to combine with their scientific and technology interests. Mattfeld reassured doubters that a greater feminine presence would add to rather than subtract from MIT's excellence: "There would be no genuine threat of a great landslide of qualified [female] applicants at this time, no danger of the ladies 'taking over' by outnumbering the men."[24]

Female students still complained about many large and small injustices in campus life, including a perceived inequality in distribution of scholarship funds and a lack of athletic opportunities for women. One pointedly remarked, "You have to walk a mile to find a ladies' room" in classroom and lab buildings. As with other high-powered engineering schools, MIT took pride in high expectations and driving students hard, an impulse that was reflected in the informal motto "Tech is hell" and in slogans that compared getting a Tech education to "drinking from a fire hose." The gendered dimensions of such analogies came through in a 1967 article about the myths about Tech's toughness, which portrayed the Institute as "an 'academic dragon' which the student must slay to prove his 'manhood.'" Graduating senior women reported in a 1964 survey that gender made MIT's normally "murderous" pressure worse. As Mona Dickson explained:

Besides having to become accustomed to being the only girl in a lecture with two hundred men, a coed must learn to compete tooth and nail with those men. . . . If she makes a mistake, academic or otherwise, she can be sure that the whole school will hear about it. . . . She must also prove that, for all her supposed brains, she is still a woman, not an oversized tomboy. Hardest of all, she must be able to throw off derogatory comments from Techmen who know her only through a warped reputation.[25]

Nonetheless, the seniors who survived spoke of relishing such challenges. They described MIT as a place where they enjoyed "the *privileges* of a woman in a man's world" and the "pleasures of holding my own" in class and discovering that "I have 'brains' . . . to fulfill my dreams; that I am not 'second rate' and incompetent as claimed by my father for so many years." This new generation of MIT undergraduate women gained national visibility; the 1963–64 AWS sent delegates to the First International Conference of Women Engineers and Scientists, an event organized by the Society for Women Engineers.[26]

While glorying in their ability to compete in a masculine intellectual world, MIT women sometimes slid into the extreme of denigrating traditional gender roles. Under the caption "We've all decided to stay home on the farm with the chickens and be 'real women!,'" one coed drew a cartoon showing an aproned woman surrounded by lots of kids and "the MAN (glow, glitter) of the family." On the kitchen table, the woman's MIT textbooks, diploma, and "great ambitions" lay buried beneath a pile of *Readers' Digest, Good Housekeeping*, and a *PTA Manual for Dedicated Mothers*. Such patronizing tendencies resulted from female students' own experiences. An early 1960s student, Christina Jansen, was shaken by disapproving comments that she overheard at parties or in conversations with other women her age who felt that her choice to attend MIT was strange. In a vicious cycle, MIT women scorned contemporaries who seemed to embrace an easier but less exciting path that centered on traditional female studies and a future that that would be defined by homemaking. Jansen recalled that she and fellow MIT women composed a song with lyrics that belittled the nearby women's colleges: "To some extent, we enjoyed the eliteness of being unusual by being MIT women students. It was both a necessary and healthy thing, and in some ways, it's very unhealthy. We looked down on all other women . . . non-MIT women and women secretaries, women housewives . . . just to justify our position, just because there was so much negative in the society."[27]

Graduating women encountered numerous incidents of discrimination during job searches. Companies questioned how long a woman scientist would remain on the job, doubted whether a female engineer could think about mechanical details "like a man," and offered salaries that were substantially lower than men's. To expose such issues, the AWS helped organize a "Symposium on American Women in Science and Engineering" at MIT in 1964. Organizers wanted to encourage young women to consider those careers by telling them about "the mythical and actual difficulties they may . . . encounter, to convey that these are not insurmountable, and to assure that the satisfaction and rewards are high." Planners hoped to attract national media coverage so that industry and the public could appreciate women's talents in science and engineering and recognize that such professionals could remain female while working in a man's world. Sensing that some potential students were scared away from entering technical work because they feared it meant sacrificing normal feminine life, organizers wanted a discussion panel with female engineers and scientists to include "one preferably young, successful and with a family; the other under 50, respected in her field and unmarried (but not the kind of person 'no one would marry anyway')."[28]

MIT's October 1964 symposium attracted college administrators and faculty, high school students and guidance counselors, and more than 250 delegates from Smith, Radcliffe, Wellesley, the University of California, Georgia Tech, Northwestern, Purdue, Bryn Mawr, Michigan, and dozens of other institutions. The novel gathering of a large group of women who worked in science and technology served an important purpose. One mechanical engineering major from Michigan State University said she found it "reassuring to see so many other women in the same situation."[29]

Among other topics, speakers devoted special attention to the challenge of juggling family and career. Psychologist Bruno Bettelheim declared, "As much as women want to be good scientists and engineers, women want first and foremost to be womanly companions of men and to be mothers." To help women balance motherhood and work, both Bettelheim and Radcliffe's president, Mary Bunting, called for development of better child-care options and flexible job scheduling. But a few female delegates objected to the tone of such conversations, worried that an emphasis on accommodations to assist women professionals might overshadow their dedication to research or call into question whether their commitment to completing a job equaled that of male coworkers.[30]

Efforts to draw more college women into science and engineering would not succeed, University of Chicago professor Alice Rossi argued, until society began rewarding girls for displaying curiosity, independence, and analytic reasoning. Psychologist Erik

Erikson encouraged women to stop depending on men for approval and to envision a future beyond being a husband's domestic helpmate. Such sentiments led to controversy. In preparing to publish symposium papers, MIT Press apparently objected that organizers had become "embroiled in 'a save-the-woman mania.'" Mattfeld retorted that far from being man-hating radicals, many AWS members were engaged to be married and thus wanted to hear speakers such as Bettelheim and Erikson, who supported their desire to combine marriage and profession: "Perhaps this was naive, but it was scarcely militantly feminist on the part of a group of 18–22 year olds."[31]

The women's symposium fulfilled organizers' aims by attracting substantial publicity, and many participants praised the experience as both professionally and personally rewarding. Yet behind the scenes, all was not well. Female undergraduates who worked on the conference had clashed with older female advisers and ignored suggestions from advisers, and many MIT women reportedly boycotted the event out of resentment, feeling that their involvement was unwelcome. Mattfeld attributed such episodes to a larger problem among the current crop of female students, what she called "a disturbingly widespread malaise . . . a massive all-inclusive suspicion and hostility . . . a sense of anonymity and futility and disillusionment."[32]

Looking at academic data in the mid-1960s, Mattfeld and others found evidence that moving numbers of female undergraduates closer to a critical mass had already generated some positive effects. Between 1961 and 1965, women's total enrollment had almost doubled from 155 to 291 (with 143 undergraduates). Figures from recent years showed that undergraduate women had the same attrition and degree completion rates as men. Graduating women earned higher grade-point averages than men did, and more of the women (over 70 percent) continued into advanced studies.[33]

Although increasing undergraduate women's enrollment to a critical mass had been a necessary and valuable first step, many observers felt that female students' apparent ongoing unhappiness proved that the Institute urgently needed to take further measures to create a healthier environment. Over previous years, female students had little chance to build up confidence or community spirit, and simply getting more bed space in McCormick Hall had not automatically generated a positive environment. Just the opposite; McCormick seemed to act as an emotional pressure-cooker (at least over the short run), fostering what Mattfeld described as "an exaggerated sense of competition among . . . the girls—a quickness to imagine or exaggerate slights, to gossip irresponsibly, to resent request for help with assignments from other students, to flare up in anger over small differences." Suddenly, observers were paying far more attention to MIT women than ever before, and many were uncomfortable with what they seemed to find—intensified tensions between students and advisers, between graduates and undergraduates, upper-classmen and newer students, married and single women, commuters and residents, and AWS leaders and other women. Frustrated female students complained to the dean about

the demoralizing effect of the early dismissal of one freshman, the long and drawn out struggles of living with two girls known to be suicidal, the feeling of being trapped in an elegant hotel with no place of get away to when one's own room felt oppressive, the sense that there is "no older person you can trust when you just want to talk or let off steam" and that "no faculty member really cares if you ever get to be a scientist or not."[34]

Mattfeld and many other MIT leaders hoped that creation of a second residential tower for McCormick Hall could address such problems by providing rooms for one to two hundred more women, social and athletic facilities, and accommodations for more adult counselors and on-site mentors. But negotiating those plans with the donor immediately grew complicated by diverging priorities. Katherine McCormick rejected the plans that were drawn up by the student affairs office, physical-plant administrators, and other campus leaders as being too expensive and elaborate, especially with regard to recreation rooms. More than that, McCormick wanted the next phase of construction to focus on providing housing for female graduate students, whom she reportedly considered a "better investment" than undergraduates with regard to their "professional motivation." MIT regarded graduate women's housing demands as less urgent, given that they were older, often married, and more independent than undergraduates. Mattfeld grew increasingly frustrated as she attempted to explain that the Institute had not yet fulfilled its obligations to undergraduate women. Mattfeld resented that some members of the Academic Council refused to invest MIT's money in expanding women's education and instead made those plans contingent on obtaining philanthropy or other external funding. Mattfeld denounced the situation as a "remnant of the historically entrenched self-image of Tech as a professional school whose only genuine responsibility is for the technical training of male students." She warned that "If the entire project continues to exist only because of the despotic benevolence of an alumna whose unchallenged orientation is that of an MIT of fifty years ago, McCormick will continue to be a large handsome women's dormitory thrust up on a men's campus whose residents continue to be beset by problems which an alien environment intensifies."[35]

Indeed, as MIT's Academic Council discussed women's future at the Institute, some still regarded the training of female undergraduates as a risky investment, given their "lower professional yield." Defenders countered that "so long as the number remains relatively small," MIT could afford to spend some time educating women who would use their background merely as "preparation for life." Admissions staff confirmed that they could find another fifty or sixty qualified female candidates each year, raising the number of women per class to about a hundred. The Council endorsed this prospect of a female undergraduate population totaling four hundred, but Mattfeld remained unsatisfied. Such plans would raise the quota of women at MIT from roughly 3 percent to 8 or 9 percent, a figure that was "scarcely representative of an ideal coeducational balance of the sexes," she objected. Leaders had "again sidestepped the possibility of deciding to admit a class on merit without regard for sex, though they are proud of their lack of discrimination on the basis of religion, race or financial status." Mattfeld concluded forcefully that a "conservative Madison Avenue or Wall Street attitude toward women still runs through MIT's veins . . . as dated as Victorian bathing suits and strangely out of keeping with MIT's liberal concepts of education."[36]

Even once MIT managed to finalize expansion plans, groundbreaking could not occur until the Institute relocated people from several existing buildings on the site, which would potentially result in a five-year delay. Instead of jumping to a hundred women per class, officials suddenly were talking about temporarily cutting back to thirty-five. Mattfeld protested vigorously that such a move would run "counter to all present governmental and social pressure for education . . . to be freed of restriction by race, creed or sex." It would convey an impression that MIT did not truly believe in training women, "a poor example for the nation's leading scientifically-based university to set." Reduced female enrollment "can only lower the morale of those presently studying here and intensify the already exaggerated pressures of being 'women in a men's school.'" Admissions officers warned that even temporary cuts would scare away good candidates, posing "a real obstacle to the recruiting of top-notch coeds for some years to come." Medical staff commented, "There is an optimal 'critical mass' for effective interaction and thus change in the educational milieu. . . . The difference between twenty and forty freshman . . . women is both socially and psychologically significant." MIT ultimately avoided cutbacks by placing coeds in married-student apartments and squeezing extra beds into McCormick Hall.[37]

Another decision soon multiplied aggravation in female students' lives. In 1967, MIT and Wellesley College initiated an exchange program that allowed undergraduates to take courses at either school. Wellesley women could take advantage of MIT's depth in math, science, and engineering classes, and MIT students gained access to offerings such as art, anthropology, and Chinese. The schools painted the plan as an experimental dialog between "two cultures" and an exciting convergence of stereotypically rational, analytical MIT students and intuitive, human-oriented Wellesley women. But students themselves did not immediately appreciate the academic possibilities. MIT undergraduates asked, "What does Wellesley have to offer us?" and assumed that Wellesley women would flounder in MIT coursework, completely out of their depth. MIT students assessed the program in social rather than intellectual terms, and many female undergraduates disdained it as a "very artificial means to get more girls on campus." MIT women regarded themselves as studying for professional aims and considered it "degrading" to be connected to what they perceived as a socially oriented women's college. Adding insult to injury, Institute men hailed the exchange as a terrific social innovation that would give them an opportunity at last "to see some pretty girls" in the classroom. A student cartoon showed two men staring as a woman in miniskirt and high-heeled boots walked by. The caption read "Coeducation comes to the 'tute"—ignoring the fact that MIT's own female students had been present all along. MIT's newspaper commented that Wellesley "girls" would "brighte[n] up the halls . . . definitely an improvement over the normal scenery found around campus." After two MIT women sarcastically suggested that MIT could also gain aesthetically by starting an exchange with Harvard men, *Tech* editors denied having any intent to slander coeds and added, "There are simply not enough of them to go around." One letter to the editor, however, meanly suggested that as part of the exchange, "MIT girls should be required to take the Wellesley classes in figure control."[38]

Some MIT officials hoped that the exchange would inspire Wellesley women to consider specialized science and engineering options, but many potential students were intimidated or discouraged by advisers who were conscious of math and physics prerequisites. In practical terms, the timing of laboratory sessions complicated scheduling, and Wellesley did not offer credit for some advanced MIT courses. The majority of Wellesley participants chose MIT classes in humanities, architecture, or city and regional planning, which helped a number to enter MIT's graduate architecture program. In fall 1968, at least three Wellesley women signed up for MIT classes in digital computer systems programming. In fall 1969, four took electrical engineering, two took civil engineering, and one took aeronautics. The following year, that total increased significantly; ten Wellesley cross-registrants enrolled in civil engineering, six in electrical, four in metallurgy and materials science, and one in mechanical engineering. In 1971, two signed up for naval architecture, and fourteen for electrical engineering. By spring 1970, six Wellesley women transferred to MIT's regular degree programs in civil and electrical engineering, math, biology, and the humanities. Overall, the exchange program was popular. In 1974–75, about 10 percent of MIT undergraduates and a quarter of Wellesley women participated.[39]

FIGURE 6.1

Although MIT had allowed female students to pursue degrees since the 1870s, young women often felt a lack of intellectual and personal respect at the school. The enthusiasm which undergraduate men expressed for the late 1960s exchange arrangement with Wellesley, welcoming it as a chance for real coeducation at last, seemed to further belittle their female classmates. *The Tech*, March 5, 1968, 4. Used with permission.

Wellesley women sought MIT classes primarily for intellectual reasons, a 1969 internal analysis concluded—to obtain "different educational perspectives," "interaction in an intellectual aggressive environment," and "a reference for future graduate study." By contrast, MIT students were interested in the exchange mainly for social ends, seeking "improved social conditions" and "an easy A." Many at MIT dismissed Wellesley as academically soft (in an era when Wellesley graduated Hillary Rodham Clinton, Madeleine Korbel Albright, Nora Ephron, Cokie Roberts, and Diane Sawyer). Contrasting MIT to what they perceived to be Wellesley's lax standards, Tech students took "extraordinary pride" in their school's rigor and competitiveness. Administrators acknowledged that the experience perpetuated "patronizing" attitudes toward the nonurban women's college. Institute students "often regard Wellesley as American tourists regard Italy: scenic, quaint, somewhat backward, and the natives too responsive to authority."[40]

Given such attitudes, tensions over the exchange continued on the personal and social side. MIT women complained about experiencing a "loss of identity" and of being mistaken on campus for Wellesley students. MIT needed to pay attention to the needs of its own coeds, they said, and if officials wanted a better gender balance in classes, they should increase female admission. Tech women feared that the exchange might lead some at MIT to take female students less seriously because more Wellesley women enrolled in "easier" humanities courses instead of science or engineering. Institute women saw little academic value in the program and resented the sense that it had been created mainly to provide social opportunities for MIT men and Wellesley women. Exemplifying MIT men's date-chasing approach, one man praised the fact that his Wellesley "basketweaving" class for once brought a sex ratio in his favor:

Alone with all those girls! Just like a mixer with no competition. Of course I realized that every once in a while I might have to read a book or two, but I figured that being an undergraduate physicist at MIT, I could handle almost anything they could give me. . . . My wildest dreams were fulfilled. I was outnumbered 23 to 1! Oh happy day! As I nestled into my seat between a sophisticated blond and a friendly redhead, . . . I really felt that I was the center of attention.[41]

MIT women had long resented stereotypes that portrayed them as less than feminine, as in the rude jingle describing "a girl five feet tall and equally wide, a slide rule hanging at her belt, who can speak only in differential equations." MIT's *Social Beaver* student handbook of the 1960s went into raptures over the beauties found at Boston women's colleges (while conveniently listing the curfew rules of each) and then added almost as an afterthought, "The story that most MIT coeds are dull and ugly is one of the first that an entering freshman encounters. With exceptions, they are that and more." Some MIT men regarded classmates as expedient if not ideal dates because the end of an evening out did not require rushing a woman back across the city.[42]

Observers pointed out that a female student's romantic partners could condition her academic well-being, influencing (for better or for worse) her psychological welfare in a masculine environment. One senior commented, "How well or how badly a girl does at MIT often depends a lot on what kind of a guy she gets to date her first year. If he . . . says things like, 'Women haven't any business in science,' . . . she usually ends up in trouble. If he builds her up and lets her know he thinks she's wonderful and that it's great that she's going into science or engineering, she does well." MIT's advisers noted that some relatively mature male graduate students seemed particularly receptive to dating and marrying women on a basis of intellectual respect. Overlooking potential complications in relationships between female undergraduates and male teaching assistants,

FIGURE 6.2                                                                                     ▶
MIT's dating culture denigrated Institute women as undesirable, with the putdown, "five feet tall, five feet wide, with a slide rule at her side." Female students sought to counter that offensive stereotype by portraying themselves as normal American women who were intelligent and attractive. From "The Coed Mystique," by Gail Halpern, *MIT Technique*, 1967, p 97. Courtesy of the MIT Museum.

Diane Mechler '68

the dean of students office concluded, "The importance of the encouragement they give our women . . . is a positive force . . . which we should foster."[43]

Female students often went to great lengths to emphasize their femininity, both as a matter of self-esteem and as a public relations move to show male classmates, critical outsiders, and younger women that academic seriousness did not turn technical women into drones. One mid-1960s student wrote, "Despite the fact that they are familiar with the intricacies of slide-rules, lathes and computers, McCormick girls are also familiar with the whims of fashion and the dictates of the clothing industry. The drab halls of MIT have been brightened lately by patterned stockings, miniskirts, Courreges boots . . . and coeds speak as easily of style, fit and fabric as they do of Maxwell's equations." Her article was illustrated with a vicious caricature of the worst stereotypes, a heavyset MIT woman with hairy legs, chin stubble, clunky sandals, muscled arms, and no real hairdo. Next to that cartoon was a drawing of the equally stereotypical sexualized ideal, a slender young woman who carried a slide rule, had a pretty hairstyle, and wore a tight sweater, accentuating her well-endowed figure. Photographs accompanying the piece showed actual MIT women posing under trees and in dorm rooms wearing fashionable sweaters, skirts, and pantsuits.[44]

## ACTIVISM, IDEALISM, AND DISPUTE IN THE LATE 1960S AND EARLY 1970S

In the early 1970s, a new sense of crisis for women at MIT led to a consolidation of gains and fostered real organizational mobilization. MIT had made a few key faculty appointments, promoting some promising women from within its small but growing group of female doctorates. In 1963, Emily Wick, who had earned her Ph.D. in chemistry at MIT in 1951, became the first woman to be granted tenure. Sheila Widnall completed her bachelor's, master's, and doctoral degrees at MIT in aeronautics and astronautics in the early 1960s and continued as assistant professor. She was promoted to associate professor in 1970 and full professor in 1974. During this era, MIT made some other influential hires. In 1963, physicist Vera Kistiakowsky left her position at Brandeis to move to Cambridge. Mildred Dresselhaus, a University of Chicago physics Ph.D. who had been on staff at MIT's Lincoln Laboratory, joined the department of electrical engineering and computer science in 1967. Those four faculty members would join MIT's existing staff and students in systematically pushing the school to improve conditions for women's education and employment in engineering and science.

Following Jacqueline Mattfeld, Emily Wick took over serving as an administrative voice for female students. Looking over applications from high school seniors, Wick and

Dresselhaus felt dismayed that MIT still seemed to impose a double standard; because housing remained tighter for women than for men, female applicants needed better test scores and stronger recommendations to qualify. In 1970, Wick formally recommended that MIT evaluate applications from high school men and women on an equivalent basis of academic and personal characteristics without artificially capping female acceptances based on the limits of campus dormitories. When McCormick Hall opened in 1963, MIT required all single, noncommuter undergraduates to room there; at the time, Wick supported this decision, believing that this would foster women's sense of community identity. But like other colleges in the late 1960s, MIT responded to student demands by loosening its *in loco parentis* control over housing arrangements. The Institute began to allow senior women new freedoms, then loosened regulations on all female upperclassmen. By fall 1969, MIT no longer had separate housing rules for women; identical regulations applied to men and women, and only first-year students were required to live on campus. At the same time, coeducational group living for unmarried students had become trendy, and the dean of student affairs agreed to experiment with letting six women move into the MIT Student House cooperative. Reactions from the twenty-four male residents ranged from the enlightened ("it's more of a home . . . we're learning to live with girls and not for them . . . you learn to think about girls as human beings—a worthwhile adjustment") to the sexist ("you may get someone to darn your socks"). In any case, Wick anticipated continued demand for coeducational residences, which would reduce pressure on McCormick space and allow MIT new flexibility in its admissions policies. Significantly, Wick's proposal did not generate any immediate dispute. Dresselhaus recalled that "changes were made in a very amicable and agreeable fashion. Everybody agreed that this was the direction to move." Starting in 1971, the Institute began assessing male and female applicants without explicit gender quotas.[45]

In 1971, when Wick wanted to return to full-time research and teaching, her move precipitated intense debate. Wick emphasized that MIT should keep a person in the dean of students' office to focus specifically on coeds' well-being. She wrote, "As the number of women students increases (and it cannot fail to do so if admissions criteria are the same for all applicants), it is essential that MIT be sensitive to their needs . . . prepared to assist women students as they make their way through this very male institution." Precisely because of their small numbers, advocates maintained, "women are treated differently from men in MIT classes." Working from her open-door office conveniently located near the women's path from McCormick Hall, Wick had stepped in to mediate when coeds encountered trouble dealing with advisers, professors, and teaching assistants.[46]

The sudden vacancy in the post of associate dean for women alarmed female students, many of whom "felt kind of dependent on that office . . . for social adjustment and for reinforcement," professor Mildred Dresselhaus later recalled. After MIT's main dean of students apparently suggested that any female students who wanted to discuss concerns could turn to the school's numerous secretaries, a "semi-riot" resulted, according to Dresselhaus. To address the matter, she and Wick convened a meeting in January 1972 that attracted two men and at least a hundred women. Most were administrative staff and secretaries, but about six professors came, plus about thirty undergraduates, grad students, and postdocs; plus research associates, alumnae, and wives of male faculty and grad students. Although Dresselhaus had originally intended to focus on the needs of undergraduate women, that initial gathering immediately broadened the discussion to cover child care, lack of opportunity for upward mobility, and the resentment felt by professional women whose male colleagues asked them to make coffee or type. Under the name of the Women's Forum, the group continued meeting twice a week, pressing upper administration to establish a thorough examination of women's issues at MIT.[47]

Heeding that suggestion, MIT leaders set up an ad hoc committee in January 1972 "to review the environment . . . for women students." Cochaired by Dresselhaus and engineering major Paula Stone, the group investigated eleven areas—undergraduate admissions and financial aid, graduate admission, academic life, the dean's office, extra-curricular activities, athletics, housing, the Wellesley exchange, medical care, child care, and employment. The final report reflected a fundamental feminist awareness, stating, "A discriminatory attitude against women is so institutionalized in American universities as to be out of the awareness of many of those contributing to it." The Institute's female students faced both outright hostility and more subtle expressions of prejudice, the group wrote:

If many people (professors, staff, male students) at the Institute persist in feeling that women jeopardize the quality of MIT's education, that women do not belong in traditionally male engineering and management fields, that women cannot be expected to make serious commitments to scientific pursuits, that women lack academic motivation, that women can only serve as distractions in a classroom, then MIT will never . . . be a coed institution with equal opportunities for all.[48]

The ad hoc committee's report received serious attention from both MIT's administrators and those at other schools. To meet the most immediate challenge, providing Institute women with a sympathetic representative, the Institute in 1973 appointed Mary Rowe as the new women's advocate with the title of special assistant to the president and chancellor for women and work.

The 1971–72 controversy significantly affected many on campus, particularly the women who helped drive developments. Dresselhaus recalled that before creating and participating in the Women's Forum, she had "very conservative" opinions on gender politics: "My thinking about women's issues evolved. . . . Up to the time of the ad hoc committee report and the Forum, I always figured that women could take care of themselves. I overlooked, in a way, all the help that I got along the way." Dresselhaus increasingly saw women signing up for her classes or even just stopping by to get a psychological boost and to find out "*how* (in very specific terms) I manage to maintain an active professional life together with a happy marriage and family." In an article for the *IEEE Transactions on Education*, Dresselhaus also noted that her male colleagues at MIT often requested advice on how to handle female undergraduates in their classrooms. Precisely because there were so few of them, these key female faculty members considered it their responsibility, as successful professionals, to lobby on behalf of other women on campus. Widnall said flatly, "You need . . . women in the community that can be counted on at critical times." She regarded women's activism as a "very exciting" force that could open wonderful opportunities for a new generation of girls: "Everybody, mothers in particular . . . , are much more aware of the importance of encouraging their daughters to take life seriously," she commented in 1976: "There's obviously a direct connection between militant feminism in the junior highs and the ultimate enrollment of women in engineering."[49]

The ad hoc committee's document represented a self-directed rallying cry, telling MIT women that academia's entrenched gender discrimination would change only when female students, faculty, and staff organized to demand improvement. The early 1970s brought a burst of activism as MIT women drew strength from the national feminist movement to assert their presence physically, intellectually, socially, and politically. Listing all the awards that coeds received, Rowe and other advocates documented that women could lead and succeed in science. The AWS produced pamphlets aimed at encouraging high school women to apply. *How MIT Looks to Its Women* did not pretend that MIT was a feminist utopia but emphasized that "there is an enormous pride in being a 'tech coed,' . . . great satisfaction in having done something difficult and worthwhile."[50]

To help MIT women establish a positive sense of identity within a male-dominated atmosphere, campus women's groups in the early 1970s initiated numerous events. AWS hosted monthly colloquia addressing wide-ranging feminist subjects such as the nature of androgyny, sexism in advertising and popular culture, and the strengths and challenges of two-career marriages. In its weekly meetings, the Women's Forum assessed topics that were central to MIT, such as recruiting more female faculty and providing better athletic facilities. Beyond that, Forum concerns reflected a broad feminist agenda; its discussions, speeches, and events covered national issues, including employment discrimination, legal inequities, women's health care, and feminist television programs and movies. The group created a "women's kiosk" in one MIT building lobby where they posted event announcements and relevant news clippings. In developing "consciousness-raising skits" and promoting women's assertiveness training, Forum leaders found it impossible to please everyone. Some women found the group (which encouraged men in the MIT community to attend meetings) too conservative and wanted more aggressive demands for change. Others judged the Forum too radical and worried about perceptions of "women's lib." As participants noted, merely attending such meetings branded them as "troublemakers" and nonconformists: "Men feel that women should want to deal with them more than with other women and accuse (subtly and not so subtly) women who seek out other women of being gay." Advocates sometimes deliberately avoided using the "f-word," afraid that its negative connotations might cause more harm than good. Nevertheless, female students, faculty, and staff spoke openly about their pursuit of equal opportunity.[51]

For advocates such as Dresselhaus, both formal and informal gatherings of MIT's female students, especially those who were heavily outnumbered in engineering, served the larger psychological purpose of giving them greater self-confidence: "I have heard many top-level women students . . . exclaim, 'I am not really smart enough to be at MIT. The only reason I got in is that I was a girl interested in engineering and therefore a curiosity,'" Dresselhaus observed. "This situation is often aggravated by the practical jokes and teasing campaigns of male classmates."[52]

It was hard to quantify issues of self-image and ambition, but MIT women also continued worrying about basic enrollment numbers. The admissions office had revised photographs and text in the Institute's catalog to highlight coeds and sent special recruiting material to all female national merit and national achievement semifinalists. AWS feared that such measures would not suffice to overcome the social forces that pushed women away from science and engineering. It would take "high-powered" efforts to increase female enrollment and "to de-mythify incorrect assumptions about women at MIT." Women's advocates worried that MIT's educational counselors—a network of almost one thousand male alumni who spoke to potential applicants—would not encourage high school women to enter nontraditional fields or address those students' concerns about coming to MIT. About a dozen alumnae also served as Educational Counselors, but since there were so few, AWS urged members to step in and contact hometown seniors themselves over the Thanksgiving and Christmas holidays: "The women in particular may just need an encouraging word from you before taking the plunge." Members also volunteered to sit in the admissions office during peak interview period to chat with interested young women. AWS members wrote personal letters to all accepted female applicants and held telethons to contact them, answering questions and providing reassurance. Internal studies in 1977 and 1978 suggested that yield rates increased noticeably among women who had more direct contact with current students.[53]

Such recruiting campaigns dismayed some engineering faculty, who worried that increased female admissions might decrease engineering enrollment, given women students' general preference for science over technology. Their suggestions that MIT risked diluting its quality by courting more female students appalled the few women professors who taught engineering, such as Sheila Widnall, who worked behind the scenes on convincing their dean to back women's education. In 1973, MIT's School of Engineering contributed $10,000 to develop a film that would combat the field's macho image among high school students, parents, guidance counselors, and the public. *Engineering: Women's Work* followed "real-life" female students and professionals through their daily routines to show that modern engineering was defined more by computers and problem solving than by heavy lifting and rough fieldwork. The organizers wanted to "focus on individuals who have come to terms with the issues of tokenism, fear of success, and overt discrimination, but still have a positive view of themselves and of their futures." The film showed female engineers who succeeded in ambitious careers while remaining overtly feminine, juggling meetings and business travel with happy marriages and children. This message that professional women could "have it all" dismayed at least one female researcher, who complained that the film "stereotypes women engineers as being overwhelming, overeducated and anonymous." But after presenting the film to female engineering freshmen at Oklahoma State and to an engineering conference for high school women at the University of Illinois, Sheila Widnall declared that the response was "overwhelmingly positive . . . the questions that concerned *them* were the questions addressed by the film." The movie's release drew national attention, and administrators considered it a major contribution to promoting the cause of women in engineering.[54]

Meanwhile, in 1973, MIT convened another workshop on women in science and engineering, this one sponsored by the Carnegie Corporation, General Electric, and the Alfred P. Sloan Foundation. The program featured panels on women's professional status and workshops on career planning. Organizers hoped to convert parents, schools, and the national media into agents of change, helping break down outdated sex role stereotypes that steered women into low-paid, shrinking occupations such as teaching: "Enlarging opportunities for women must include not only opening all doors, but also helping women to have the motivation and the courage as well as the educational preparation for walking through them."[55]

Embracing feminist language and principles, speakers such as MIT's president Jerome Wiesner spoke about a need "to encourage women's participation in every aspect of our technological society. This is another front in the almost universal battle for equality of opportunity." Workshop leaders called for revising lower-school curricula to attract girls to nontraditional fields, to sensitize parents to girls' ambitions, and to teach boys to "understand the importance of eliminating sex barriers." Although the conference pushed equal rights, MIT people worried about the controversial overtones of the word *feminism*. In a memo inviting representatives from major corporations to attend sessions on "women in industry," organizers wrote, "This will not be an ardent feminist production, so please don't hesitate to include the names of men who would otherwise be terrified by an onslaught of screaming females." That reluctance to challenge conservative business mores offended some women, who denounced the conference as "a 'whitewash' of existing conditions." Rather than telling women how to "make it in the 'system,'" those critics wanted to demand a radical restructuring of corporate culture.[56]

Despite such disagreements, women's advocates considered 1973 occasion for celebration. In June, the Association of MIT Alumnae (AMITA) commemorated the one hundredth anniversary of MIT's women graduates. AMITA's centennial hailed tangible gains. Female enrollment had tripled since 1963, reaching 816 (roughly 13 percent of the total student body). In the freshman class, women went from forty-eight out of 958 students in 1965 (5 percent) to 211 out of 1,036 in 1974 (20 percent). The proportion of women who completed degrees on time had risen from 33 to 64 percent (which was equivalent to male students' performance). Women graduated with higher GPAs than men, and a larger proportion moved on to graduate studies. By academic year 1974–75, MIT had a total of 1,111 female students. Out of the 414 sophomores, juniors, and seniors, the largest number (190) majored in the school of science, and 107 women majored in engineering.[57]

By 1974, female faculty and staff had started getting together for monthly lunches, hoping to offset their small numbers by multiplying their impact on Institute policy. Pursuing an activist stance inside the engineering school, Dresselhaus and Widnall inaugurated a new freshman seminar titled "What Is Engineering?" Although not restricted to MIT's female students, the course was aimed primarily at them, starting from an assumption that women often avoided technical subjects because they sounded unfamiliar.

Researchers from various engineering fields visited class to explain their work, and the syllabus included lab work in electronics, welding, soldering, drafting, and building Heath Kits, projects meant to help women feel comfortable with male-associated manual skills. To assist MIT women in moving confidently from college into employment, Dresselhaus helped to organize meetings entitled "Let's Talk about Your Career" where female students could consult faculty, staff, and guest lecturers for advice on graduate school, employment, and the eternal question of how best to combine marriage with work. MIT brought female visiting professors, such as neurobiologist Rita Levi-Montalcini, to spend weeks in residence at McCormick Hall to talk to students about their experiences as female professionals. Arguing that male students' familiarity with the business world gave them a competitive advantage, AMITA started an annual seminar called "Getting the Job You Want in Industry: A Woman's Guerrilla Guide to the Pin-Striped World." By advising women on resume writing and interview techniques, alumnae hoped to level the playing field.[58]

MIT women engaged with the national women's movement, thinking about how their immediate interest in supporting women's science and engineering education connected with broader feminist issues. In 1974, the Women's Forum invited Gloria Steinem to speak, explaining to her, "Although MIT is usually considered a male institution, there are substantial numbers of women on campus." Addressing a packed auditorium in January 1975, Steinem told an audience that feminism aimed to give women political, social, and educational autonomy, not dominance over men.[59]

Women's advocates found much to contemplate in Steinem's message, given that female MIT students still frequently felt like second-class citizens. Male undergraduates continued spreading stereotypes of MIT women as "stupid," "ugly," and "stuck-up." Female undergraduates complained that much of the "faculty has trouble accepting women students as women. They expect them to behave exactly like men; to laugh at the same jokes, not to be offended by sexist comments and to have the same background; e.g. be as familiar with electronics, engineers, shops etc. as men." Undergrads saw no place to register their grievances and feared that complaining to professors might hurt their grades. MIT also made a number of its male students unhappy, over-stressed, and feeling mistreated, but women's groups contended that "women need support *more* than men at MIT, as they are so isolated and have so few role models." The system had been designed by and for men, "who therefore get support from it, by its very maleness."[60]

Women's representatives saw danger ahead. By 1976, budget cuts led the admissions office to reduce targeted mailings and other "extras" that were needed to draw female applications. Although MIT had once led efforts to recruit high school women who were talented in math and science, other colleges such as Cornell, Caltech, and the University of Chicago had since launched their own campaigns. AWS called for MIT to redouble efforts at competing for that small pool of potential students, suggesting that all admissions committees should include members with an explicitly feminist outlook. That campaign raised some difficult questions, as even some insiders who felt sympathetic to women's causes worried that overly ambitious recruiting had already brought in some female undergraduates who were unprepared to handle technical courses. "I am strongly opposed to recruiting measures that will increase the number of women who have trouble in their freshman year," science professor Vera Kistiakowsky wrote, "not just because it will give grounds for increased prejudice against women, but also because those women will be unhappy at MIT. I have already taught a number who said 'why on earth did I come here,' and see no reason to increase their number."[61]

Seeking methods of attracting more top-level women students, an ad hoc faculty and staff committee found that negative images of MIT still discouraged well-qualified female candidates from seeking entrance. In 1978, 1,122 high school women had filed preliminary papers but failed to send in final applications. At least 212 of those had scored over six hundred points on the math SAT exam. Admissions officers sent out questionnaires to discover why only 40 percent of the women who expressed initial interest in MIT chose to follow through (versus 55 percent of men). Some women reported that they had perceived a distressing "lack of warmth" in the campus climate: "I saw most students walking alone and eating alone during my summer visit. They appeared alienated from each other." Others feared they might not be able to keep up with MIT's "killer" courses or worried that the Institute's scientific and technical emphasis might prevent them from exploring liberal arts. A number voiced reservations about the life of MIT women: "I felt the male-female ratio was unhealthy. I have heard unfavorable stereotypes of both the MIT male and the MIT female," one explained. Another wrote, "My decision not to return the application . . . to MIT was my fear of going to a predominantly male school. However, I did decide to go to RPI because, although predominantly male, they did quite a bit to assure me of the interest the school has in their female population."[62]

Spurred by such comments, female undergraduates, faculty, and staff renewed efforts to welcome potential coeds. During a spring-vacation telethon, volunteers called 172 high school women who had been accepted for the next year. Of those who were contacted, 114 (two-thirds) decided to attend MIT; among those not reached, only 43 percent chose the Institute. The sense that this personal touch made a difference in

raising the "yield" convinced a few undergraduates to undertake a more intensive project in 1979. Noticing that women made up just seven out of forty-one students who
had been accepted from their home state of Michigan in 1978, this small group set a goal
of having women comprise 50 percent of the Michigan contingent that would arrive at
MIT in 1980. To reach interested candidates, this Michigan Project Committee sent out
hundreds of newsletters to try to combat the stubborn "perception among most . . . girls
that science and technology are not appropriate or desirable fields of study or work for
them." At symposia in Southfield and Kalamazoo, Michigan, MIT professors and recent
graduates encouraged over 150 high school women to keep their educational and career
prospects open by staying in math and science classes.[63]

The "Michigan project" symbolized where MIT women stood at decade's end. By
the late 1970s, female students made up 17 percent of undergraduates, 16 percent of the
graduate body, and 12 percent of engineering majors (up from just 4 percent in academic year 1971–72). The increase in population mattered because as women became
more of a presence on campus, activists gained a critical mass for organization. In 1979,
the Society of Women Engineers opened a chartered chapter at MIT, which promptly
held a series of job programs and student workshops that won the Institute's Karl Taylor
Compton prize to honor good MIT citizenship. Female graduate students formed their
own society, as did women in architecture, women in chemistry, and women at Lincoln
Lab. Such groups kept women's issues on the front burner, providing a sense of visibility, an identity, and a cause for many individuals (which became especially valuable
to some female faculty and graduate students who were based in departments with few
other women).[64]

Many of MIT's female faculty members repeatedly praised the commitment that
Institute leaders had demonstrated in supporting women's interests. Assistant chemistry
professor Ellen Henderson commented:

MIT has done more for women than any other major first-rank educational institute. . . . That
doesn't mean that sometimes the nitty-gritty of day-to-day life at MIT is any easier because you
know that Jerry Wiesner is behind you. Well, it does make it easier, in fact. Those kinds of attitudes
percolate down through all the ranks. Now when . . . someone uses a racist or a sexist term, and
some other member of the faculty corrects it, and they realize that they were wrong in saying it,
then we're a long way toward change. That comes, in part, from the grass roots pressure of women,
but . . . in large part also, from the top-level people.[65]

Kistiakowsky wrote in 1979 that "the Johnson and Wiesner administrations have
supported what appears to be the country's best record in terms of improvement in

equality for women faculty and students." MIT had achieved "what we always wanted: an enrichment of talent and no loss in scientific creativity or excellence."[66]

Yet MIT women did not consider their gains necessarily safe. Mary Rowe noted that in the mid-1970s, the cause of equal rights had come under national attack as women's discrimination cases gathered dust at the laggard Equal Employment Opportunity Commission, as Washington politicians felt safe to ignore the demands of feminist organizations. Rowe praised MIT administrators for their dedication to women's well-being even as other schools attacked affirmative action. Yet Kistiakowsky warned that even with favorable institution presidents, women's progress at MIT remained "extremely vulnerable to loss of support." Reflecting that sense of fragility, those looking at numbers worried that in the late 1970s, the previous decade's rise in interest among female potential students had leveled off. Back in 1965, 202 high school women had applied to MIT for admission, and that number jumped to 411 by 1970; but from 1976 through 1979, numbers had plateaued around 750. Observers noted that MIT faced challenges in competing against other schools to recruit top candidates but worried that the Institute was still handicapped by perceptions of it as too masculine-dominated.[67]

Pockets of feminist awareness did not transform MIT into a woman's paradise. In 1982, all the female graduate students and technical staff in electrical engineering and computer science prepared a lengthy list of problems that they had encountered. Too many male coworkers refused to take women seriously or made them feel invisible, "excluded from discussions . . . ignored in meetings, constantly interrupted and talked over as if I wasn't even there." Official department letters all opened with, "Dear Sir." Female students had been mistaken for secretaries and literally pushed away from equipment. Patronizing supervisors devalued women's technical knowledge and gave them make-work tasks such as proofreading papers. Women heard comments such as, "You [only] got into grad school because the department needs more women" or "because professor X is in love with you." They heard male faculty embrace the stereotype that women were too nurturing to succeed as engineers. One reportedly said, "I don't like to take on female graduate students. . . . I can't stand it when they start to cry if you criticize their work." Male professors and grad students alike declared that women could not make it through MIT since they lacked competitive spirit, yet in a classic double-bind, women who were perceived as aggressive were nicknamed "Mrs. Attila the Hun" or told, "'You sure are bitchy today; must be your period." Pointing to the insidious effect of such repeated comments, the women wrote, "We hear about the lack of qualifications so much that we are led to believe it, undermining any semblance of self-confidence we have, and putting that much more pressure on us to have to prove ourselves." They related numerous sexist incidents, including receiving obscene mail sent through the computer system and having Playboy-type pictures posted on their office doors. Many women reported being constantly stared at by both male faculty and fellow grad students or dismissively addressed as "honey" and "dear." While one female student was working on her computer, a male professor approached her from behind and gave her an unwanted backrub. Women heard staffers say, "you are the student I would most want to lose my job over" and announce that there was "a sweepstakes going on to see who can get into [another woman graduate student's] pants first." Sexual tensions inhibited normal professional exchanges. Women sensed that simply smiling at a male professor or asking fellow students for help might be interpreted as a "come-on," and eating lunch with a male colleague opened floodgates of gossip about dating.[68]

The women explained that such specific incidents added up to a damning indictment that risked driving away current female students and staff and confirmed outsiders' existing suspicions that it was unwise for women to work at MIT. They praised the "positive examples" of "many" colleagues but added that some who considered themselves allies exhibited protective or chivalrous behavior that more subtly but ironically deepened women's feeling of being "undervalued." The women asked MIT's research group leaders to set good examples, examining their laboratories to ensure that they welcomed, supported, and credited women's professional participation on equal terms. Faced with such extensive documentation of women's discomfort within his program, acting associate head of computer science Peter Elias wrote a memo directing all faculty members to take seriously these problems of bias. In 1980, MIT finished raising funds for its new Ellen Swallow Richards Professorship to honor distinguished female faculty, a plan that had been announced seven years earlier to coincide with the centennial of her graduation. Nancy Lynch, who had earned her MIT Ph.D. in 1972, joined the department of electrical engineering and computer science in 1982 as the first Richards chair. Although the presence of one more female professor in engineering could not change anything by itself, it symbolized and encapsulated advocates' hopes that conditions for MIT women would continue to improve.[69]

## CONCLUSION

Officially, women won the right to attend MIT decades before they claimed the right to suffrage in the United States. Yet in practice, that longer history of coeducation did not mean that MIT women escaped the tensions and challenges that paralleled in many ways the struggles at Georgia Tech and Caltech. Precisely because the numbers of female students were so small over many decades, especially in engineering, the Institution's image, both inside and outside, effectively remained an all-male school. Technology defined MIT as the heart of its name, and gendered assumptions separated women from that elite purpose of advanced technical training. The battle throughout the twentieth century, then, was not to convince or force a reluctant institution to begin admitting women. Instead, the campaign shifted to a different but equally intense discussion about whether, why, and how to make MIT's existing women more visible and more welcome.

For many years, the debate revolved around numbers. Women's advocates focused on the goal of attaining a "critical mass," a tipping point where female students could naturally evolve into a real community that would provide mutual support and an increased presence that could command respect. At the same time, they realized that campaigns to recruit, admit, and enroll more women at MIT were necessary but not sufficient. Women's success would ultimately turn on wider issues of making the campus climate less antagonistic, discouraging pervasive hostility, and bringing more mentors into faculty ranks and top administration. The situation involved delicate psychological strategies to build women's self-confidence before they attempted to break into the competitive employment race. Even more challenging, the MIT community recognized that efforts to attract more women must ultimately stall unless society embraced the goal of convincing young girls to pursue serious interests in technology and science as worthwhile careers. In negotiating those bigger questions, many individual MIT women and their multiplying organizations drew inspiration from the national feminist movement, even though they sometimes shunned the more extreme manifestations of radical sentiment.

By the 1960s and 1970s, this debate extended well beyond MIT to other colleges, SWE, and other groups of professional women. Those concerned with the well-being of female engineers were bringing issues of harassment and discrimination to the forefront. The late 1970s and 1980s would bring new energy and visibility to the discussion of women's engineering education and add new momentum to the demands for change.

At MIT, Caltech, Georgia Tech, and other institutions in the postwar period, female engineering students found both the best and the worst environments. Allies stepped forward to offer vital emotional, professional, and practical support, and critics fought against the idea of letting women share center stage in the discipline. But gender did not automatically make allies or enemies. Male students, faculty, administrators, family members, and members of the engineering profession often provided special encouragement, while fellow female undergraduates, faculty, administrators, or others sometimes remained skeptical. Looking back on their early days in the field, some female engineers recalled a few women professors and professionals who seemed to be "queen bees" who rejoiced in their unique status and rigidly demanded that younger women endure the same tough rites of passage that they had survived. But on balance, the greater impact came from supportive women, who volunteered countless hours to assist other women and young girls in pursuing the dream of an engineering education. In individual efforts, female engineers mentored others, taught special classes, and offered informal advice on both career and personal questions. This tradition of help extended across generations. Well into her eighties, Lillian Gilbreth traveled around the country to meet with female engineering undergraduates, who in turn hosted outreach programs for girls in elementary, junior high, and high schools. At a group level, female engineering students at dozens of colleges leveraged their numbers by organizing support networks and numerous activities. At an institutional level, the Society of Women Engineers (SWE) established many different forms of support mechanisms that expanded over the years. The net effect contributed significantly toward making the intellectual, social, and personal atmosphere for women in engineering somewhat more welcoming from the 1970s into the twenty-first century.

ONGOING FRUSTRATIONS AND DANGERS: ISOLATION, SEXISM, AND HARASSMENT

One commonality typified student life at MIT, Georgia Tech, Caltech, and many other public and private colleges for much of the twentieth century. An undergraduate climate that assessed women primarily in terms of their appeal as potential girlfriends or sex objects made it hard for women to be treated as intellectual equals, future professional colleagues, or even normal fellow students. These persistent social tensions had particularly tangible repercussions at technical schools and in engineering departments, where female students entered an atmosphere set up for male students to interact with women not as classmates but as girlfriends. Annually, engineering colleges sponsored St. Patrick's balls or "lipstick and slipstick" dances for which men imported dates from nearby women's colleges. Engineering magazines published at Iowa State, Penn State, Purdue, and elsewhere by engineering students for fellow students, faculty, alumni, and the engineering community frequently objectified women. In 1944, under the headline "There's Glam in Engineering!," the *Michigan Technic* depicted four female engineering students as queens on a set of playing cards, showing their flowing hair in dramatically lit photographs.[1]

Alongside technical articles, updates on campus events, alumni news, corporate advertisements, and pages of sexist jokes, school engineering magazines for years included a "girl of the month" or "girl of the year" feature. The March 1956 issue of the *Penn State Engineer* devoted a full page to "our little Engineer Girl this month," art education major Annette Bair. An accompanying photo of Bair in a bathing suit showed her in a sexy (although awkward) pose, legs stretched out, head tilted back, arms pulled back to display her chest better. The comments enthused, "Her hobby, swimming, is suited to her streamlined 6 ft. 37-25-37 figure." That year, the *Penn State Engineer* published a "girl of the month" full-year calendar that was illustrated by photographs of women in evening dresses, tight sweaters and shorts, and formal wear, accompanied by trite sayings such as "A pretty girl is like a melody." Similar features continued for years. In 1967, the *Penn State Spectrum* selected its homecoming queen, Susan Politylo, as its Miss November. The magazine commented, "Miss Spectrum's major is Fashion Merchandising, and it's no wonder—she's got the fashion and the merchandise! When asked what she thinks of engineers, she replied, 'I admire them for their ambition to be an engineer.'"[2]

To modern eyes, such features might appear sexist but innocent or even romantic in a corny manner. But on a number of occasions, such semi-naked features underlined a naked anti-woman attitude. In its 1956 April Fool's issue, Purdue's engineering magazine ran a photo of a pig as its "co-ed of the month," calling the animal "a very good average representative for the campus . . . 50-85-80 . . . eyes: brown, hair: blond. Sorry,

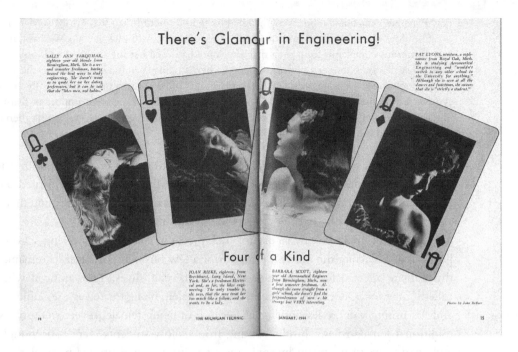

FIGURE 7.1

Engineering magazines from a number of schools presented women primarily not in terms of intellectual respect but in sexual terms, either as targets of chauvinist jokes or as desirable sex objects. The University of Michigan highlighted the "glam" side of some female engineering students in 1944, even as other publications underlined the patriotic dimension of women's engineering work. *Michigan Technic* (January 1944): 14–15, thanks to the Bentley Historical Library, University of Michigan.

fellows, she's taken." Like engineering magazines at other schools, Purdue's publication routinely recycled stale sexist jokes: "Girls are like cigarettes—a fact you must admit. You can't enjoy them fully until you get them lit." But over the years, Purdue's engineering magazine stood out from the others as it became increasingly crude in its tone. Its annual "*Playboy* parody" issue was an excuse to run drawings of scantily clad or naked women posing with slide rules and compasses. One drawing from 1970 showed a stereotypical male nerd (complete with thick glasses) in bed with a sexpot, fondling her bare breasts with one hand and clutching a slide rule with the other. The caption read, "Cool and calculating, this engineer uses his slipstick to its best advantage." A March 1968 article full of double entendres named the slide rule itself as Purdue engineers' "girl of the month".

Few Playmates can boast about being handled by the number of men that our Miss March claims. . . . She does seem to enjoy being worked by anyone who will pick her up. We have, however, had numerous complaints from men that after they had used her for a while, she loosened up so much that it was no longer a pleasure to handle her. . . . She spends a lot of time in the sun sans-a-top much to the pleasure of the male population. . . . She confesses that she actually enjoys the aggressive way men take her out of her clothing to make use of her most important feature. . . . She has even taken to working late at night with some of the more resourceful young men just so that they could learn more about her abilities.[3]

Another year's *Playboy* issue used ones and zeros to form a computer printout image of an electrical engineering major's ideal woman: "Miss March is easily turned on by a [computer science] man. . . . She gets a real thrill out of helping all the young men learn . . . practical application of their courses in control stimulation."[4]

Although it might be easy to dismiss such pieces in campus engineering-school publications as reflecting the immature humor of a few boys, important facts remain. First, this climate affected female engineering students' daily lives. They noticed what one called "rude jokes, derisive comments about women; being treated as a sex object only" and "assuming you are less intelligent because of being female, severe stereotypes into traditional female roles." Insults coming from fellow students were bad enough, but awkward or sexist treatment by some male faculty members proved particularly upsetting. Women noted that teaching assistants and professors often were particularly willing to help girls but worried that such favoritism would lead male students to say "girls have it easy here" or "you just got in because you're a woman." By the early 1970s, women at MIT began complaining about sexual harassment, such as entering computer labs filled with pinup pictures and experiencing frequent sexual remarks and pressures.

In a campus engineering climate that too often focused on women's bodies rather than their brains, engineering humor remained a double-edged sword that reduced women to sex-object status but often singled out female engineering majors as uniquely undesirable. The cruel stereotype of MIT engineering women as "five feet tall, five feet wide, with a slide rule at her side" exploded into the gendered controversy of the Wellesley exchange program. Caltech men's snide references to "'real women' vs. 'Tech women'" were possibly fanned by a "sour grapes" feeling among those who had trouble attracting a girlfriend because of the extensive competition for dates. In addition to making social and emotional life harder for female students, such negative images both reflected and perpetuated the impression that normal women did not want to pursue engineering, a message that often was blamed for discouraging girls at the elementary and secondary school levels. Female engineers in the 1950s and onward often emphasized their femininity by stressing how many of them were married and had children. In the 1960s and 1970s, female engineering students dressed in fashionable outfits and mentioned that they enjoyed sewing and cooking. Yet this emphasis on standard femininity served only to underline women's continued position as outsiders in both engineering and femininity. A 1971 poem at Purdue told the story of "Brainzella Bold," a plump 1920s young woman with nerd glasses who won a prize at her junior high science fair and felt "determined to make her stand, in a world that was traditionally known as a man's":

When other little girls played dress up like Mom
With frilly white dresses and pink pom poms,
Brainzella was busy building tinker toy houses,
Fighting with boys and ripping her blouses. . . .
It was while a senior in high school that she decided to be
A girl engineer—at Purdue—OH, NO, a C.E.!
Her parents, they shuddered, they pleaded, they begged—
"Home Ec, or nursing, but NOT engineering!" they said.
But Brainzella was stubborn, determined to be,
The first bridge building woman C.E. . . .
Infamous for her many "A"s
She became the first woman Chi Ep in her days.
Brainzella graduated magna cum laude
For her it was victory, she'd not been downtrod. . . .
About a year to the day, that she was graduated from Ol' Purdue,
Brainzella was married, as most of them do.
But let us not hide our heads in disgust—
Face it, girls, for Brainzella Bold that was sheer luck![5]

More seriously, the campus climate that appealed to male engineering students by portraying women students as sex objects was carried into the professional world. Advertisements placed in campus engineering magazines by major corporations in the 1950s and 1960s contained photographs of engineering workspaces that were filled with men, which reinforced and normalized the demographic reality that women remained scarce (although present) in the profession. One ad for Kodak said it all: under the heading "This is the image of a Kodak mechanical engineer," a man in nerd glasses, white shirt, and tie worked on a camera. Advertisements projected assumptions that men were the world's technical doers. One drawing that promoted ITT's avionics division showed a young grinning boy who was feverishly building his own computer as a young girl looked up at him admiringly, clutching her doll and asking, "Have you always been a genius?"

Other advertisements featured sexist and sexual references to catch the eye of male undergraduates. Under the headline "Want the opportunity to explore your discipline?," a 1966 ad for Douglas aircraft showed two women on a tropical beach who gazed admiringly at a young man as he lounged offshore in a flotation tube. Another 1966 ad from Douglas (billing itself "an equal opportunity employer") showed a man who was enthusiastically sketching a naked flying woman on his drafting board and calculating formulas for the aerodynamics of her full breasts. The headline read, "Intrigued by exotic designs?" An advertisement for Vought aircraft showed a young man courting a female robot with metallic conelike breasts and pouting lips. The ad's text asked, "Do mechanical brains intrigue you? Do these intellectual vamps arouse your engineering instincts? Then why go on ogling? Especially if you're an electronics or mechanics major! Plan to enjoy the company of the best mechanical computers. Create your own electronic brains for missile guidance. . . . Our representative will be in your placement office."[6]

By the mid-1970s, the tone of corporate engineering recruiting had changed radically, almost overnight, as companies faced civil rights pressure to hire female engineers. Leading firms began courting women engineering majors (at least on paper) by touting exciting job opportunities and emphasizing how much their executives valued female engineers. In a 1978 advertisement for IBM, staff member Evelyn Gratrix, a University of Washington electrical engineering graduate, praised IBM as a place where each employee's career path was limited only by ambition and ability. Kodak upped the ante by showing advertisements featuring not one, but groups of female engineering employees, highlighting the wide range of responsibilities within the company. General Electric promised female engineers a chance to advance to management status. Under the headline "We're looking for engineers who were born to lead," one ad showed a woman leaning back in an executive chair, enjoying high-level job security. GE boasted that "We make products other than appliances—and hire people other than men. . . . We want to add still more trained and talented women to work with us in creating, manufacturing and marketing the advanced, high technology jet aircraft engines we build. . . . If you'd welcome the chance to work equally with men . . . you'll have the same opportunities for professional advancement as your male counterparts, the same pay and the same status." Satirizing the sudden ubiquity of ads recruiting certain demographics,

FIGURE 7.2                                                                              ▶
Some corporate recruiting advertisements of the 1960s used sexist imagery to catch the attention of potential employees, while conveying a less than subtle message that the typical engineer was a heterosexual male who expected to encounter women in erotic dreams, not as coworkers. *Tech Engineering News*, February 1966, p. 20.

## Intrigued by exotic designs?

Come to Douglas. We have a lot to intrigue you: extensive and exceptional Southern California facilities, where there are many independent research and development programs underway; engineering design problems to challenge the most creative minds; encouragement to publish. Why not find out about Douglas? Contact your placement office or send a resume to L. P. Kilgore, Box 701-I , Corporate Offices, Douglas Aircraft Co., Inc., Santa Monica, Calif.

**DOUGLAS**
An equal opportunity employer

the telephone-equipment firm Stromberg-Carlson ran its own ad saying: "What! No photographs of women and minorities in a recruiting ad? No. We think today's college graduates have got the message that industry wants to add women and minorities." Many ads sought to counter suspicions of tokenism. Harris Communications ran one that said: "We do not hire 'women engineers.' (We do hire outstanding able engineers, some of whom happen to be female.)" Most notably, some companies capitalized on feminist language and sentiment, as in one Omaha firm's ad: "Women engineers: you've come a long way, but you'll be surprised how much farther you can go, with Gibbs and Hill." Semiconductor maker Inmos ran a drawing of a boy in baseball uniform under the heading, "When I grow up, I want to be an engineer, like my Mom."[7]

In reality, of course, many female engineering students still encountered both subtle biases and overt discrimination in actual corporate employment practices. Assurances of fair treatment and equal pay for women engineers often proved all too hollow and left some professionals only more frustrated and demoralized. But in contrast to earlier decades, when images of engineering students and professionals either kept women invisible or denigrated them, the 1970s trend of highlighting their presence and talent presented a striking new note. It was no coincidence that corporate advertisements praising and seeking female engineering students ran not only in campus engineering magazines but also in the *SWE Newsletter*. By the mid-1970s, SWE had proven itself to be a powerful force in promoting women's interests in the discipline.

## THE SOCIETY OF WOMEN ENGINEERS: MULTIPLYING ADVOCACY IN THE 1960S AND 1970S

Feminist groups on campus and in the wider community provided vital support for female engineering students by helping them to vocalize concerns about issues such as harassment, to strategize and learn techniques to counter such problems, and to develop self-confidence and positive feminine awareness. But many young women remained hesitant about openly identifying themselves as feminists. More than that, the intensive nature of engineering education limited time for extracurricular activities. Many female engineering students, whether or not they embraced a political and social feminist

FIGURE 7.3                                                                              ▶
The tone of engineering recruiting ads that ran in magazines changed radically with the passage of civil rights laws and the mobilization of feminist pressure, as companies competed with each other to demonstrate how much they appreciated the talents of women in engineering and how well they treated female employees, *Tech Engineering News*, January 1975, back cover.

consciousness, remained more comfortable with groups that were embedded in their engineering subculture—the various specialized professional societies and SWE.

After the official incorporation of the Society of Women Engineers in 1952, its leaders dedicated enormous effort to creating a new organization with a crusading vision, SWE's leaders systematically approached the cause. They poured personal attention into reaching potential converts; members of SWE's professional guidance and education committee personally wrote to dozens of high school women, sending pamphlets and replying to questions. In 1954, four SWE members had lunch with one William and Mary first-year woman who was looking at engineering as a way of using her talent for math. Elsie Eaves wrote, "Roslyn Gitlin, Althea Thornton and myself . . . and Betty Mills . . . gave her a pretty well rounded picture of civil, chemical and mechanical engineering and suggestions of how she could check with Columbia for planning her liberal arts work so that she could transfer to engineering if she wishes."[8]

Throughout the 1960s, the number of SWE student chapters multiplied on college and university campuses across the country. Established members offered support; for instance, the Los Angeles section of SWE provided speakers and counselors to student sections at USC, UCLA, Loyola Marymount, Harvey Mudd, Cal State Long Beach, Pomona, Fullerton, and Cal Poly San Luis Obispo. Overall, campus SWE groups fulfilled vital intellectual, social, and psychological support roles for female engineering majors. Karen Lafferty Instedt, a student at Ohio State from 1968 to 1971, later wrote that SWE gave her "an opportunity to meet the other female engineers who, like me, were isolated in their respective fields and classrooms. The SWE section functioned as a refuge of sorts—where one could find an understanding ear from a peer or a kindhearted, encouraging professor or dean." By the end of the 1970s, student sections had been chartered in over 170 colleges, universities, and technical institutes. SWE held an annual national student conference featuring technical sessions and exhibits, professional workshops, industrial tours, and sessions on career planning, power dynamics, management, personal assertiveness training, and "dressing for success." To make everyday campus life more manageable, the SWE chapter of the University of Missouri at Rolla compiled a 1974 "Dear John" pamphlet that listed locations of all women's restrooms on campus (and warned that "female johns" were "non-existent" in one building). The same concern had plagued female students at MIT and other historically male-dominated schools whose older architecture did not respect female convenience.[9]

By the late 1970s, SWE's overall membership totaled over ten thousand women and men. As SWE grew, its leaders drew on their expanding membership resources and mobilized the political clout to attract outside support. In the most obvious manifestation of this, SWE philanthropists donated and collected money to help young women finance their higher education. Starting in 1958, SWE had instituted the Lillian Moller Gilbreth scholarship for a woman in her junior or senior year of engineering school. Local chapters in the Southwest, in Kentucky, and elsewhere soon created their own scholarship funds. The Pittsburgh section offered awards to women engineering students who were Pennsylvania residents or had finished freshman year in a Pennsylvania university. By the end of the 1970s, SWE administered nineteen annual scholarship competitions that were worth more than $27,000. RCA supported SWE scholarships for third- or fourth-year women who were enrolled in electrical engineering, and the Westinghouse Educational Foundation funded Bertha Lamme-Westinghouse Scholarships (named in honor of that company's pioneering woman engineer) for first-year women.[10]

SWE activities at the college, regional, and national levels exploded in the 1970s, driven by members' enthusiasm and dedication, by the feminist movement, by government equal opportunity laws, and by university public relations needs. One of the most energetic programs came at Purdue University's engineering school, which had created a special staff position in 1968 to increase its female enrollment and promote retention. That intensive campaign paid off; Purdue's women engineering students rose from 46 in 1968 to 280 in 1974 to more than a thousand in 1979, which represented the nation's largest female engineering enrollment. Female enrollment soared during the 1970s, even as ongoing elements of masculine engineering-campus raunch culture continued

FIGURE 7.4
Sponsored by Corning Glass, Purdue's chapter of the Society of Women Engineers built a go-kart and entered it in the 1974 campus championships, with an all-female pit crew and driving team. *Purdue Engineer* (November 1974): 13. Coutesy of Purdue University Libraries, Karnes Archives and Special Collections.

# 1972 placement brochure

Purdue student section
SOCIETY OF WOMEN ENGINEERS

to denigrate female engineers. It was not coincidental that within this climate, Purdue developed one of the country's most active student SWE chapters. Among other activities, engineering coeds published their own newsletter, ran a "big sister" program pairing entering women with upper-class mentors, offered help in locating summer jobs, and produced an annual members' "resume book" for sale to potential employers. Each weekday, SWE "hostesses" volunteered to talk to prospective engineering students, take them to lunch, and lead tours of residence halls.[11]

The 1970s witnessed the organization of dozens of conferences, open houses, and other public events in many states to celebrate and assist women pursuing engineering. Some meetings were organized by and for women who were already out in the work world to give each other encouragement and suggestions for promotion. For example, a 1974 "Women in Engineering" conference that was jointly sponsored by SWE, the Engineering Foundation, and the Engineers' Council for Professional Development focused on advising women how to update their skills (especially after temporary child-rearing leave) and advance into other areas, including management. Other conferences were designed for women still in college, to bring them together with each other and with older mentors who might help undergraduates succeed in their studies and prepare to enter the professional world. For instance, the University of Washington (with 230 women engineering students in 1975 and 445 in 1977) hosted an annual conference where those women met with working professionals such as Bonnie Dunbar, a Rockwell ceramics engineer. The SWE section at the University of North Dakota sponsored a 1979 conference titled "Transitions: College to Careers" that brought in corporate representatives (many of them alumni) to talk about how to project a professional image, have a successful interview, handle postcollege finances, set career goals, and balance work and marriage. Speakers offered practical advice, suggesting, for example, that women make an effort to communicate with the boss, describe their career goals, and draft a projected schedule for accomplishing them.[12]

Still other conferences were organized by college engineering women for younger girls as a way of encouraging potential interest. SWE analysts assumed that girls and boys

◀ FIGURE 7.5

In the 1970s, the Society of Women Engineers brought women in engineering together for various events that were designed to promote their mutual support and multiply their chances for success. To maximize the visibility of female engineering students, SWE chapters at Purdue and other schools around the country produced resume books that highlighted their members' qualifications and then sent that material to corporations. Society of Women Engineers, National Records Collection, Walter P. Reuther Library and Archives of Labor and Urban Affairs, Wayne State University.

essentially possessed a similar ability to excel in math and science and that girls potentially could find as much interest in technology as did boys. They blamed girls' relative
lack of interest in engineering on socialization patterns that handed dolls to girls and toy
tools to boys, which put girls in home economics and boys in shop class. SWE further
attributed girls' underrepresentation in engineering to failures of the school system, especially the problem of guidance counselors who refused to take girls' career ambitions
seriously and who let them drop math and science classes. To counter such problems,
in 1974, the University of Iowa hosted a meeting titled "Women in Engineering: Why
Not You?" The conference brochure was distributed to high school women and read:
"Right now you're probably going through the list of things you do and don't want to
do with your life. College, teaching, the Peace Corps, marriage, or just getting a job
are a few of the things you may have considered. Well, if you're looking into a career,
we'd like you to think of one more possibility—engineering. While engineering has
always been thought of as a man's profession, it is no more masculine than cooking is
feminine. All you need to be a good engineer is an interest in math and science, and
the desire to plan and solve problems. In fact, most engineering students are a lot like
you." At the conference, current and former coeds spoke about "student life: trials and
tribulations, joys and expectations"; industry representatives and educators discussed
coursework and careers.[13]

Similar events occurred across the country to familiarize young women with engineering, employment opportunities, and the most exciting sides of technical work and
to allow girls to meet role models. In 1976, the New Jersey Institute of Technology
hosted an all-day program for three hundred young women. Organizers had received
more than six hundred attendance requests, far beyond their capacity. Faculty member
Marion Spector said, "Typically women just let things happen, they float along with
the current, not making any effort to set career goals. What we are trying to do is to
give them an introduction to personal direction and to introduce alternatives while they
are young enough to make strong changes." A 1973 University of Illinois conference,
"Women in Engineering: It's Your Turn Now," gave high school juniors and seniors a
chance to participate in "rap sessions," informal conversations with college SWE members and older women engineers. A 1974 symposium sponsored by SWE sections at the
Universities of Florida and South Florida featured a tour of the Kennedy Space Center,
plus discussions of student financial aid, co-op programs, career problems and openings,
and men's reaction to women engineers. Promotional material read, "As an engineering student you'll gain something most women don't get in college, a professional skill
which can be used immediately upon graduation . . . [with] the highest starting salary

bracket of the major professional job categories for women holding a bachelor's degree. . . . You owe it to yourself to look into the possibilities and opportunities offered by engineering."[14]

Other SWE chapters went directly into high schools as self-described "missionaries," seeking to spread the gospel. Starting in 1976, Berkeley's SWE section sent teams of three or four students and engineers to visit local junior high school classes. In 1980, members gave presentations to about a thousand students in ten Bay Area schools. Presenters described how they became interested in engineering and sought "to dispel myths about women in engineering . . . and give special encouragement to girls who are interested in math and science." One mechanical engineering major prepared posters showing how an engineer might design a pair of skis, and another team brought slides illustrating the construction of a hydropower plant. One organizer commented, "We discovered that women engineering students can be excellent role-models for girls in grades 7–12. A practicing engineer or scientist may be inspiring, but her achievement may seem unattainable to students who have not even started college. Junior high students, in particular, are more willing to take advice from those closer to their own age. 'I was happy to find out that there are women engineers!' said one enthusiastic student. . . . 'It showed me another kind of work I might be interested in.'" Berkeley noted that running this community outreach program benefited SWE members themselves by giving them experience in public speaking, introducing them to useful professional contacts, and providing favorable publicity. Berkeley members compiled a handbook for other SWE chapters, containing advice on how to start a similar outreach program. Taking outreach further, in the 1980s, SWE's San Francisco section hosted a program titled "Tinker . . . Toys . . . Technology," in which seventy-one teenage Girl Scouts spent two weeks learning computer programming, running physics experiments, touring Silicon Valley companies, and talking with women engineers and astronauts.[15]

Other SWE members hoped to influence even younger girls, those still in elementary school. In the mid-1970s, the Boston section sought to "infuse a seven- or eight-year-old" with enthusiasm and curiosity about how and why things worked. They wrote and published a coloring book, *Terry's Trip*, which told the story of a girl visiting her aunt, a mechanical engineer who worked in a toy factory. The heroine, Terry, talked to industrial engineers supervising the production line, chemical engineers mixing polystyrene, electrical engineers working with fancy calculators, and concluded, "Maybe someday I'll be an engineer like Aunt Jennifer and her friends at the factory." In a similar project, the North Carolina section of SWE put out a 1983 booklet titled *Betsy and Robbie*, which described a girl's visit to her cousin at a university engineering fair, where she became fascinated with Robbie, a computer-controlled robot designed by a female student. SWE materials emphasized that women were just as qualified as men for engineering, a discipline that required creativity and logic more than physical strength. Illustrations and photographs documented the daily activities of women who worked as a university professor, a safety engineer for General Motors, and a government environmental engineer. By making such role models visible and attractive, SWE and other advocates sought to win young women's interest and public confidence.[16]

Some SWE experts admitted that in the end, it was virtually impossible to find a direct causal correlation between advocacy efforts and changing patterns of women's

FIGURE 7.6

The Boston section of the Society of Women Engineers produced the children's book *Terry's Trip* in the 1970s, which encouraged young girls to visualize engineering as a possible and exciting career option for them. Society of Women Engineers, National Records Collection, Walter P. Reuther Library and Archives of Labor and Urban Affairs, Wayne State University.

engineering education. Taken in isolation, a child's coloring book, a conference for high school women, or even a new dormitory would seemingly do little to affect such big decisions as where to attend college, what major to choose, or which career to follow. Yet as a whole, the multidimensional set of actions started in the 1950s, 1960s, 1970s, and 1980s by the national Society of Women Engineers, local chapters, student sections, and individual women at places such as MIT and Purdue added up to a substantial force. SWE set up a social and professional bridge between generations that was rich with meaning for those who were both giving and receiving support. College women benefited from the guidance of older SWE members while they simultaneously served as outreach ambassadors to younger girls. Coming out of the Vietnam era, when protesters linked technical research and corporate employment to the horrors of Agent Orange, napalm, and impersonal computer-aided weapons proliferation, advocates sought to counter the masculine, militarized image of engineering. Some observers looked to female students to resuscitate the profession, arguing that by nature, women were perfectly suited to develop connections to the burgeoning environmental movement and Great Society–era community problems.

## STUDIES OF WOMEN ENGINEERING STUDENTS, SEX DISCRIMINATION, AND SUPPORT

In the 1960s and early 1970s, many observers optimistically believed that simply convening a "critical mass" of female engineering students would create the necessary conditions for their advancement in the discipline. The messages of women's success would trickle down throughout American culture, they hoped, creating a snowball effect to interest more young girls in entering engineering. But by the 1990s, it was apparent that although the numbers of women entering engineering study had indeed increased dramatically and although female enrollment in higher education overall had surpassed men's, women still claimed far fewer engineering degrees than men did. At the end of the twentieth century and beginning of the twenty-first, engineers, sociologists, psychologists, and other academics and policymakers began undertaking a wealth of studies to tease out the complicated intersection of multiple factors contributing to that underrepresentation.[17]

Some studies maintained that consciously or unconsciously, many white male engineers still insisted on defining themselves as an elite fraternity who were inherently more able than women at mathematical and technical matters. Even (or perhaps especially) in a feminist era, some men seemed heedless of whether behavior that felt comfortable

BETSY and ROBBIE

WRITTEN BY CAY POSEY
DESIGNED/ILLUSTRATED BY SUE HALL

FIGURE 7.7

In 1983, SWE members in North Carolina created *Betsy and Robbie*, the story of a girl who became intrigued with the robotics work being performed by female engineering students at her local college. Such efforts aimed to cultivate technical interests among elementary- and secondary-school girls so that they did not rule out engineering before even reaching college. Society of Women Engineers, National Records Collection, Walter P. Reuther Library and Archives of Labor and Urban Affairs, Wayne State University.

to them might upset or exclude some women. Day-to-day observations of one soph-omore-level engineering design class in the 1990s revealed that male students relished profanity and double-entendre asides as accepted discourse, which female classmates overheard but studiously ignored. The male professor casually deployed violent meta-phors to explain technical work and mocked his female colleagues who were coleading the class; he delighted in saying that his "job is to put on stress" and in making engi-neering education into ordeal tests designed to push students to the breaking point.[18] On team projects, female students often wound up taking notes; out in the workplace, observers assumed that females in the office must be secretaries rather than engineers.

Such examples fed into what became an extensive discussion by multiple scholars, such as Sue Rosser, Judith McIlwee, Gregg Robinson, and a host of others, about whether engineering remains a "chilly climate" for women where group norms and institutional operations systematically send coded messages that make females feel like outsiders.[19]

Other investigators in the 1990s put female engineering majors themselves under the microscope, trying to identify the personal, social, and academic characteristics that either contributed to their success or hampered it. Numerous studies focused on how gender, race, and class influenced students' construction of identity. Several studies monitored the seemingly crucial measure of self-confidence; one researcher suggested that over months of study, female engineering majors gained assurance as they mastered an engineer's problem-solving discourse. Those women who survived learned to play the "weeding-out" "game" of academic success, relishing their ability to survive and even thrive under rapid-fire academic immersion and highly competitive coursework.[20] But overall, a number of longitudinal studies suggested that even female students who exhibited high achievement test scores and good grades often disproportionately showed lower self-confidence than male peers.[21]

The concept of "stereotype threat," which attracted substantial traction in the early twenty-first century, suggests that the self-assurance of women (or minorities) can be undermined by negative comments about their group's potential.[22] In a set of studies of computer science majors at Carnegie-Mellon, Jane Margolis and her coauthors noted that women in the program felt outclassed from the start, after recognizing that their male classmates entered with substantially more programming experience, although not necessarily more talent. Female students feared that men "know so much more, do the work so much more easily, with less effort. These comparisons can be the kiss of death," the researchers wrote. The reality, however, was that "men tend to exaggerate their achievements more than women, and that men are more likely to attribute their successes and failures in a self-affirming way than are women." Even as female majors earned good grades, they became "socially alienated," discouraged by male peers' extreme competitiveness and obsession with computers. Explicit sexist "taunting" also undermined women's confidence; 22 percent had heard someone claim that program leaders had accepted female students only because of their gender.[23]

As academic investigators extended analysis of stereotype threat, many mainstream media reporters noted a fall 2011 *American Sociological Review* article documenting how a group of 288 female engineering students expressed ongoing uncertainty that their chosen major matched their personal values and felt unsure whether they could succeed in the profession. Lead author Erin Cech commented, "It stems from very subtle differences in the way that men and women are treated in engineering programs and from cultural ideologies about what it means to be a competent engineer. Often, competence in engineering is associated in people's minds with men and masculinity. . . . So, there are these micro-biases that . . . add up [and] they result in women being less confident in their expertise and their career fit." On the positive side, investigations concluded that stereotype threat does not have to pose a permanent, irredeemable problem, however. Researchers have suggested that engineering professors and other actors can head off or counter the loss of confidence by creating an inclusive classroom atmosphere that affirms the academic potential of everyone but especially women and minorities.[24]

Many observers focused on the fact that although elementary school women often express enthusiasm for science and math at levels parallel to those among boys, the female presence in high school and college science classes drops and in graduate school and the workplace falls even more dramatically. Phrases such as "the leaky pipeline" became common currency among analysts and activists, who extended the conviction of observers in earlier decades that the problem of women's relatively low representation in the field originated long before college. Local and national intervention efforts aimed at bringing female engineers into the classroom, to help girls visualize their own options in nontraditional fields. The influential Girl Scouts organization began offering young women the chance to earn multiple badges by exploring technology and science, while troops hosted science and engineering day camps and similar programs.[25]

In 1990, college advocates for women in engineering formed the group Women in Engineering ProActive Network (WEPAN), which joined SWE and other efforts in both research and action.[26] Numerous campuses around the country began to institutionalize support mechanisms for female engineering majors, which experimented with widely ranging academic and social interventions that were designed to boost retention rates and female students' well-being. In the 1990s, officials at Penn State, troubled by women's 13 percent level of enrollment in engineering at its Altoona campus, designed a special computer diagnostics class to create a "safe environment" for women to have fun exploring technical work. By promoting small-group "hands-on" work and by linking classroom theory to "real-world" project-based learning, advocates believed that they could improve both technical knowledge and positive thinking among female students. With funding from the National Science Foundation and similar support, other schools also experimented with creating new courses to familiarize first-year female students with hardware, hand tools, and spatial visualization, areas in which many experts believed that boys typically gained more informal knowledge than girls long before arriving at college. In one course at UC Davis in the 1990s, instructors and female engineering students alike recognized that childhood socialization had not, in fact, made every single college man fully comfortable with technical skills. Nonetheless, evaluators decided that "an all-women classroom environment is best" for creating a "non-threatening, non-defensive" space that honored women's distinctive communication styles and preference for cooperation over competition: "[S]tudents unanimously agree that the course would not be as successful if men were a part of the class." They feared that male classmates might isolate women and take over the lab benches and that their presence might constrain women from asking beginners' technical questions, discussing their concerns openly, and forming friendships with each other. In the UC Davis class, female engineering students disassembled cars, built robots, and kept journals to stimulate their "self-discovery" and reflections on gender equity as addressed by guest speakers and reading assignments. Such pedagogical experiments, combined with personal-development seminars and community-building exercises, sought to foster a sense of cohesion and confidence.[27]

As in earlier decades, a number of individual female engineering students avoided such optional programs. Some worried that there were already too many demands on their time, and others preferred not to self-identify with women's groups, especially if that definition carried feminist overtones. Several researchers have documented the way that some female engineering students in recent years almost recoiled in horror from feminist critiques of engineering and shut down counterparts who spoke about ongoing discrimination.[28] Of course, women in fields other than engineering have also rejected the word *feminism* or found it daunting to confront men about obnoxious and unfair conduct. Still, one 2009 British study concluded pessimistically that engineering represented a particularly dramatic case where some women internalized and even colluded in maintaining an environment that favored masculine values. For self-protection and security or by conservative personal preference, female technical students "performed their gender in a particular way in order to gain male acceptance. . . . [W]hilst there are multiple masculinities and femininities that can be performed by anyone, only 'traditional' masculinity performed by men is valued in engineering cultures specifically." The researchers saw the women seeking to "blend in" or even to outdo colleagues in machismo, ignoring or making excuses for sexist conduct in a way that ended up normalizing discrimination. Female engineering students joked around to position themselves as "one of the boys" and put down other women as weak and even "irritating." The authors warned, "In 'doing' engineering, women often 'undo' their gender. Such gender performance does nothing to challenge the gendered culture of engineering, and in many ways contributes to maintaining an environment that is hostile to women."[29]

Going even further, some female engineering students worried that targeted programs that were meant to assist women in the twenty-first century ended up stigmatizing and hurting them by inviting a resentful male backlash against perceived favoritism. In earlier decades, at schools such as MIT, Georgia Tech, Caltech, the land-grants, and many others, men had expressed similar suspicions that professors and teaching assistants "went easy" on women (especially if they were attractive or flirty) and that college admissions committees bent over backward to give women preference. Detailing the consequences of special mentoring and similar arrangements, one researcher wrote in 2005, "[T]he process of singling women out marks them simultaneously as different than male students, and in need of additional help and, therefore, less capable. This induces self-doubt in women students, and . . . spawns criticism from male students, who form all sorts of misconceptions about their female peers' scholastic ability and their 'privileged' status within the college, thus bolstering sexist pre-conceptions, conditions, and behavior."[30] Advocates maintained that schools could avoid such a trap by promising that male as well as female students would benefit from institutional changes and curriculum revisions to make engineering study less pressured, more user-friendly, and more effective. Others insisted that women who sought sisterhood and support to succeed in engineering were demonstrating strength, not weakness. They saw no need for women to apologize for honest efforts that aimed at helping them fight through a difficult climate.

From the historical perspective, these modern analyses show just how deeply rooted the issues of gender in engineering education are. Recent questions about critical mass, mentoring, role models, stereotypes, socialization, and self-confidence all have precedents in discussions of engineering coeducation during earlier periods. Although no one used the phrases "leaky pipeline" or "chilly climate" decades ago, the underlying concerns would have been familiar to observers of female education at Georgia Tech in the 1960s, Caltech in the 1970s, and MIT over many years. By the twenty-first century, comments painting women as "invaders" had generally disappeared, but many female engineering students and some of their faculty role models still often felt like outsiders in what remained, in many eyes, masculine territory by default.

## NATIONAL MOBILIZATION EFFORTS INTO THE TWENTY-FIRST CENTURY

By the first decade of the twenty-first century, programs to support women's success in engineering almost amounted to an industry in their own right, characterized by substantial funding, intensive research, multiplying publications, and symbolic weight. Professional development workshops and other programs highlighted an awareness of diversity profiles and sought to engage faculty and administrators in supporting women students. Accompanying long-standing efforts by the Society of Women Engineers to make technical studies more female-friendly, the National Science Foundation (NSF) threw its unique clout behind the cause. In 2001, the NSF created the influential AD-VANCE grant program (ADVANCE: Increasing the Participation and Advancement of Women in Academic Science and Engineering Careers) that over the following decade distributed over $130 million to more than a hundred schools. The list included pub-lic universities such as Michigan, Wisconsin, Nebraska, Colorado-Boulder, UC Irvine, Iowa State, and Ohio State, plus private schools such as Brown, RPI, and Case Western Reserve. Georgia Tech was among the first recipients of "institutional transformation" funding, which called for "innovative and systemic organizational approaches to trans-form institutions of higher education in ways that will increase the participation and advancement of women in STEM academic careers." The Tech program collected data and tracked trends in hiring, retention, and promotions of female and minority faculty and offered workshops and individualized career coaching to support junior candidates. Specially designated ADVANCE professors worked with department chairs to dissemi-nate information about "best practices" in equity and issues, such as stopping the tenure clock for family leave. The school created an online program titled Awareness of Deci-sions in Evaluating Promotion and Tenure (ADEPT) that was meant to help faculty pre-pare strong portfolios for consideration and to alert campus leaders to insidious sources of assessment bias. Although such initiatives primarily targeted faculty welfare, they helped ensure that undergraduate and graduate women in science and engineering could find role models and mentors on campus, while promoting discussions about work-life bal-ance and similar challenges that faced women in nontraditional fields.[31]

Other national and international organizations joined NSF and SWE in mobilizing to address the overall climate for women in engineering. In 1999, the National Academy of Engineering (NAE) organized a "Summit on Women in Engineering" in Washing-ton, D.C., where panelists and attendees evaluated the state of existing gender-equity efforts and highlighted the advantages that a more diverse technical workforce could offer American business and government. The NAE convened follow-up workshops, surveyed diversity-management policies at firms recognized as the best employers for women, and underlined corporate interests in improving elementary, secondary, and college education. NAE president William Wulf wrote, "There is a real economic cost to our lack of diversity. . . . Every time we approach an engineering problem with a 'pale male' team, we do so with a set of potential solutions excluded, including perhaps the most efficient and elegant." In its 2007 report Beyond Bias and Barriers, the NAE collaborated with the National Academy of Science and Institute of Medicine to address underrepresentation of female faculty in engineering and science institutions. Surveying years of research, the high-profile document concluded that the academic system could hold back talented women in multiple ways, including gender stereotypes, unconscious biases and subjective standards in grant and tenure evaluations, and organizational obstructions (such as the assumptions that professors had wives who would handle household matters).[32]

Beyond spreading diversity awareness, the NAE capitalized on the unprecedented reach of the World Wide Web to begin directly encouraging girls to consider technical studies. In 2001, the NAE funded creation of the *EngineerGirl* Web site, aimed primarily at middle-school women. One of the site's first pages asked young women, "Why be an engineer? . . . You'll have the power to make a difference . . . money and job security . . . lots of options . . . [and] get to do cool stuff!" Panels of established female professionals offered online advice about which classes aspiring engineers should take and gave personal feedback about educational and career choices. Young women from around the United States submitted questions, as did others from Ireland, Nigeria, Namibia, India, Malaysia, the Philippines, and many different countries. Although the NAE intended *EngineerGirl* to serve a junior high school audience, numerous inquiries came from high school students, college women already majoring in engineering, and even graduates. Responding to this evident demand, the NAE soon teamed up with the WGBH Educational Foundation and over a hundred educational and engineering partners to create *Engineer Your Life*—a Web site, a video series, brochures, a Facebook page, and more—to serve high school women (plus their parents, teachers, and guidance counselors). Together, *EngineerGirl* and *Engineer Your Life* aimed to help female students think of engineering as a career path that was accessible to them. Both sites presented profiles of role models who were doing exciting jobs, women who seemed both personally and professionally admirable. The Web sites sought to make engineering compelling for young women by emphasizing the opportunities for creativity, travel, and job flexibility. Many analysts suggested that young women were particularly interested in the social value of engineering and the altruistic pleasure of creating technological innovations that could clean up the environment, promote better health care, and raise living standards worldwide. Accordingly, both NAE Web sites defined engineering as a way to make a difference in the world, highlighting projects such as improving water supplies in developing communities. *Engineer Your Life* told site visitors, "[W]omen who become engineers save lives, prevent disease, reduce poverty, and protect our planet. Dream Big. Love what you do. Become an engineer."[33]

Other high-profile groups hoped to appeal to girls by portraying creative problem solving as cool and intelligent female engineers as hot. Most notably, in 2000, Karen Panetta, a faculty member in electrical and computer engineering at Tufts University, created the "Nerd Girls" group that "celebrates smart-girl individuality . . . to encourage other girls to change their world through Science, Technology, Engineering and Math, while embracing their feminine power." Tufts undergraduate women gained leadership experience outside the classroom by building a solar-powered race car, and over subsequent years, teams (including men) also developed solar home systems, other renewable-energy projects, and technical aids for the disabled. Through their Web site, members of *Nerd Girls* introduced themselves as team leaders who designed the solar car's suspension, braking, and other technical systems while pursuing outside interests such as robotics, math games, yoga, rollerblading, art, violin, parties, and high-fashion Manolo Blahnik shoes. Accounts indicated that about 98 percent of undergraduate Nerd Girls moved on to graduate school within three years, far above the overall rate for women earning bachelor's degrees in engineering. In 2011, President Barack Obama presented Panetta with a Presidential Award for Excellence in Science, Mathematics, and Engineering Mentoring, accompanied by $25,000 from the National Science Foundation. With support from the IEEE and companies such as Verizon, the group spread across the United States and abroad, running mentoring and educational programs that reportedly had reached over 25,000 American youngsters, parents, and teachers by 2012.[34]

The Nerd Girls movement cultivated a substantial media presence, making videos for the IEEE Web site and YouTube that intercut images of the "knockout brainiacs" at work on their race car and solar-lighthouse projects with shots of them flirtatiously posing in stilettos, superhero capes, and a geek's stereotypical black-framed eyeglasses. After members appeared on the *Today Show*, NBC commented that any of them "could find work as a model." *Newsweek* linked real-life members such as biomedical engineering grad student Cristina Sanchez, "a former cheerleader . . . who can talk for hours about aerodynamics," to pop-culture icons Tina Fey and Danica McKellar. To counteract decades of depressing stereotypes of engineering as an inherently unfeminine retreat for unattractive women, the Nerd Girls updated their predecessors' efforts to present female engineers as well-rounded, happy individuals. The Nerd Girls philosophy—"Brains are beautiful. Geek is chic. Smart is sexy"—echoed third-wave feminist boldness in urging women to seize ownership of their sexuality and relish cultivating their attractiveness. Panetta reveled in pink high heels and a crystal-covered "Hello Kitty" laptop computer, a playful style that resonated with some observers but discomfited others who worried about existing problems of objectification and harassment. Panetta dismissed accusations of "using sex to sell science," saying, "The most damaging thing for women is to com-partmentalize and limit themselves. . . . [A]fter decades of trying to be more 'male,' the new generation of women is comfortable in embracing all aspects of who they are and celebrating it."[35]

National organizations such as the NSF, NAE, and IEEE were able to direct sub-stantial funding and create high-profile initiatives, but beyond that, many other factors continued to promote transformation in the gendered practices and principles of engi-neering in the first years of the twenty-first century. Media outlets, such as the *New York Times* and *PBS NewsHour*, monitored efforts to close the engineering gender gap, as did a multiplying host of Web sites and blogs. Female students themselves both reflected and drove changes in the social and college climate, even as some maintained (especially during their early college days) that they had never been treated any differently than men in engineering. Although some young women rejected the label *feminist* or chose not to become involved with gender-specific engineering programs, second-wave and third-wave feminism had nonetheless left an impact. Progressive parents, educators, media outlets, and organizations such as the Girl Scouts and Nerd Girls helped create a conscious embrace of women's potential and a mentality of empowerment. Many young women faced skeptics or continued to question themselves periodically, but oth-ers carried a sense of confidence and belonging into their college years, having internal-ized bold visions for their education, future careers, and personal lives. In 2007, Caltech freshman Elizabeth Mak said simply, "[O]pportunity and success should be equal for both sexes."[36]

As visible role models for women undergraduate and graduate students, female en-gineering faculty members, department chairs, and top administrators remained under-represented across American academia but were becoming less rare by the decade. In 1984, Pratt Institute selected electrical engineering professor Eleanor Baum as the first woman in the country to head a school of engineering. Baum subsequently became the American Society for Engineering Education's first woman president and, in 1997, dean of Cooper Union's engineering school. When she initially expressed a youthful interest in engineering in the 1950s, Baum recalled, her mother had tried to dissuade her by warning, "You can't do that. People will think you're weird, and no one will marry you." One engineering school rejected her application with the excuse that it did not have bathroom facilities to accommodate women. Baum became the sole female engineering student of her day at the City College of New York, where, she remem-bered, "A lot of people did indeed think that I was a very strange person, and that was not very comfortable. . . . In lab classes, I was expected to be the data taker rather than the person who did things. . . . [I]t was not a wonderful experience." As her own career advanced, Baum monitored the gendered state of engineering life. In 1989, she led a survey of SWE members that documented how many more women had entered the field since the late 1970s. On the positive side, more than 90 percent of respondents said that entering technical work had not compromised their femininity, and 82 percent felt satisfied with their salary levels. On a more discouraging note, roughly 70 percent of the women believed that they had to work harder than men to succeed, and over 50 percent reported encountering sexual harassment in the workplace.[37]

In the 1990s, a few other women joined Baum as engineering school deans, including Cynthia Hertzel at Temple University and Ilene Busch-Vishniac at Johns Hopkins. In 1996, Denice Denton became the first female engineering leader of a NRC-Research One category institution, at the University of Washington. In 2005, the University of California at Santa Cruz installed Denton as chancellor, where she planned bold steps to promote diversity and social justice on campus. An MIT graduate, Denton had survived what a coworker called a "hostile climate" in her first faculty post, at the University of Wisconsin at Madison, and won national recognition as a role model, mentor, and advocate for women in engineering. At UC Santa Cruz, Denton stepped into a hotbed of pre-existing labor trouble, concerns about racism, and disputes over military recruiting on campus. Controversies over her compensation and other perks, including allegations of making elaborate renovations on her official university residence, drew statewide attention. Denton's status as one of the few openly gay university leaders in the United States generated further volatility, apparently including some antilesbian slanders and physical threats (including broken windows and protesters who blocked her door and her car on campus). Reportedly depressed by medical issues, personal stress, and the highly public backlash, Denton evidently committed suicide in summer 2006. Friends and students memorialized her as something of "a rock star" idol "for women in the sciences and engineering" and as "a trailblazer in pursuit of equity and multiculturalism." Two former colleagues wrote that many "wished we had her passion and strength while slogging through the detritus of the entrenched 'old boys' network." They concluded poignantly, "Perhaps the most plausible speculation is that those who break through the glass ceiling may be wounded—even destroyed—by the shards."[38]

In the twenty-first century, it remained newsworthy but no longer seemed revolutionary for institutions such as Tufts University, Purdue, Texas A&M, Yale, and Harvard to name a woman to lead the school of engineering.[39] Those professional women and their male allies often reminded other institutional players to maintain awareness of gender issues, both within their own schools and across the discipline as a whole. They created local programs, served on national committees, and implemented initiatives to make engineering study less intimidating and more compelling for women.

As in earlier years, some individuals volunteered exceptional efforts and organized programs with striking, if localized, effects. Enrollment at Harvey Mudd College, a southern California private school with strengths in engineering and science, was less than 10 percent female in 1970, with twenty-nine women in its total of 387 undergraduates. Its numbers hovered around 15 percent female in the early 1980s, grew to around 25 percent female by the mid-1990s, and hit one third of total enrollment in 2002.[40] But when former Princeton University engineering dean Maria Klawe became Harvey Mudd's president in 2006, she observed with disappointment that women comprised only about 10 percent of majors in her home discipline, computer science. Program leaders worried that the tough Java programming required for every freshman signaled that only nerds (male by default) were welcome in computer science and chased off potentially excellent female students. Klawe and other observers linked the daunting effect to what psychologists nicknamed "the impostor syndrome," a severe perfectionism and insecurity that led some very bright women to consider themselves frauds who did not deserve to succeed and feared being found inadequate.[41] To create a more comfortable climate, Harvey Mudd's computer science department split its introductory class into two tracks, separating novices from more experienced classmates. Faculty members demystified programming by embracing alternative, less hard-core languages while revising the course to emphasize interdisciplinary problem solving and highlight applications of computer science to biology, health care, anthropology, and the arts. Klawe claimed, "Our introductory course went from drudgery to outrageous fun." Follow-up surveys of students documented that both men and women found the revamped class far more positive but that it proved especially influential in exciting women about computer science and reinforcing their confidence. Klawe raised funds on campus and from tech companies to offer first-year women free trips to each year's Grace Hopper Celebration of Women Conference to deepen their professional interest and help them think about careers and graduate school. Such initiatives more than tripled female representation in computer science at Harvey Mudd and took women up to 40 percent of the major by 2011, the highest in all U.S. coeducational schools. In that year, women also made up 37 percent of Harvey Mudd engineering students and 42 percent of its overall enrollment. To increase the yield of female applicants who actually chose to attend Harvey Mudd, Klawe sent personalized e-mail messages and, for several years, even sent handwritten notes to all those accepted. Not coincidentally, she also emphasized a gender-inclusive tone in all the college promotional material and touted the importance of female science and engineering faculty as influential role models.[42]

Meanwhile, prescriptions for engineering education have continued to evolve, reflecting ongoing social and academic discussions of what future professionals should ideally learn. Most notably, as the twentieth century came to a close, the Accreditation Board for Engineering and Technology (ABET) published "Engineering Criteria 2000," a new set of assessment and outcomes standards that reviewers could use in judging the quality of engineering programs and their graduates. Sue Rosser has suggested that in emphasizing factors such as communication skills, cooperative teamwork, and an interdisciplinary understanding of the way that technological development connects to the human social context, the ABET measures might inspire some curriculum reforms in directions that experts believe may help cultivate diversity and "make engineering more female-friendly."[43]

The very end of the twentieth century brought one more telling development in women's engineering education. As noted at the start of this book, women made substantial gains in many areas of science and medicine in the late 1800s and early 1900s in part because women's colleges had made concerted efforts at preparing students to enter such fields. Although female scientists and physicians still generally faced extreme challenges in securing employment, recognition, and fair treatment, institutions such as Wellesley, Vassar, Smith, Mount Holyoke, Barnard, and Bryn Mawr opened the gateway, graduating a small but visible cohort of women in botany, zoology, psychology, math, chemistry, geology, physics, and astronomy before World War II.[44] Although the Seven Sisters and other single-sex institutions could justify scientific studies as a legitimate component of proper modern education for women, those arguments did not stretch to encompass engineering education, which remained too closely identified with a male domain and therefore seemed impractical and undesirable for female students. Hence, the women's colleges that made such a difference in nurturing American female scientists simply did not play any direct role in graduating female engineers—until 1999. One hundred and twenty-eight years after its founding, Smith College created the country's first permanent degree program in engineering at a women's school.

Earlier in the twentieth century, Smith had joined other single-sex and coeducational institutions in creating temporary wartime programs to train women in a few technical subjects and mechanical skills such as ambulance driving. As described above, the Wellesley-MIT exchange program, started in the late 1960s, also attempted to establish a path for drawing women into areas of study that were not otherwise available at their home institution. Around 1976, Smith's first female president, Jill Ker Conway, cooperated with the University of Massachusetts to open a dual-degree program in engineering, and Smith opened an engineering minor. Over the years, other women's colleges, such as Mills, Agnes Scott, and Spelman, also instituted dual-degree options, representing efforts to rethink the relationship between gender and technical training.[45] On balance, however, such initiatives often only underlined how much the traditionally male engineering curriculum remained physically and psychologically distant from the world of the all-female school.

Smith College's decision in 1999 to create an engineering degree made front-page headlines in the *New York Times*. In defining their ambitions, campus leaders and trustees sought to prove that single-sex schools had left behind their "white-glove image" to become modern, relevant, and appealing. Smith administrators hoped to show that well-designed educational efforts could help remedy engineering's lack of gender, ethnic, and racial diversity, a pattern that represented a real concern if American economic leaders wanted to remain globally competitive. Finally, they believed that twenty-first-century engineering education should avoid overspecialization and (as ABET had indicated) should extend beyond a mastery of essential technical knowledge to encompass humanistic learning and messages about the social and ethical responsibilities of engineering. Advocates argued that Smith—as a liberal arts institution and as a female-friendly environment where roughly 30 percent of students majored in the sciences, which far surpassed the national average for college women—could contribute in tangible and unique ways to women's engineering education. Pointing to ongoing unease about retention issues for well-qualified women in coeducational environments, Smith's provost, John Connolly, said:

A student's chances of leaving Smith with an engineering degree are likely to be much greater than they would be at a university. . . . If today five out of six engineering students are male, even while medicine, law and business are approaching gender parity, then clearly there is great need for another educational paradigm. . . . Smith will be a leader in educating women engineers, not by virtue of numbers—we are never going to have a very large program—but by virtue of having an exemplary approach.[46]

In September 2000, twenty Smith students entered the Picker Program in Engineering and Technology, named after a 1942 graduate whose family contributed $5 million to launch a new endowment. The Ford Motor Company financed several full scholarships and faculty research and became the lead donor for a $73 million building designed to promote collaboration by locating teaching and laboratory facilities for science and engineering under one roof (which was LEED environmentally certified). Engineering enrollment rose faster than the founders had anticipated, and in 2004, twenty

Smith students received the first bachelor's degrees in engineering issued by an American women's college. Those graduates said that the level of public interest made them "feel like 'rock stars.'" By 2010, Smith had granted engineering science degrees to 160 women, about 60 percent of whom entered corporate employment and 40 percent of whom pursued advanced studies, some with prestigious fellowships from the National Science Foundation. Certain engineering graduate programs at Johns Hopkins, Tufts, Dartmouth, and elsewhere agreed to offer automatic acceptance to Smith alumnae who recorded at least a 3.5 grade-point average.[47]

National leaders such as NAE president William Wulf and Hewlett-Packard CEO Carly Fiorina hailed Smith's engineering graduates as worthy pioneers and praised the school's innovative pedagogy. Critics, however, publicly jeered at the program, voicing fears that in cultivating gender diversity and interdisciplinarity, engineering educators undercut the importance of technical knowledge as the core of engineering. *National Review*'s Robert VerBruggen commented that if Smith's program were really good, its graduates "wouldn't need affirmative action deals" to get into graduate school. He scoffed at the sociology, philosophy, and other liberal arts elements of Smith's training as "distractions," and readers added their own putdowns about not wanting to fly in any airplane designed by such pseudo-engineers. To his credit, VerBruggen later posted responses from engineers who wrote him to "*heartily* endorse" widening the definition of a good professional to include skills such as communication. Princeton University engineering administrator Peter Bogucki confirmed to VerBruggen that Smith students who took engineering classes at Princeton under an undergraduate exchange program were "very well prepared . . . able to take the really tough Princeton junior-year engineering courses without a problem. . . . [T]heir academic content is equal to other top engineering programs, since you can't repeal the second law of thermodynamics, and the level of rigor is very high."[48]

Over its first decade in existence, Smith deliberately shaped its engineering to be distinctive in many ways, a program that "fosters an intellectual and emotional balance," language that was strikingly absent from traditional disciplinary pedagogy. National-award-winning professor Glen Ellis said that after arriving at Smith, he had to break himself of "intimidating" classroom habits, such as "cold-calling" students and making exams excessively brutal. The program's learner-centered methods, such as portfolio work, led to a 90 percent student-retention rate, and roughly 60 percent of its faculty and staff were female, data points that contrasted sharply with national trends.[49] Given that Smith would inevitably remain smaller than traditional engineering-education powerhouses such as MIT and many state land-grant institutions, the school did not subdivide its program into customary specialties. Instead, Smith developed two "adaptable expertise" degree offerings by 2012, an ABET-certified B.S. in engineering science (which also required a double major, minor, or substantial coursework in social sciences or humanities) and a B.A. in engineering arts, which linked technical understanding with public policy, global development, education studies, landscape and sustainability studies, or similar fields. The introductory course, significantly named "Engineering for Everyone," emphasized design-related teamwork experience and asked students to "critically analyze contemporary issues related to the interaction of technology and society." Advanced courses covered standard topics such as systems, strength of materials, circuit theory, engineering mechanics and biomechanics, thermodynamics, modeling, and year-long design–clinic practice.[50]

Conferences on topics such as "Engineering, Social Justice, and Peace" extended Smith's emphasis on the interdisciplinarity and the social, ethical, and emotional accountability of technical work. Beyond that, the women's-college experience overlapped in substantial ways with coeducational engineering, with parallel experiences such as engineering internships and field trips. Although Smith's classroom conversations might have foregrounded gender-equity conversations more often and more overtly than other schools did, its programs to promote women's place in engineering echoed those that had been developing at coeducational schools since the 1960s. Like dozens of other schools, Smith acquired its own chapter of the Society of Women Engineers, which hosted both social and professional-development events. Its outreach efforts gave college women the chance to mentor younger counterparts. Smith hosted both an "Introduce a Girl to Engineering" day to reach young women in middle school and a summer program to bring high school women to campus for research experience. In a further effort to help attract young teenage girls to engineering, Smith developed a Web site called *Talk to Me*. That 2011 project featured an online story about a young teenager who used her problem-solving skills and collaborated with friends to develop a robot cat programmed for speech recognition and artificial-intelligence abilities. The girl then deployed the robot to save her mother from arrest under false charges of embezzling funds from her cell-phone engineering job. The Web site allowed users to play games that explained engineering ethics, design, and sustainability and invited visitors to read a blog with "an insider's look at what it's like to be an engineering student at Smith College!"[51] Beneath such interactive elements, this twenty-first-century project shared the same faith that had gone into SWE's 1970s booklet *Terry's Trip* and 1983's *Betsy and Robbie*—the faith that female engineering students and established professionals could and should use their visibility to nurture the next generation.

CONCLUSION: UPDATING CASE-STUDY INSTITUTIONS INTO THE TWENTY-FIRST
CENTURY

Although feminism and broader cultural changes made the case that female engineering students were entitled to intellectual respect and educational opportunity, other factors also continued to drive interest in women's engineering enrollment in the early twenty-first century. There remained a social expectation that science and technology-oriented colleges could best serve students by providing gender balances that maximized opportunities for heterosexual pairings. In 2007 Caltech senior Michael Woods hailed rising female enrollment as crucial to the "emotional learning" of male "nerds," something that he called "a vital aspect of college that isn't represented in tuition and lecture halls." In more tactful wording, Caltech's assistant vice president for student affairs, Erica O'Neal, declared, "The more women we have on this campus, the better it is for everybody. . . . It is better for women to not feel so isolated. And it is better for the guys to learn how not to be awkward with the opposite sex."[52]

Today, many female engineering students, alumnae, faculty, and professionals can and do disagree about the gendered status of their chosen field, as do men in engineering. Some highlight the continued obstacles that face women, citing a continued tacit assumption (or even less subtle indications) among male colleagues and the American people in general that engineering by definition represents a masculine domain. Others relish the advances that have been made by female engineers, both individually and as a group, and consider gender discrimination primarily a relic of the past.

Issues of sexism and personal experiences can be subjective, but statistics remain undeniable proof of women's ongoing underrepresentation in engineering. Of all engineering bachelor's degrees that were issued in the United States in academic year 2010–11, 18.4 percent went to women. The figure had been higher just six years before: women took home 20.9 percent of the country's undergraduate engineering diplomas in 2002, 20.4 percent in 2003, and 20.3 percent in 2004, before sliding a bit each year between 2005 and 2009. Underneath the wide umbrella of engineering education, specialties varied, sometimes dramatically, in their appeal to female students. In academic year 2010–11, women earned 44.3 percent of bachelor's degrees in environmental engineering, 39.1 percent in biomedical engineering, and 33.1 percent in chemical engineering, versus just 9.4 percent in computer engineering, 11.5 percent in electrical, and 11.7 percent in mechanical engineering.[53]

To put that in perspective, by the twenty-first century, women comprised a sizeable majority of total nationwide undergraduate enrollment and completed a majority of all

American bachelor's degrees. Women's figures in both categories stood at around 57 percent in 2008. Across all the disciplines ranked by the U.S. Department of Education, engineering still graduated the lowest share of women. Women's representation in many pure science fields offered some striking contrasts. In academic year 2007–08, women earned 59.4 percent of all bachelor's degrees awarded in biological and biomedical sciences; their numbers remained around or just under half of all chemistry and mathematics undergraduate degrees. Numbers remained far lower for women in physics and computer science, where the economic downturn, the bursting of the technology bubble, and other factors reversed earlier progress. In 1985, women claimed 37 percent of computer science degrees, but by 1997–98, women earned just 26.8 percent of all computer and information science bachelor's degrees, and ten years later, their share had fallen to 17.6 percent.[54]

It is not inevitable or mandatory that the female presence in undergraduate engineering work rise to 50 percent. Yet just when many women and male allies hope at least that the age when critics openly questioned female intellectual ability has passed, speculations still reappear about whether women are biologically inferior to men in mathematical ability and inherently less suited to pursue "hard" fields. Most notably, at a 2005 conference on increasing diversity, Harvard's then-president, Larry Summers, sought to "provoke" listeners by saying that perhaps women possessed less "intrinsic aptitude" for science and engineering than men did and were less ambitious, letting "legitimate family desires" keep them away from "high intensity" positions. Summers also suggested that women might be opting to stay away from science and engineering by free choice, a perspective that minimized issues of socialization and discrimination.[55] His comments appalled many in his audience, which included prominent female scientists, female engineers, and researchers in the specialty of gender, education, and professionalization. In one among many responses, Stanford-affiliated electrical engineer Carol Muller, former astronaut Sally Ride, and seventy-seven other women and men coauthored a letter to *Science,* saying:

Well-accepted, pathbreaking research on learning . . . shows that . . . [i]f society, institutions, teachers, and leaders like President Summers expect (overtly or subconsciously) that girls and women will not perform as well as boys and men, there is a good chance many will indeed not perform as well. . . . [W]ell-documented evidence demonstrates that women's efforts and achievements are not valued, recognized, and rewarded to the same extent as those of their male counterparts.[56]

Subsequent conversations in the popular media and academic analysis, following up on Summers's explosive remarks, drew increased attention to what some critics had been saying for years about the persistent chilly climate for women in engineering and science disciplines. In 1994, tenured women in MIT science programs began comparing notes on disturbing experiences and persistently frustrating conditions that hindered their worklife. Those professors collected data documenting that female science faculty received less financial assistance, less laboratory space, and fewer rewards than male colleagues obtained. Their 1999 "Study on the Status of Women at MIT" commented, "Each generation of young women, including those who are currently senior faculty, began by believing that gender discrimination was 'solved' in the previous generation and would not touch them. Gradually however, their eyes were opened to the realization that the playing field is not level after all, and that they had paid a high price both personally and professionally as a result." Engineering professor Sallie Chisholm described the phenomenon as "microdiscrimination"—the subtle, mostly nondeliberate biases and marginalizations that ultimately added up to serious assaults on women's careers.[57] Modern female engineering students' situation may well be considered in the same light, continued cumulative difficulties that prove harder to combat as they become relatively less blatant.

Indisputably, some ongoing gender conflicts surrounding engineering-student life remain anything but subtle. The uncensored corners of the Internet offer sites for anonymous posters to denounce the entire category of female engineers as repulsive beings who enjoy unfair preferences, who somehow simultaneously lack all sex appeal but also use feminine wiles to get ahead of men. Extending the battle of the sexes, other anonymous commenters retort by attacking male engineers as hopelessly immature geeks (as in the saying deployed at many tech schools to describe the availability of men: "The odds are good, but the goods are odd").

By the twenty-first century, gendered resentments and frustrations had become ingrained in engineering and science school slang. At Georgia Tech, the phrase "the ratio" instantly summed up male students' vocal unhappiness with the quantity and quality of potential dates.[58] In a 2005 insiders' guide to campus life, one male sophomore at Georgia Tech wrote, "The women at Tech are also, generally, brilliant, but also either unattractive or stuck up. We have a term for their behavior . . . Tech Bitch Syndrome," where "a moderately attractive girl will come to Tech and will instantly get more attention from guys than she's ever had in her life. The women get spoiled by the attention, and before long they're stuck-up and full of themselves."[59] The bitter sentiments behind the "TBS" label occasionally flared into flame wars, as in an extended exchange of letters in the *Technique* preceding Valentine's Day 2001, where one fellow wrote, "Why can't any GT girls find quality guys? Because the quality guys don't want to have anything to do with them. Why can't GT guys find quality girls? Because there are none at Tech, or at least very few." Follow-up letters from other Tech men echoed indignant stories about being unfairly rejected by snobbish women. Some female students leapt to defend themselves as a group, linking the dating debate to deeper issues, such as women's intellectual seriousness. Katie O'Connor wrote, "Women have worked hard for years to eradicate the stereotype of the infamous MRS degree, the notion that women attend college for no other reason than to find a husband. Now we're attending college in larger numbers than ever, and we have BETTER THINGS TO DO." On the positive side, the dialog did not devolve into a pure insult exchange between men and women. Some female and male students alike wrote thoughtful comments on potential links between dating attitudes and serious problems such as date rape. Several letters from men encouraged their fellows to try "building a normal, healthy friendship" with women rather than merely shooting for hookups. In one piece with the headline "Women are people, not simply a thing to 'date,'" Alex Salazar described his pleasure in platonic companionships with female classmates, enjoying them as "people you can talk to . . . not simply 'girlfriend material.'"[60]

The improved but still uneven male-female ratio also continued to cause stress at Caltech throughout the 1980s and beyond. Campus slang encapsulated gender tension in the term "glomming," referring to the phenomenon of multiple men competing for a woman's attention and following her incessantly, in person or electronically. At its worst, glomming might tiptoe dangerously near harassment or cross the borderline into stalking, but according to Caltech graduate student Kathleen Richter, at least some male students considered it "normal and natural" and even a "tradition." Describing the problem for *Ms.* magazine in 2001, Richter linked the "culture of disrespect" to episodes of vandalism at Caltech's Women's Center, to the appearance of pornographic images and vulgar sexual references in student publications and Web sites, and to a campus life where "women seem to become invisible, and everything centers around what men want to say." Some female students who complained were consequently snubbed or effectively punished, Richter wrote, and as a result, others tend to "play down glomming and make excuses for it . . . unwilling to rock the boat because they are afraid to be seen as man-haters." Some at Caltech consciously fostered that climate of "denial," Richter charged, fearing that openly addressing glomming risked calling attention to it and deterring female applicants, thereby reversing gains in the school's gender balance. Richter warned that ignoring the issue would not work and that glomming had forced some women to leave the school. In one high-profile example, neuromorphic engineering specialist Misha Mahowald publicly recounted how in the 1980s, the fraternity mentality, verbal harassment, and constant pestering temporarily drove her away from

Caltech. After electrical engineering professor Carver Meade helped arrange off-campus housing for her, Mahowald returned to finish her bachelor's degree in biology and earn a doctorate in computational neuroscience.[61]

Sexist episodes, discriminatory thinking, and obstacles to women's success in engineering studies have by no means been eradicated, but equally indisputably, another factor has changed since the late 1800s and early 1900s. Institutions that once equated technical and scientific excellence with an all-male identity seek the status in the twenty-first century of appearing female-friendly. Before 1952, Georgia Tech had no undergraduate women; in 2011, Tech's undergraduate population was almost one third female (4,490 women out of 13,948 total). Among that year's entering students, women comprised 50 percent of the biomedical engineering majors, 45 percent of chemical engineering majors, and 40 percent of industrial and systems engineering majors. Since other engineering programs had less gender diversity, women ended up representing 24 percent of all students in Georgia Tech's college of engineering in 2010, substantially higher than the national 2010 figure of 18.1 percent. According to the American Society for Engineering Education, Georgia Tech produced more female undergraduate engineers than anywhere else in the nation; in academic year 2010–11, Georgia Tech granted 387 engineering bachelor's degrees to women.[62]

Over the preceding two decades, Georgia Tech had not ignored women's issues. With funding from the National Science Foundation in the 1990s, Georgia Tech headed up a statewide program, "Integrating Gender Equity and Reform" (InGear), that was intended to make K–12 science and math education equally welcoming to girls and boys. In a required self-study of their own institution's data and gender climate, members of a Georgia Tech team assessed their school's areas of progress and weakness in women's education and faculty employment. Their 1998 "Report on the Status of Women" collected "pipeline" data showing that 29 percent of all the undergraduate admission requests to Georgia Tech came from women in 1996 and that acceptance rates across the two sexes were basically equal, at around 57 percent for both Georgia Tech overall and the College of Engineering. Statistics covering various years after 1986 indicated that female students changed majors more often than male students (45.0 percent versus 39.7 percent) but had a higher retention rate (75.5 percent versus 68.5 percent) and an overall grade-point average that was 0.06 higher than men's. The report declared:

Within the College of Engineering, women who were enrolled in 1997 earned an equivalent or higher GPA at Georgia Tech than their male classmates in every engineering major except Computer Engineering, where men outscored women by 0.15 grade points. . . . [W]omen outperform men to the greatest degree in fields where they are the most out-numbered. In fields such as biology, chemistry, and psychology, where female students are commonplace, it is acceptable for a woman to be merely average. However, it appears only women with outstanding academic abilities choose to major in fields such as physics, or in many of the engineering disciplines.[63]

Looking beyond such indicators, the report saw trouble remaining in the way that some "difficult . . . attitudes held by male faculty and administrators about gender" made for frustrating experiences that belied the warm welcome implied in the school's recruiting messages. "[S]ubtle "put-downs" and patronizing "comments" still made everyday campus interactions tense for many women: "One of the characteristics of the climate that exists at Georgia Tech is an understanding that jokes and snide remarks about women are a common mode of banter. . . . Occasionally the comments are more directly sexually demeaning or express some sort of sexual innuendo, thereby becoming part of the more serious problem of sexual harassment." In surveys, female students reported feeling discomfort when male counterparts failed to respect their questions or comments in the classroom and when faculty excluded women from discussions. Tech's cutthroat competitiveness took a disproportionate toll on women: "While male students at Georgia Tech also report that the classroom climate is alienating and off-putting, women are more socialized to interpret negative comments and behaviors as a comment on their worth and capabilities." Undergraduate, graduate-student, and junior staff women also faced challenges in finding mentors and role models, given the school's "glass ceiling" that limited the advance of female faculty and administrators and sent discouraging signals on gender progress. Georgia Tech kept pace with its peer-group schools in hiring women but fell behind in female representation among associate and full professors, apparently due in part to the absence of policies to promote work-life balance and provide more flexibility in tenure and promotion paths.[64]

In a revealing note, the 1998 InGear report confirmed that while the total number of female engineering undergraduates at Georgia Tech had ballooned from 312 in 1975 to 1,119 in 1981, the gain in female majors had then essentially halted. Looking at the same question about a decade later, some observers worried that Georgia Tech still found it easier to attract more women students in the liberal arts, sciences, and architecture, while their representation in engineering effectively remained flat. Christine Valle, head of the Women in Engineering program and a Tech mechanical engineering Ph.D. herself, commented, "Research estimates you need 30 percent to change the dynamic. At 30 percent a minority group will not feel singled out. Some majors at Georgia Tech, like biomedical engineering, are well above that mark but others haven't come close." Addressing the sense that gains in overall female enrollment had leveled off, undergraduate admissions director Ingrid Hayes commented, "We sit right on the edge of the tipping point . . . but it's very difficult for us to get beyond that."[65]

Accompanying the InGear report, the engineering college at Georgia Tech convened its own task force, whose 1998 report, "Enhancing the Environment for Success," echoed the need to create better mechanisms to address sexual harassment, make promotion and tenure evaluations more fair, and provide much-needed family leave and day-care policies. Such campus-climate improvements would help Tech retain and cultivate female senior faculty in engineering, the task force concluded, without seeming to demand special women's accommodations. Three women who were involved with Tech's self-assessment later told the American Society for Engineering Education, "Women faculty and students do not want gender-specific programs for women because they fear that the existence of such programs will create the perception that women cannot be successful without special assistance."[66]

Tech administrators made some moves to address faculty well-being and family-balance issues, while also expanding programs directly aimed at recruitment and retention of female students. In that way, Georgia Tech offers an excellent case study of how progressive-minded experiments in support, as pioneered by earlier generations of female engineers, had expanded into the mainstream of engineering-school operations by the twenty-first century. Gender-equity workshops and outreach efforts had become a source of institutional pride, favorable publicity, and almost a matter of necessity for a school that wanted to appear socially conscious. Tech's Women in Engineering (WIE) program operated out of the dean's office within the College of Engineering and offered scholarship funds totaling about $165,000 in 2012. Its regular staff supervised mentoring arrangements that paired female students with upper-class partners and professional women in engineering, as well as practical advice sessions on navigating career fairs and building good resumes, an annual awards banquet, "tea with the dean," and weekly "coffee talks" in collaboration with the Women's Resource Center. To cultivate enrollment interest, WIE hosted an annual "Engineering Career Conference" that gave female high school students a chance to interact with Tech admission counselors, female faculty, and current women engineering majors. Reaching down to middle-school students, the program ran both an "Introduce a Girl to Engineering" day session and a week-long summer Technology, Engineering, and Computing Camp, where college women served as counselors and supervised hands-on experiences in Web design, chemical engineering, acoustics, and aeronautics. WIE student ambassadors volunteered to visit local K–12 schools, and student Maggie Burcham commented, "I have found that we usually leave more inspired ourselves. . . . Each time I visit a school I leave feeling more determined to be successful at Georgia Tech." Meanwhile, Tech's student-run Society of Women Engineers continued as an affiliate of the national organization, where members organized their own professional-development programs, hosted social events, and ran separate outreach programs for middle schools and high schools, some in collaboration with local Girl Scout troops. Georgia Tech also featured a Women in Electrical and Computer Engineering group, a Women@College of Computing community, a Center for the Study of Women, Science, and Technology, and an NSF-funded ADVANCE program to create a more favorable climate for female faculty.[67]

By 2012, most schools with engineering departments had created or been involved with some sort of efforts to enhance women's chances of success in nontraditional fields. Many offered multiple overlapping programs, including sustained options that were built into the academic structure (such as residential learning communities or societies for women in engineering) as well as one-time events (such as workshops or conferences). At least twenty-seven schools, including Ohio State, Michigan-Ann Arbor, Rutgers, Penn State, UCLA, Case Western Reserve, and Howard, hosted chapters of Phi Sigma Rho, a social sorority created in 1984 at Purdue for women in technical studies, whose academic demands could make it difficult for them to join the standard Greek rush.

Did the dollars and effort poured into this wide variety of support programs actually pay off, in terms of drawing more female students into engineering and helping those already there? Various observers worked to answer that broad question, looking both at specific programs within individual schools and at nationwide trends. A 2002 survey, funded by the NSF and the Alfred P. Sloan Foundation, studied twenty-six institutions that offered official programs for Women in Engineering, as well as twenty-seven comparison schools lacking such formal structures. Through surveys, site visits, and data analysis, those researchers concluded that almost half the women in their sample who left engineering for other majors had averaged A or B grades, confirming conclusions from other studies that the discipline was losing female students who were fully able to handle the work. Causes of attrition seemed to lie deeper than academic failure, with some women "negatively interpreting grades that may actually be quite good," experiencing "diminished self-confidence," or not feeling part of the "community in engineering." The NSF/Sloan study saw institutional programs as valuable in both raising female students' sense of self-efficacy and providing a sense of belonging. Engineering schools were more likely to retain those female majors who took part in study groups, engineering social events, field trips, or workshops with guest speakers, while tending to lose the women who did not participate in such "support activities." After accounting for students' self-analysis of their confidence levels and their satisfaction with their departments' atmosphere, "there is a unique quality of Social Enrichment activity participation that makes women want to stay in engineering," the research concluded: "[S]tudents who participate in support activities are more positive about their department and their classes than students who do not participate."[68]

In each month, from coast to coast, dozens of large and small campuses hosted outreach efforts that mobilized current undergraduate women to excite, encourage, and educate younger students about engineering, from Iowa State University's "Taking the Road Less Traveled" career conferences for girls in secondary school to Worcester Polytechnic Institute's "Camp Reach." Such programs expanded but did not deviate radically from the principles that had galvanized SWE leaders and other activists during the post–World War II decades—the philosophy of helping professional women, college women, and young girls to connect one-on-one, cultivating success in engineering through sisterhood and support. Other activities sought to accomplish similar mentoring through virtual interactions. A fall 2012 online community, "Women in Technology Sharing Online" (WitsOn), recruited an initial three hundred academics and professionals (including female executives from companies such as Microsoft and Cisco) to answer e-mail questions from undergraduate women.[69]

As a more sustained institutional investment, a number of schools, including Wisconsin-Madison, Texas-Austin, Iowa State, and the University of Minnesota, developed residential learning communities for women in nontraditional fields. Such commitment reflected research that suggested that individuals became more comfortable and hence more likely to persist when they developed a "psychological sense of community" through shared experiences, common affiliations, emotional safety, interpersonal connections, and positive reinforcement. Nilanjana Dasgupta wrote, "Individuals' choice to pursue one academic or professional path over another may feel like a free choice but is often constrained by subtle cues in achievement environments that signal who naturally belongs there and who does not." Dasgupta cited evidence that when young women worked with female peers, they gained motivation and "self-efficacy," a confidence of being able to accomplish their goals. More than that, successful female role models served as "social vaccines" for fragile novices by helping "inoculate" them against self-doubt and the perniciously nagging damage that could be inflicted by negative stereotypes.[70] Watching established professionals did not always reinforce young women's desire to emulate them, however. As a number of researchers have pointed out, seeing female professors who spent extended hours in the laboratory only reinforced alarm among some young women that choosing a competitive career entailed sacrificing a "normal" devotion to family and personal life.

Learning communities sought to foster a sense of "belonging" among female science and engineering majors through such mentorship and shared social activities, ranging from chocolate engineering to ropes-course team-building exercises. Programs also gave women students tangible assets, including access to special computer labs, study groups, focused seminars, and academic tutoring. For undergraduates, those practical advantages did not always seem to offer sufficient incentives to join a distinctively set-apart group. A self-study at Virginia Tech noted that almost 40 percent of residents had been pushed into its "Hypatia" community by their parents and initially resisted joining because "they did not see themselves as being similar to the women who would choose to participate. In their vernacular, they viewed Hypatia as being for 'geeky girls. . . .'" In a dramatic turnaround, many reluctant students described the learning-community experience as being highly rewarding and came to recognize their fellows as "'normal' girls with far ranging interests from rugby to dance but [who] have one specific goal in common: that of becoming [an] engineer." Over the learning community's first three years, about 90 percent of residents chose to remain engineering majors, as compared to a retention rate of just 73 percent among those female Virginia Tech freshman engineers outside Hypatia. Between 2005 and 2011, women's representation among Virginia Tech's first-year engineering students rose from 15.6 percent to 20.1 percent, and the learning community came to serve over a third of the entering female majors.[71]

Engineering-oriented schools today keep a close eye on their gender balance, bragging publicly whenever female enrollment grows. At the start of 1970, Caltech's undergraduate school included no women at all, and that year's entering class was 14 percent female. In academic year 2011–12, women comprised 39 percent of undergraduate enrollment (381 out of 978 students).[72] There was no historical law that women's presence would rise in a steady path of unbroken progress; as in previous decades, numbers often fluctuated year to year. In 2001, Caltech's freshman class surged to roughly 36 percent female, but in 2006, women represented just 28.5 percent of first-year students.[73]

Looking over the total list of educational institutions that offered engineering bachelor's degrees, MIT stood second only to Georgia Tech in handing out the most diplomas to female students, graduating 289 women in 2011. That comprised 43.4 percent of all engineering bachelor's degrees awarded at MIT in 2011, ranking MIT second in the country (behind only the Franklin W. Olin College of Engineering) in terms of graduating the highest percentage of women.[74] The demographic shifts at MIT over the postwar era were substantial and no accident. Starting in the late 1990s, after leaders acknowledged the justice of female science professors' complaints, MIT devoted substantive formal and well-publicized attention to equity issues. Focusing on conditions for faculty women in engineering, a 2002 follow-up report documented that although the representation of female faculty in MIT's School of Engineering had doubled from 5 percent in 1990 to 10 percent by 2001, the two largest departments together gained only a net two female faculty members in ten years. Forty percent of female job candidates rejected MIT offers (versus 14 percent of men), and many engineering areas also had a poor record of retaining female faculty. Interviews revealed what the report called "a consistent pattern of marginalization for many of the women faculty," involving less favorable teaching assignments, less support from mentors, discrepancies in past compensation, and family or child-care stress. Although official promotion and tenure practices treated women well, male faculty generally possessed greater access to influential contacts, informal networks, collaborative research grants, influential department positions, and pipeline administration posts. Engineering dean Thomas Magnanti wrote, "Barriers persist and all too many of us remain oblivious to them. . . . MIT is not a hospitable environment for many women faculty. Simply put, this situation is unacceptable." Although MIT's high-profile reports focused on evaluating conditions for faculty, the situation there carried clear implications for female engineering students, both in practical terms (access to women mentors and role models) and symbolically (signals regarding the field's hospitability).[75]

Backing up the engineering school's vows of redemption, the following decade did create momentum for change. New hiring policies designed to ensure broader search pools meant that between 2001 and 2011, the number of female faculty almost doubled again, from thirty-two (10 percent of the total) to sixty-two (17 percent). Some at the junior level experienced problems of "being bullied by some senior male faculty" or encountered insensitive "disparaging remarks." But female faculty overall observed a marked improvement in MIT engineering programs' climate over the decade, and many expressed enthusiasm about their access to resources for research, institutional support, and professional collegiality.[76]

In fact, female faculty, students, and high school women who were seeking role-model inspiration at MIT could look right to the top after 2004, when MIT garnered international attention by naming brain researcher Susan Hockfield as its new president. Hockfield reportedly accepted the post against the advice of her mentor, James Watson, who worried that she might not win respect as a woman leader at a male-dominated school and as a biologist at an institute traditionally more prominent in engineering and physical sciences. By 2012, as Hockfield finished her eighth year in the presidency, many observers both on and off campus lauded her for leading MIT to path-breaking energy research and interdisciplinary cancer studies, physical expansion of campus facilities, improved student life, and curriculum updates.[77]

Since the late 1800s, female engineering students at MIT, state land-grant schools, and other institutions have gone from exclusion to inclusion, from invisibility to a high-profile presence featured on school Web sites and in school publications. At Georgia Tech and Caltech, women moved from a tense and tenuous acceptance to today's welcome and encouragement (at least officially). They no longer need to fight to "invade" formally all-male strongholds, and the most overt forms of discrimination are no longer common or even permissible in most educational and professional circles. At schools that formerly were monopolized by men, support mechanisms for female undergraduates, graduate students, faculty, and staff in engineering have become high-profile features. Campus leaders increasingly view learning communities and administrative advocacy as investments that can help them compete to attract young women as applicants and retain female engineering majors.

It remains vital to recognize and combat the biases and sexual harassment that linger, along with the challenge of addressing society's more deep-seated gender issues that wind up discouraging some talented girls from pursuing nontraditional interests and contribute to female underrepresentation at higher professional levels. But the history of women's engineering studies underlines just how dramatic a revolution in American institutions, social assumptions, and individual lives has already occurred. Its most essential lesson lies in the simple fact that today's young women take it for granted that they have the right to explore technical and mechanical interests, the right to enroll in even the nation's most prestigious engineering and science schools.

# NOTES

## INTRODUCTION

1. Brian Yoder, "Engineering by the Numbers," American Society for Engineering Education, 2012, http://www.asee.org/papers-and-publications/publications/college-profiles/2011-profile-engineering-statistics.pdf, accessed October 14, 2012.

2. Amy Slaton, *Race, Rigor and Selectivity in U.S. Engineering: The History of an Occupational Color Line* (Cambridge, MA: Harvard University Press, 2010).

3. Marion Hirsh, "The Changing Position of Women in Engineering Worldwide," *IEEE Transactions of Engineering Management* 47 (3) (2000): 345–359.

4. Margaret Rossiter, *Women Scientists in America: Struggles and Strategies to 1940* (Baltimore: Johns Hopkins University Press, 1984), xv.

5. Suzanne Le-May Sheffield, *Women and Science: Social Impact and Interaction* (New Brunswick, NJ: Rutgers University Press, 2005); Margaret Alic, *Hypatia's Heritage* (Boston, MA: Beacon Press, 1986); Ann Shteir, *Cultivating Women, Cultivating Science: Flora's Daughters and Botany in England 1760 to 1860* (Baltimore: Johns Hopkins University Press, 1999); M. Susan Lindee, "The American Career of Jane Marcet's *Conversations on Chemistry*, 1806–1853," *Isis* 82 (1991): 8–23.

6. Benjamin Rush, *Thoughts upon Female Education, Accommodated to the Present State of Society, Manners, and Government, in the United States of America, Addressed to the Visitors of the Young Ladies' Academy in Philadelphia, 28th July, 1787, at the Close of the Quarterly Examination* (Philadelphia, 1787), http://www.swarthmore.edu/SocSci/rbannis1/AIH19th/female.html, accessed October 7, 2012; Linda K. Kerber, *Women of the Republic: Intellect and Ideology in Revolutionary America* (Chapel Hill: University of North Carolina Press, 1980); Mary Beth Norton, *Liberty's Daughters: The Revolutionary Experience of American Women, 1750–1800*, rev. ed. (Ithaca, NY: Cornell University Press, 1996); Nancy F. Cott, *The Bonds of Womanhood: "Woman's Sphere" in New England, 1780–1835*, 2nd ed. (New Haven, CT: Yale University Press, 1997).

7. Kerber, *Women of the Republic*; Cott, *The Bonds of Womanhood*.

8. Kathryn Kish Sklar, *Catharine Beecher: A Study in American Domesticity* (New Haven, CT: Yale University Press, 1973); Mary Kelley, *Learning to Stand and Speak: Women, Education, and Public Life in America's Republic* (Chapel Hill: University of North Carolina Press, 2008).

9. Ercel Sherman Eppright and Elizabeth Storm Ferguson, *A Century of Home Economics at Iowa State University* (Ames, IA: College of Home Economics, 1971).

10. Andrea Radke-Moss, *Bright Epoch: Women and Coeducation in the American West* (Lincoln: University of Nebraska Press, 2008).

11. Patricia Albjerg Graham, "Expansion and Exclusion: A History of Women in American Higher Education," *Signs* 3 (4) (Summer 1978): 759–773; Leslie Miller-Bernal, *Separate by Degree: Women Students' Experiences in Single-Sex and Coeducational Colleges* (New York: Peter Lang, 2000); Lynn Gordon, *Gender and Higher Education in the Progressive Era* (New Haven, CT: Yale University Press, 1990).

12. Rosalind Rosenberg, "The Limits of Access: The History of Coeducation in America," in *Women and Higher Education in American History,* ed. John Mack Faragher and Florence Howe, 107–129 (New York: Norton, 1988); Graham, "Expansion and Exclusion."

13. Edward H. Clarke, *Sex in Education, or, A Fair Chance for the Girls* (Boston, MA: Osgood, 1873); Margaret Lowe, *Looking Good: College Women and Body Image, 1875–1930* (Baltimore: Johns Hopkins University Press, 2005); Martha Verbrugge, *Able-Bodied Womanhood: Personal Health and Social Change in Nineteenth-Century Boston* (New York: Oxford University Press, 1988); Patricia Palmieri, "From Republican Motherhood to Race Suicide: Arguments on the Higher Education of Women in the United States, 1820–1920," in *Educating Men and Women Together: Coeducation in a Changing World,* ed. Carol Lasser, 49–64 (Urbana: University of Illinois Press, 1987); Barbara Miller Solomon, *In the Company of Educated Women* (New Haven, CT: Yale University Press, 1985).

14. Helen Lefkowitz Horowitz, *Alma Mater* (New York: Knopf, 1984); Pamela Mack, "Straying from Their Orbits: Women in Astronomy in America," in *Women of Science: Righting the Record,* ed. Gabriele Kass-Simon and Patricia Farnes (Bloomington: Indiana University Press, 1990); Linda Eisenmann, ed., *Historical Dictionary of Women's Education in the United States* (New York: Greenwood, 1998).

15. Rossiter, *Women Scientists in America*; Julie Des Jardins, *The Madame Curie Complex: The Hidden History of Women in Science* (New York: Feminist Press at CUNY, 2010); George Johnson, *Miss Leavitt's Stars* (New York: Norton, 2005).

16. Naomi Rogers, "Women and Sectarian Medicine," in *Women, Health, and Medicine in America,* ed. Rima D. Apple, 273–302 (New Brunswick, NJ: Rutgers University Press, 1990); Regina Morantz-Sanchez, *Sympathy and Science: Women Physicians in American Medicine* (New York: Oxford University Press, 1985).

17. Steven Peitzman, *A New and Untried Course: Women's Medical College and Medical College of Pennsylvania, 1850–1998* (New Brunswick, NJ: Rutgers University Press, 2000); Virginia

Drachman, *Hospital with a Heart: Women Doctors and the Paradox of Separatism at the New England Hospital, 1862–1969* (Ithaca, NY: Cornell University Press, 1984).

18. Ellen More, *Restoring the Balance: Women Physicians and the Profession of Medicine, 1850–1995* (Cambridge, MA: Harvard University Press, 1995); Thomas Neville Bonner, *To the Ends of the Earth: Women's Search for Education in Medicine* (Cambridge, MA: Harvard University Press, 1992); Mary Roth Walsh, *Doctors Wanted: No Women Need Apply* (New Haven, CT: Yale University Press, 1977).

19. Daryl Hafter, *Women at Work in Preindustrial France* (Philadelphia: Pennsylvania State University Press, 2007); Daryl Hafter, *European Women and Preindustrial Craft* (Bloomington: Indiana University Press, 1995); Francesca Bray, *Technology and Gender: Fabrics of Power in Late Imperial China* (Berkeley: University of California Press, 1997); Rachel Maines, *Hedonizing Technologies: Paths to Pleasure in Hobbies and Leisure* (Baltimore: Johns Hopkins University Press, 2009); Chandra Mukerji, *Impossible Engineering: Technology and Territoriality on the Canal du Midi* (Princeton, NJ: Princeton University Press, 2009).

20. Bertrand Gille, *The Renaissance Engineers* (London: Lund Humphries, 1966); John Hubbel Weiss, *The Making of Technological Man: The Social Origins of French Engineering Education* (Cambridge, MA: MIT Press, 1982); Lawrence Grayson, *The Making of an Engineer* (New York: Wiley, 1993); Peter Lundgreen, "Engineering Education in Europe and the U.S.A., 1750–1930," *Annals of Science* 47 (1) (1990): 33–75.

21. Daniel Calhoun, *The American Civil Engineer: Origins and Conflict* (Cambridge, MA: MIT Press, 1960); Grayson, *The Making of an Engineer*; Lundgreen, "Engineering Education in Europe and the U.S.A., 1750–1930"; Paul Nienkamp, "A Culture of Technical Knowledge: Professionalizing Science and Engineering Education in Late Nineteenth-Century America," Ph.D. dissertation, Iowa State University, 2008; Terry Reynolds, "The Education of Engineers in America before the Morrill Act," *History of Education Quarterly* 32 (4) (1992): 459–482; James Brittain and Robert McMath Jr., "Engineers and the New South Creed: The Formation and Early Development of Georgia Tech," *Technology and Culture* 18 (2) (1977): 175–201.

22. Monte Calvert, *The Mechanical Engineer in America* (Baltimore: Johns Hopkins University Press, 1967); Daniel Calhoun, *The American Civil Engineer: Origins and Conflict* (Cambridge, MA: MIT Press, 1960).

23. Gary Lee Downey, "Low Cost, Mass Use: American Engineers and the Metrics of Progress," *History and Technology* 23 (3) (2007): 289–308; Bruce Sinclair, *A Centennial History of the American Society of Mechanical Engineers, 1880–1980* (Toronto: American Society of Mechanical Engineers, 1980); Terry Reynolds, *Seventy-five Years of Progress: A History of the American Institute of Chemical Engineers* (New York: American Institute of Chemical Engineers, 1983); Edwin Layton Jr., *The Revolt of the Engineers: Social Responsibility and the American Engineering Profession* (Cleveland, OH: Case Western University Press, 1971); A. Michel McMahon, *The Making of a Profession: A Century of Electrical Engineering in America* (New York: Institute of Electrical and Electronics Engineers,

1984); David Noble, *America by Design: Science, Technology, and the Rise of Corporate Capitalism* (New York: Knopf, 1977).

24. Carroll Pursell, "Toys, Technology and Sex Roles in America," in *Dynamos and Virgins Revisited: Women and Technological Change*, ed. Martha Moore Trescott, 252–267 (Metuchen, NJ: Scarecrow Press, 1979); Sally L. Hacker, *"Doing It the Hard Way": Investigations of Gender and Technology*, ed. Dorothy Smith and Susan Turner (Boston, MA: Unwin Hyman, 1990); Cynthia Cockburn, *Brothers: Male Dominance and Technological Change* (London: Pluto Press, 1991); Judy Wajcman, *Feminism Confronts Technology* (University Park: Pennsylvania State University Press, 1991); Roger Horowitz, ed., *Boys and Their Toys? Masculinity, Class, and Technology* (New York: Routledge, 2001); Nina Lerman, Ruth Oldenziel, and Arwen Mohun, eds., *Gender and Technology: A Reader* (Baltimore: Johns Hopkins University Press, 2003); see also Joan Rothschild, ed., *Machina Ex Dea: Feminist Perspectives on Technology* (New York: Pergamon Press, 1983).

25. Ruth Oldenziel, *Making Technology Masculine: Men, Women, and Modern Machines in America, 1870–1945* (Amsterdam: Amsterdam University Press, 1999), 10.

26. Karen Tonso, "The Impact of Cultural Norms on Women," *Journal of Engineering Education* (July 1996): 217–225.

27. Betty Reynolds and Jill Tietjen, *Setting the Record Straight: The History and Evolution of Women's Professional Achievement in Engineering* (Denver, CO: White Apple Press, 2001); Frank B. Gilbreth Jr. and Ernestine Gilbreth Carey, *Cheaper by the Dozen* (New York: Crowell, 1949); Frank B. Gilbreth Jr. and Ernestine Gilbreth Carey, *Belles on Their Toes* (New York: Crowell, 1951); Jane Lancaster, *Making Time: Lillian Moller Gilbreth—A Life beyond "Cheaper by the Dozen"* (Boston, MA: Northeastern University Press, 2006); Laurel Graham, *Managing on Her Own: Dr. Lillian Gilbreth and Women's Work in the Interwar Era* (Norcross, GA: Engineering and Management Press, 1998); Margaret E. Layne, *Women in Engineering: Pioneers and Trailblazers* (New York: American Society of Civil Engineers, 2009). The American Society of Civil Engineers has published *Women in Engineering: Pioneers and Trailblazers*, a collection of articles focused on the lives on individual women in engineering, including Emily Roebling, Kate Gleason, Edith Clarke, and Emma Barth.

28. Annie Canel, Ruth Oldenziel, and Karin Zachmann, eds., *Crossing Boundaries, Building Bridges* (Amsterdam: Harwood, 2000).

29. Robert McMath, *Engineering the New South: Georgia Tech, 1885–1985* (Athens: University of Georgia Press, 1985). There are numerous books on the history of MIT and specific programs there, including David Kaiser, ed., *Becoming MIT: Moments of Decision* (Cambridge, MA: MIT Press, 2010); Philip Alexander, *A Widening Sphere: Evolving Cultures at MIT* (Cambridge, MA: MIT Press, 2011); Julius Stratton and Loretta Mannix, *Mind and Hand: The Birth of MIT* (Cambridge, MA: MIT Press, 2005); Karl Wildes and Nilo Lindgren, *A Century of Electrical Engineering and Computer Science at MIT, 1882–1982* (Cambridge, MA: MIT Press, 1985).

30. For instance, see Judith McIlwee and J. Gregg Robinson, *Women in Engineering: Gender, Power, and Workplace Culture* (Albany, NY: SUNY Press, 1992). There are also more specialized

explorations, such as those on gender and computing, including Jane Margolis and Allan Fisher, *Unlocking the Clubhouse: Women in Computing* (Cambridge, MA: MIT Press, 2003); Thomas Misa, ed., *Gender Codes: Why Women Are Leaving Computing* (Hoboken, NJ: Wiley-IEEE, 2010); Joanne Cohoon and William Aspray, eds., *Women and Information Technology: Research on Underrepresentation* (Cambridge, MA: MIT Press, 2008).

31. Examples of this genre include Jill Bystydzienski and Sharon Bird, eds., *Removing Barriers: Women in Academic Science, Technology, Engineering, and Mathematics* (Bloomington: Indiana University Press, 2006); National Academy of Sciences, *Beyond Bias and Barriers: Fulfilling the Potential of Women in Academic Science and Engineering* (Washington, DC: National Academies Press, 2007); Ronald J. Burke and Mary Mattis, eds., *Women and Minorities in Science, Technology, Engineering, and Mathematics* (New York: Edward Elgar, 2007); Henry Etzkowitz, Carol Kemelgor, and Brian Uzzi, *Athena Unbound: The Advancement of Women in Science and Technology* (Cambridge: Cambridge University Press, 2000); Susan Ambrose et al., *Journeys of Women in Science and Engineering: No Universal Constants* (Philadelphia, PA: Temple University Press, 1999). See also Sue Rosser, *Breaking into the Lab: Engineering Progress for Women in Science* (New York: New York University Press, 2012); Sue Rosser, *The Science Glass Ceiling: Academic Women Scientists and the Struggle to Succeed* (New York: Routledge, 2004); Sue Rosser, *Re-Engineering Female Friendly Science* (New York: Teachers College Press, 1997); Mary Frank Fox, Gerhard Sonnert, and Irina Nikiforova, "Programs for Undergraduate Women in Science and Engineering: Issues, Problems, and Solutions," *Gender and Society* 25 (October 2011): 589–615; Gerhard Sonnert, Mary Frank Fox, and Kristen Adkins, "Undergraduate Women in Science and Engineering: Effects of Faculty, Fields and Institutions over Time," *Social Science Quarterly* 88 (December 2007): 1333–1356.

32. Evelyn Fox Keller, *Reflections on Gender and Science* (New Haven, CT: Yale University Press, 1985); Sandra Harding, *The Science Question in Feminism* (Ithaca, NY: Cornell University Press, 1986); Sandra Harding, *Whose Science? Whose Knowledge? Thinking from Women's Lives* (Ithaca, NY: Cornell University Press, 1991); Londa Schiebinger, ed., *Gendered Innovations in Science and Engineering* (Stanford, CA: Stanford University Press, 2008).

33. Linda Eisenmann, *Higher Education for Women in Postwar America, 1945–1965* (Baltimore: Johns Hopkins University Press, 2006).

34. Studies from the 1990s suggested that although American men had begun putting more time and effort into housework, women continued to perform two or three times as much of the routine chores as did men. Scott Coltrane, "Research on Household Labor: Modeling and Measuring the Social Embeddedness of Routine Family Work," *Journal of Marriage and the Family* 62 (November 2000): 1208–1233; see also Arlie Hochschild, *The Second Shift* (New York: Avon, 1989).

35. American Association for University Women, *How Schools Shortchange Girls: The AAUW Report* (New York: Marlowe, 1992).

36. Psychological, sociological, and education research has generally confirmed that it is overly simplistic to say that all men always prove superior to all women with regard to all aspects of

mathematical ability, visual-spatial perception, and related skills. The supposed gaps between men and women have often been exaggerated, especially in comparison to performance variations within each sex, and may well be traced to gender differences in socialization and childhood experiences rather than biological inevitability. See, for example, Robyn Scali, Sheila Brownlow, and Jennifer Hicks, "Gender Differences in Spatial Task Performance as a Function of Speed or Accuracy Orientation," *Sex Roles* 43 (5–6) (2000): 359–376.

## CHAPTER 1

1. Alva Matthews, "Emily W. Roebling," in *Women in Engineering: Pioneers and Trailblazers*, ed. Margaret Layne (Reston, VA: ASCE, 2009), 123–130; David McCullough, *The Great Bridge: The Epic Story of the Building of the Brooklyn Bridge* (New York: Simon & Schuster, 1972).

2. Virginia Fortiner, "Women Storm Field of Engineering: Engineers of Fair Sex Tell How Barriers Are Swept Down," *Newark Sunday Call* (New Jersey), April 13, 1930; "One Woman Is Recognized as Automotive Engineer," *The Independent* (St. Petersburg), September 5, 1925, 14.

3. Melanie McCulley, "The History of Women at Rensselaer," 1991, RPI archives, subject file "women—institute policies"; "School letters," no date, ca. 1992, RPI archives, subject file "women—articles 1/2"; Samuel Rezneck, *Education for a Technological Society: A Sesquicentennial History of Rensselaer Polytechnic Institute* (Troy, NY: RPI, 1968), 180, 320–321, 347, 392.

4. Ercel Sherman Eppright and Elizabeth Storm Ferguson, *A Century of Home Economics at Iowa State University* (Ames: Iowa State University Press, 1971); see also Sarah Stage and Virginia B. Vincenti, eds., *Rethinking Home Economics: Women and the History of a Profession* (Ithaca, NY: Cornell University Press, 1997).

5. U.S. Department of Transportation, *Women in Transportation: Changing America's History* (Washington, DC: U.S. Government Printing Office, 1998), 11.

6. Richard Weingardt, "Elmina and Alda Wilson," *Leadership and Management in Engineering* 10 (4) (2010): 192–196; Elmina Wilson, *Modern Conveniences for the Farm Home*, USDA Farmers' Bulletin no. 270 (Washington, DC: U.S. Government Printing Office, 1906); U.S. Department of Transportation, *Women in Transportation: Changing America's History* (Washington, DC: U.S. Government Printing Office, 1998), 11; "Elmina Wilson, Alda Wilson," *The Arrow of Pi Beta Phi* 21 (1904): 93; Iowa State University Web sites at http://www.lib.iastate.edu/arch/rgrp/21-7-24.html and http://www-archive.ccee.iastate.edu/who-we-are/department-history/marston-water-tower .html, accessed January 3, 2012; Amy Slaton, *Reinforced Concrete and the Modernization of American Building, 1900–1930* (Baltimore: Johns Hopkins University Press, 2001).

7. Richard Weingardt, "Elmina and Alda Wilson," *Leadership and Management in Engineering* 10 (4) (2010): 192–196; "Elmina Wilson, Alda Wilson," *The Arrow of Pi Beta Phi* 21 (1904–1905): 93; Iowa State University Web sites at http://www.lib.iastate.edu/arch/rgrp/21-7-24.html, accessed January 3, 2012.

8. Margaret Ingels, "Petticoats and Slide Rules" (1952), in *Women in Engineering: Pioneers and Trailblazers*, ed. Margaret Layne (Reston, VA: ASCE, 2009): 85–97; Jeff Meade, "Ahead of Their Time" (1993), in *Women in Engineering: Pioneers and Trailblazers*, ed. Margaret Layne (Reston, VA: ASCE, 2009): 137–140; Martha Moore Trescott, "Women in the Intellectual Development of Engineering," in *Women of Science: Righting the Record,* ed. Gabriele Kass-Simon and Patricia Farnes (Bloomington: Indiana University Press, 1990); Martha Moore Trescott, *New Images, New Paths: A History of Women in Engineering in the United States, 1850–1980* (Dallas: T&L Enterprises, 1996); Ruth Bordin, *Women at Michigan: The Dangerous Experiment, 1870s to the Present* (Ann Arbor: University of Michigan, 2001).

9. Margaret Ingels, "Petticoats and Slide Rules" (1952), in *Women in Engineering: Pioneers and Trailblazers*, ed. Margaret Layne (Reston, VA: ASCE, 2009), 85–97.

10. Laurel Sheppard, "Kate Gleason Leaves Her Legacy on RIT's College of Engineering," *SWE* (September–October 1999): 34–40; Henry Petroski, *An Engineer's Alphabet: Gleanings from the Softer Side of a Profession* (Cambridge: Cambridge University Press, 2011), 329; http://www.rit .edu/kgcoe/about/aboutkategleason.htm and http://www.winningthevote.org/F-KGleason .html, accessed January 3, 2012; Eve Chappel, "Kate Gleason's Careers," in *Women in Engineering: Pioneers and Trailblazers*, ed. Margaret Layne (Reston, VA: ASCE, 2009): 131–135; Charlottee Williams Conable, *Women at Cornell: The Myth of Equal Education* (Ithaca, NY: Cornell University Press, 1977).

11. *College of Engineering Alumni Register,* University of Illinois archives.

12. Lillian Moller Gilbreth, "Opportunities for Women in Industrial Engineering," *Woman Engineer* (December 1924): 12; Lillian Gilbreth, draft of article, ca. 1927, file "Industrial Engineering as a Career for Women," Purdue University archives; Lillian Gilbreth, "Technical Education for women" notes, ca. 1940, file "Institute of Women's Professional Relations," box 139, Gilbreth papers, Purdue University archives.

13. Lillian Moller Gilbreth to C. R. Place, August 13, 1925; Frederick Waldron to Gilbreth, August 15, 1925; August Hosea Webster to Gilbreth, August 21, 1925; Gilbreth to James Hartness, September 4, 1925; Gilbreth to L. P. Alford, September 2, 1925; R. A. Wentworth to Gilbreth, September 9, 1925; Fred Low to Gilbreth, September 9, 1925; Charles Main to Gilbreth, September 15, 1925; John Freeman to Gilbreth, April 28, 1926; Freeman to Gilbreth, September 10, 1926; all in box 131, file "Dr LMG, membership work, ASME 1940 NHX 0830-1," Gilbreth papers, Purdue University archives.

14. Margaret Ingels, "Petticoats and Slide Rules" (1952), in *Women in Engineering: Pioneers and Trailblazers*, ed. Margaret Layne (Reston, VA: ASCE, 2009): 85–97; Virginia Fortiner, "Women Storm Field of Engineering: Engineers of Fair Sex Tell How Barriers Are Swept Down," *Newark Sunday Call* (New Jersey), April 13, 1930; Alice Goff, "Women Can Be Engineers" (1946), in *Women in Engineering: Pioneers and Trailblazers*, ed. Margaret Layne (Reston, VA: ASCE, 2009): 155–177; see also the University of Kentucky College of Engineering Web site at http://www .engr.uky.edu/alumni/hod/margaret-ingels, accessed January 3, 2012.

15. Fortiner, "Women Storm Field of Engineering: Engineers of Fair Sex Tell How Barriers Are Swept Down"; National Society of Professional Engineers, press release, June 18, 1959, Potter papers: box 1, unlabeled folder, Purdue University archives; Susan Ware, ed., *Notable American Women* (Cambridge, MA: Harvard University Press, 2005), 295–297.

16. Ingels, "Petticoats and Slide Rules"; University of Colorado, *Journal of Engineering* 2 (1905–1906): 94; "She's a Miner!," *San Jose News*, April 13, 1928, 8; "Girl Has Unique Degree," *Spokesman-Review*, August 5, 1928; "Beauty Meets Resistance," *Penn State Engineer*, October 1934.

17. Margaret Ingels, "Petticoats and Slide Rules"; "The T-Square Society," *Michigan Technic* 28–29 (1915): 75; Sarah Allaback, *The First American Women Architects* (Urbana: University of Illinois Press, 2008).

18. Allaback, *The First American Women Architects*; "Bertha Louise Yerex Whitman," http://www.findagrave.com/cgi-bin/fg.cgi?page=gr&GRid=60398180, accessed January 17, 2012; Nancy Ruth Bartlett, *More Than a Handsome Box: Education in Architecture at the University of Michigan, 1976–1986* (Ann Arbor: University of Michigan College of Architecture, 1995).

19. "Women Organize National Engineering Society," *Engineering News-Record* 85 (4) (July 22, 1920): 190; *Colorado Engineer* 15–17 (1918–1920); "Personal," *Electrical Review* 72 (January 12, 1918): 92; Lauren Kata, "The Shortage of Engineers Is No Laughing Matter" (2003), in *Women in Engineering: Professional Life*, ed. Margaret Layne (Reston, VA: ASCE: 2009): 91–100.

20. Marilyn Ogilvie and Joy Harvey, eds., "Barney, Nora Stanton (Blatch) De Forest," *The Biographical Dictionary of Women in Science* (New York: Routledge, 2000), 82–83; Christopher Gray, "Streetscapes / 220 West 57th Street; Civil Engineers' 1897 Clubhouse," *New York Times*, November 12, 2000; Petroski, *An Engineer's Alphabet*, 328.

21. "Woman a Civil Engineer: First of Her Sex to Enter the Heretofore Masculine Profession," *Spokane Spokesman Review* (Washington), August 26, 1923, Alumni folder, "Dennis, Olive Wetzel," Cornell University archives; U.S. Department of Transportation, *Women in Transportation: Changing America's History* (Washington, DC: U.S. Government Printing Office, 1998), 15; Alice Goff, "Women Can Be Engineers" (1946), in *Women in Engineering: Pioneers and Trailblazers*, ed. Margaret Layne (Reston, VA: ASCE, 2009), 155–177.

22. Harold Peterson, "Co-ed Engineers: Man's Domains Are Again Invaded," *Minnesota Techno-log* (May 1925): 11; Kathryn Gallagher, "Wisconsin Girl Engineers Like Their Jobs," *Milwaukee Journal*, July 4, 1926, 24.

23. Pat Stenberg, "Coed Invades 'Man's World'; Finds Success in Engineering," *Purdue Exponent* (October 1941): 1; "Female Engineers Invade Chem Labs and Airplanes," *Detroit Collegian*, October 14, 1940, 3.

24. Virginia Fortiner, "Women Storm Field of Engineering: Engineers of Fair Sex Tell How Barriers Are Swept Down," *Newark Sunday Call* (New Jersey), April 13, 1930.

25. Proceedings of a Special Meeting of the Cornell Society of Engineers, November 10, 1938, folder 1938 18–23 "Cornell Society of Engineers Proceedings, NYC," Cornell Club, New York City, Hollister Papers 16/2/2077, Cornell University archives.

26. "Light Shed on Engineering Job by an Expert Feminine Touch," *New York Times*, March 31, 1940.

27. Raymond Howes, "Concerning Sibley Sues," *Cornell Alumni News* (October 13, 1938): 30.

28. Jane Lancaster, *Making Time: Lillian Moller Gilbreth—A Life beyond "Cheaper by the Dozen"* (Boston, MA: Northeastern University Press, 2006); Robert Topping, *A Century and Beyond: The History of Purdue University* (Lafayette, IN: Purdue University Press, 1988); Susan Butler, *East to the Dawn: The Life of Amelia Earhart* (Cambridge, MA: DaCapo Press, 2009); Ray Boomhower, "The Aviatrix and the University: Amelia Earhart at Purdue University," *Traces of Indiana and Midwestern History*, 6 (3) (1994): 36–41; John Norberg, *Wings of Their Dreams: Purdue in Flight* (Lafayette, IN: Purdue University Press, 2003).

29. Ellen Zeigler, "Ross Camp," *Purdue Engineer* (October 1940): 10, 14; Kathleen Lux, "Following the Civils," *Purdue Engineer* (November 1941): 42.

30. "Are Women Engineers?," *Minnesota Technolog* (December 1931): 63.

31. Charlotte Bennett, "An Opinion on Engineering for Coeds," *Purdue Engineer* (November 1935): 33.

32. Ellen Zeigler, "Move Over, Men!," *Purdue Engineer* (March 1940): 87–88.

33. "Coed Engineer," *Penn State Engineer* (March 1938); "Today of Yesteryear: Thirty Years Ago," *Observer-Reporter* (Washington, PA), April 7, 1969, A13, http://www.libraries.psu.edu/digital/findingaids/325.htm, accessed January 19, 2012.

34. Margaret Rossiter, *Women Scientists in America: Struggles and Strategies to 1940* (Baltimore: Johns Hopkins University Press, 1982).

35. Marilyn Ogilvie and Joy Harvey, eds., *The Biographical Dictionary of Women in Science, L–Z* (New York: Routledge, 2000); Elizabeth Durant, "A Lab of Their Own," *Technology Review* (May–June 2006), http://www.technologyreview.com/article/405825/a-lab-of-their-own.

36. Letter from Mrs. Frederick T. Lord to Karl Compton, January 11, 1938, and Louise P. Horwood to Mrs. Frederick Lord, January 5, 1938, both in AC 4, box 240, folder 12, Massachusetts Institute of Technology archives (all of the following archival references are from MIT). For background, see Margaret W. Rossiter, *Women Scientists in America: Before Affirmative Action, 1940–1972* (Baltimore: Johns Hopkins University Press, 1995).

37. Association of Women Students, *Cheney and All That: Freshman Coed Handbook*, no date, ca. 1956, AC 4 box 1 f7; Beth Boge to Henry Pearson, no date, MC65 box 5, file "Women at MIT statistics 1945–1973"; Marilynn Bever, "The Women of MIT, 1871 to 1941," undergraduate thesis, Massachusetts Institute of Technology, June 1976; MIT Alumni Association and War Records

Committee, *Technology's War Record* (Cambridge, MA: MIT Alumni Association, 1920); "A Record Graduating Class," *Technology Review* 16 (7) (July 1914): 441; Joan Cook, "Marion R. Hart, 98," *New York Times*, July 4, 1990; Helen Hill Miller, "Science: Careers for Women," *Atlantic* 200 (4) (October 1957): 123–128.

38. Association of Women Students, *Cheney and All That: Freshman Coed Handbook*, no date, ca. 1956, AC 4 box 1 f7; Beth Boge to Henry Pearson, no date, MC65 box 5, file "Women at MIT statistics 1945–1973"; Bever, "The Women of MIT, 1871 to 1941"; Deborah Douglas, *American Women and Flight since 1940* (Lexington: University Press of Kentucky, 2004).

39. Association of Women Students, *Cheney and All That: Freshman Coed Handbook*, no date, ca. 1956, AC 4 box 1 f7; Beth Boge to Henry Pearson, no date; MC65 box 5, file "Women at MIT statistics 1945–1973"; Bever, "The Women of MIT, 1871 to 1941."

40. "Glamor Girl MIT," *The Tech* (October 8, 1940): 1; *Massachusetts Institute of Technology Handbook* (Cambridge, MA: Massachusetts Institute of Technology, 1941).

41. "Fulton Invented Water Craft; Niece Will Design Airplane," *Iowa State Daily*, October 15, 1941, 4.

42. Eloise Davison, "A Course in Home Economics: A Report of a Successful Course Offered to Sophomore Women at Iowa State College," master's thesis, 1924. For more on the history of Iowa State's household equipment program, see Amy Sue Bix, "Equipped for Life: Gendered Technical Training and Consumerism in Home-Economics, 1920–1980," *Technology and Culture* 43 (4) (October 2002): 728–754.

43. "Brief Outlines of Courses in Household Equipment Taken by Majors," n.d., Iowa State University Archives, Ames, Louise Jenison Peet Papers, box 3, folder 14; Bessie Spratt, "Development of the Home Economics Curriculums of Iowa State College from 1923 to 1953," master's thesis, Iowa State, 1953; Faith Madden, *Household Equipment Experiments* (Ames: Iowa State College Press, 1952).

44. Eloise Davison, "Stove Efficiency Tests," Iowa State College, 1923.

45. "Miss Sater Likes Crafts," *Iowa Homemaker* (March 1932), inside cover.

46. "Another Viewpoint on Home Freezers," *Electrical Merchandising* (October 1945): 108, 110; Catherine Raymond, "Behind Closed Doors in Appliance Research," *Iowa Homemaker* (February 1941): 4.

CHAPTER 2

1. Margaret Rossiter, *Women Scientists in America: Before Affirmative Action, 1940–1972* (Baltimore: Johns Hopkins University Press, 1995); Kathleen Williams, *Improbable Warriors: Women Scientists and the U.S. Navy in World War II* (Annapolis, MD: U.S. Naval Institute Press, 2001); Jordynn Jack, *Science on the Home Front: American Women Scientists in World War II* (Urbana: University of Illinois

Press, 2009); Ruth Howes and Caroline Herzenberg, *Their Day in the Sun: Women of the Manhattan Project* (Philadelphia, PA: Temple University Press, 2003); Jennifer Light, "When Computers Were Women," *Technology and Culture* 40 (3) (July 1999): 455–483; see also Laura Puaca, "A New National Defense: Feminism, Education, and the Quest for 'Scientific Brainpower,' 1940–1965," Ph.D. dissertation, University of North Carolina at Chapel Hill, 2007.

2. Ruth Milkman, *Gender at Work: The Dynamics of Job Segregation during World War II* (Urbana: University of Illinois Press, 1987).

3. "Women Engineers in the Field," *New York Times,* October 17, 1943.

4. H. P. Hammond to R. A. Seaton, January 21, 1941, National Archives, RG12, entry 214, box 2, folder "Members—Hammond, H.P.—advisory committee"; Society for Promotion of Engineering Education, March 2, 1941, RPI archives, AC1, box 9, folder 105; J. W. Studebaker, "Proposals to Expand the Program of Training for National Defense through Schools and Colleges," July 27, 1940, and "Tentative Engineering School Intensive Course for the National Defense Program," August 15, 1940, National Archives, RG12, entry 222, box 1, folder "vol. 2."

5. U.S. Office of Education, "Answers to Questions Pertaining to ESMWT," September 1943, RPI Archives, Houston papers, box 62, folder 645; minutes, meeting of regional advisors, June 27, 1941, National Archives, RG12, entry 222, box 1, folder "vol 3."

6. Minutes of EDT regional advisors meeting, December 21, 1940, National Archives, RG12, entry 222, box 1, folder "3 EDT-misc"; "Defense Program Moves Rapidly," *Penn State Alumni News* (March 1941): 2.

7. Memo from the SPEE, December 17, 1941, National Archives, RG12, entry 214, box 2, folder "Potter, Dean A. A."; Minutes, national advisory committee meeting, December 19–20, 1941, RPI Archives, Hotchkiss Papers AC1, box 9, folder 105.

8. "War Goes to College," *Time*, February 15, 1943, 89–90; Benjamin Fine, "Educational Changes Here to Stay," *New York Times*, July 18, 1943.

9. "Guidance for Girls," *Purdue Exponent*, July 31, 1942, 2; Ed Stephenson, "'Why Do We Have Coeds,' Asks Inquiring Reporter," *Purdue Exponent*, January 21, 1943.

10. "Rumor about 'Roomers,'" *Iowa State Daily*, August 11, 1942, 4; "Rumor Epidemic Again Sweeps the Campus," *Iowa State Daily*, November 17, 1942, 4; Charles Friley, "A Message to Iowa State Students," Ames, IA, November 20, 1942, University of Illinois archives, Willard papers 2/9/1, box 77f Provost.

11. "University to Test Science Aptitudes of Coed Students," *Purdue Exponent*, December 1, 1942, 1, 4; "Coed Conclusions," *Purdue Exponent*, December 8, 1942, 2; "Coeds to Take Tests Tomorrow at Convocation," *Purdue Exponent*, December 8, 1942, 1, 4; "Prof. F. C. Hockema Hits Army Rumor," *Purdue Exponent*, December 11, 1942; "Hockema Expresses Desire for Coeds to Remain Here," *Purdue Exponent*, December 10, 1942; Elaine Ahrens, "Elliott Answers Coeds' Questions," *Purdue Exponent*, March 5, 1943, 1; "Opportunity Is Knocking, Coeds," *Purdue Exponent*, October

27, 1942; Mary Jordan, "University Offers Special Courses in Training Women for War Work," *Purdue Exponent*, January 24, 1943; American Council on Education, "College Women Students and the War," October 17, 1942, University of Minnesota archives, AB1.1, Presidential papers, Supplemental, box 8, file "women in defense"; "Woman Engineers Needed," *Science News Letter*, December 26, 1942, 407.

12. American Council on Education, "College Women Students and the War," October 17, 1942, University of Minnesota archives, AB1.1, Presidential papers, Supplemental, box 8, file "women in defense"; "Woman Engineers Needed," *Science News Letter*, December 26, 1942, 407.

13. "Women Are Needed as War Engineers," *New York Times,* January 5, 1943; Penn State ESMWT booklet and "The Pennsylvania State College Trains for War" pamphlet, no date, ca. 1944, Penn State archives, GVF/Events/War/WWII/PSC ESMWT;. George Case, "What More Can Engineering Colleges Do through ESMWT?," October 28, 1942, University of Illinois archives, Willard papers 2/9/1, box 70f, ESMWT; *Penn State Alumni News* (January 1943): 17; "Navy Girl," *Daily Collegian,* January 30, 1943, 4; "Women Study Engineering," Cornell University archives, 16/2/2077, box 18; *Cornell Alumni News*, March 11, 1943, 269; Leland Antes to Jerry Martin, August 3, 1943, Center for American History, University of Texas at Austin.

14. Henry Armsby, "Engineering, Science, and Management War Training: Final Report," U.S. Office of Education, 1946, 47; "Engineering Fundamentals for Women" advertisement, *Minnesota Alumni News,* January 16, 1943, 247; "Junior Engineer Course for Women," *Education for Victory,* November 16, 1942, 12; Maria Leonard to Illinois alumnae, February 4, 1943, 31/3/804, University Illinois archives.

15. Blythe Foote, "San Marino Woman Prepares to Enter Engineering Field," *Pasadena Post*, March 7, 1943.

16. "College Opens Map-Making Course to Coeds," *Cornell Engineer* 8 (3) (January 1943): 15; "Army Sponsors Special Course in Map Making," *Cornell Daily Sun,* February 11, 1943, 5; "College to Open Map Making Class to Senior Women," *Cornell Daily Sun,* January 5, 1943, 5; "Women Enroll for War Courses," *Cornell Alumni News*, February 25, 1943, 1.

17. Phil Ferguson to Dean W. R. Woolrich, August 3, 1943, Center for American History, University of Texas at Austin.

18. Walter Rolfe to Jerry Martin, July 29, 1943, Center for American History, University of Texas at Austin; "Woman Engineers Needed," *Science News Letter*, December 26, 1942, 407.

19. "Excerpts from Some Letters on ESMDT Courses," no date, ca. April 1942, Center for American History, University of Texas at Austin; Walter Rolfe to Jerry Martin, July 29, 1943, Center for American History, University of Texas at Austin; Worth Cottingham to Jerry Martin, July 29, 1943, Center for American History, University of Texas at Austin; "Woman Engineers Needed," *Science News Letter*, December 26, 1942, 407.

20. "Training Women in Engineering, Science and Management," *Education for Victory*, November 16, 1942, 20; "Enrollment of Women Encouraged by ESMDT," *Education for Victory*, March 16, 1942; "Excerpts from Some Letters on ESMDT Courses," no date, ca. April 1942, Center for American History, University of Texas at Austin; University of California to ESMWT, no date, ca. August 1942, National Archives, RG12, entry 222, box 13, folder "Replies to misc. 45"; ESMWT newsletter, September 30, 1942, National Archives, RG12, entry 214, box 2, folder "Potter, Dean A. A."; "WEEDS Join Women's Alphabet Armies," *Daily Illini*, October 2, 1942, 2.

21. Penn State ESMWT booklet and "The Pennsylvania State College Trains for War" pamphlet, no date, ca. 1944, Penn State archives, GVF/Events/War/WWII/PSC ESMWT.

22. Wesley Hennessy to ESMWT officials, January 18, 1944, National Archives; "105 Women to Study for Engineering Jobs," *New York Times*, June 21, 1943.

23. "Columbia Training Women Engineers," *New York Times,* September 1, 1942; Anne Peterson, "Women Fitting Themselves Fast for Skilled Jobs in War Plants," *New York Times,* September 6, 1942; "Women Aides End Engineering Study," *New York Times,* January 8, 1943; "Aircraft Plant to Open Course in Engineering," *Cornell Daily Sun*, May 4, 1943, 3; *A Career for You with Grumman Aircraft Engineering Corporation*, booklet, National Archives.

24. Pamphlet, "Grumman Engineering Aide Training for College Women," no date, ca. 1942; National Archives.

25. *A Career for You with Grumman Aircraft Engineering Corporation*, booklet, National Archives.

26. Ibid.

27. Wesley Hennessy to ESMWT officials, January 18, 1944, National Archives.

28. C. Wilson Cole, "Training of Women in Engineering," *Journal of Engineering Education* 34 (1943–1944): 167–184.

29. H. Heald, to A. C. Willard, October 12, 1942, Willard papers, box 70, folder "engineering college," 2/9/1, University of Illinois archives; Elmer Franklin Bruhn, "A History of Aeronautical Education and Research at Purdue University for 1937–1950," ca. 1968, Purdue archives; Michael Bezilla, *Engineering Education at Penn State* (University Park: Pennsylvania State University Press, 1981), 142–145.

30. Ad, *New York Times*, January 14, 1943, 38; "Education: Engineer Cadettes," *Newsweek*, December 14, 1942, 99; "Engineering Cadettes of the Curtiss-Wright Corporation," brochure, 1942, Cornell University archives.

31. "Curtiss-Wright Interviews Twenty," *Daily Texan*, December 10, 1942.

32. Surveys of former Cadettes, 1991–1992, University of Minnesota archives.

33. Ibid.

34. Joseph Corn, *The Winged Gospel: America's Romance with Aviation, 1900–1950* (New York: Oxford University Press, 1983).

35. Surveys of former Cadettes, 1991–1992, University of Minnesota archives; Marian Loofe, "Campus Interests—Curtiss-Wright Cadets [sic]," *Iowa State Daily*, February 13, 1943, 1; "Engineer Courses Draw Many Women," *New York Times*, January 14, 1943, 16.

36. Surveys of former Cadettes, 1991–1992, University of Minnesota archives; "From Colleges throughout Country," *Minneapolis Morning Tribune*, February 13, 1943; Cornell Cadette histories, Cornell University archives, 16/2/2077, box 18.

37. Surveys of former Cadettes, 1991–1992, University of Minnesota archives; Jean Curry, "St. Louis, Here We Come," *The Cadetter*, July 1943, Iowa State University archives.

38. "Janet Russell, "Train Women Here," *Iowa State Daily*, December 4, 1942, 1; "Cadettes Will Start Class Feb. 15," *Purdue Exponent*, February 7, 1943, 1S; D. G. Frier, "Cadettes Invade Campus," *Purdue Engineer* (March 1943): 78–79.

39. A. F. Grandt Jr., W. A. Gustafson, and L. T. Cargnino, *One Small Step: The History of Aerospace Engineering at Purdue University* (West Lafayette, IN: Purdue University School of Aeronautics and Astronautics, 1995), 32–39; Ed Stephenson, "Cold, Men, Impress Cadettes," *Purdue Exponent*, February 16, 1943: 1; *The Debris*, Purdue yearbook, August 1943, 13, 20, 169.

40. Melanie McCulley, "The History of Women at Rensselaer," 1991, RPI archives, subject file "women—institute policies"; "Curtiss Wright Women Enter Rensselaer to Begin Ten Month Aeronautics Course," *Rensselaer Polytechnic* (February 16, 1943): 1; Tracy Drake, "This Is Rensselaer," *Rensselaer Polytechnic* (March 20, 1943): 1, 16.

41. M. J. Thompson to C. D. Simmons, August 11, 1943, Center for American History, University of Texas at Austin.

42. W. F. Spafford to J. William Long, July 22, 1943, RPI archives, Houston collection, box 62, folder 644; D. G. Frier, "Cadettes Invade Campus," *Purdue Engineer* (March 1943): 78–79.

43. "Curriculum Meeting for Engineering Cadette Program" and "Summary of Conference on Curriculum," December 1942, Cornell University archives, 16/2/2077, box 18; Paul Hemke, "Aeronautical Engineering," *Alumni News* (June 1943): 5–6.

44. S. C. Hollister, January 31, 1944, and V. J. Norris to Hollister, November 15, 1943, Cornell University archives, 16/2/2077, box 18.

45. C. Wilson Cole, "Training of Women in Engineering," *Journal of Engineering Education* 34 (1943–1944): 167–184; "Engineer Courses Draw Many Women," *New York Times*, January 14, 1943, 16.

46. "War Training programs—WWII—Curtiss-Wright Engineering Cadette Training Program," Iowa State College, Ames, IA, 1945; "University of Minnesota, Institute of Technology: Accreditation of Courses Given in the Curtiss-Wright Engineering Cadette Program," May 15, 1944,

University of Minnesota archives; "Curriculum for Curtiss-Wright Engineering Cadette Program," Cornell University archives, 16/2/2077, box 18; "Cadettes Make A-25 Models," *Iowa State Daily*, October 8, 1943, 3.

47. "Curriculum Meeting for Engineering Cadette Program" and "Summary of Conference on Curriculum," December 1942, Cornell University archives, 16/2/2077, box 18; Fred W. Ocvirk, "Cadettes at Cornell, *Cornell Engineer* (April 1943): 8–9, 24.

48. Anonymous, *Fifty Years of Aeronautical Engineering: University of Minnesota, 1929 to 1979*. See http://www.aem.umn.edu/info/history and http://www.aem.umn.edu/info/history/03_Sources -Photos.shtml.

49. Anonymous, *Fifty Years of Aeronautical Engineering*; "'Lady Engineers Upset Tradition," *Pittsburgh Press*, May 17, 1944, 5; Warren Bruner, "A Report on the Engineering Cadette Training Program of the Curtiss-Wright Corporation," RPI archives.

50. Marjorie Allen, "Cadette Column," *Iowa State Daily*, March 5, 1943, 2; "Purdue Personalities: George," *Purdue Engineer* (April 1946): 38; Jean Kneeland Geelan, August 16, 1991, University of Minnesota archives; Surveys of former Cadettes, 1991–1992, University of Minnesota archives.

51. Rosella Katz, "Shop Talk," *The Cadetter*, April 1943, Iowa State University archives; C. Wilson Cole, "Training of Women in Engineering," *Journal of Engineering Education* 34 (1943–1944): 167–184.

52. William O. Hotchkiss, "Through the Transit," *RPI Alumni News* (April 1943): 22; "Curtiss-Wright Sponsors Strenuous Program," *Penn State Alumni News* (March 1943): 1; Isabel Lion to All Cadettes, October 19, 1943, Cornell University archives, 16/2/2077, box 19; Elmer Franklin Bruhn, "A History of Aeronautical Education and Research at Purdue University for 1937–1950," ca. 1968, Purdue University archives.

53. Elaine Garrabrant Stulb, "A Memoir of a Curtiss-Wright Cadette," RPI archives, subject file "Curtiss-Wright Cadettes"; Milton Beveridge, "Letter to the Editor," *Rensselaer* (June 1992): 3; "Remember: The Curtiss-Wright Cadettes," *At Rensselaer* (December 1981): 13; Jean (Schaaff) Cook, "Letter to the Editor," *At Rensselaer* (August 1982): 2.

54. "RPI's New Cheerleaders," *Rensselaer Polytechnic* (October 12, 1943): 5; *The Transit*, RPI yearbook, 1944, 16–18, 35–36.

55. "Cadettes Have Five-Day Vacation, Starting Today," *Cornell Daily Sun*, July 15, 1943, 9; *The Cadetter*, July 1943, Iowa State University archives; "Curtiss-Wright Sponsors Strenuous Program," *Penn State Alumni News* (March 1943): 1; Jean Kneeland Geelan, August 16, 1991, University of Minnesota archives.

56. Marguerite Leeds, "Girls Explain Cigs, Greasy Overalls," *Purdue Exponent*, June 17, 1943, 2; Surveys of former Cadettes, 1991–1992, University of Minnesota archives.

57. Sam Gadd, "Reader Declares Slacks Loom as Nation's Menace," *Iowa State Daily*, January 7, 1944, 3.

58. "Engineering Cadettes: Girls Train for Aircraft Industry," *Life*, May 10, 1943, 45, 46, 48; Frances P. Baldwin, "Heard at Texas," *The Cadetter*, September 1943, Iowa State University archives.

59. Mary Glover, "With One Purpose—Victory!," *The Cadetter*, March 1943; *Purdue Exponent* advertisement, February 12, 1943; "Minnesota Cadettes" scrapbook, University of Minnesota archives, Akerman papers; Betty Fletcher and Janet Graham, "Cadette Song for Cornell," *The Cadetter*, April 1943, Iowa State University archives; Joan Ryan, "Night and Day," *The Cadetter*, September 1943, Iowa State University archives.

60. Margaret Perkins, "Sidelights at RPI," *The Cadetter*, March 1943, Iowa State University archives; Marjorie Allen, "Cadette Column," *Iowa State Daily*, February 26, 1943, 2; *The Transit*, RPI yearbook, 1944, 16–18; Anonymous, "Harriet with the Light Blue Jeans," *The Cadetter*, March 1943, Iowa State University archives; Leona M. Cheklinski, "In the Good 'Ole' summertime," *The Cadetter*, July 1943, Iowa State University archives.

61. Helen Bachrach, "All about Us," *The Cadetter*, July 1943, Iowa State University archives; Anonymous, from University of Minnesota, "Lay That Slide Rule Down," *The Cadetter*, September 1943, Iowa State University archives; "Cadette Notes," *The Daily Collegian*," March 17, 1943, 4; "Cadette Song," *The Cadetter*, April 1943, Iowa State University archives.

62. "Engineer Courses Draw Many Women," *New York Times*, January 14, 1943, 16; Patricia Gilbert, "The Propeller with a Mind of Its Own," *The Cadetter*, September 1943, Iowa State University archives; Janet Sabine, "Crystal Gazing," *The Cadetter*, July 1943, Iowa State University archives; Jean Curry, "St Louis, Here We Come," *The Cadetter*, July 1943, Iowa State University archives.

63. Marjorie Allen, "Cadette Column," *Iowa State Daily Student*, March 5, 1943, 2; "Aero. E's Awarded," *Iowa Engineer* (March 1943): 166–168; R. M. Harnett to Anne Blitz, October 15, 1942, University of Minnesota archives, AW9.1, Dean of Women's Papers, folder 27, "Women in War."

64. Virginia Allen, "Purdue Chapter," *The Cadetter*, May 1943, Iowa State University archives; Betty Marshall, "Penn State Chapter," *The Cadetter*, July 1943, Iowa State University archives.

65. Press release, ca. early December 1943, Cornell University archives, 16/2/2077, box 18; "RPI 'Graduates' Cadettes," *Alumni News*, January 1944, 23; Grandt, Gustafson, and Cargnino, *One Small Step: The History of Aerospace Engineering at Purdue University*, 32–39.

66. Marjorie Allen, "Women Find Their Place in the Home of the Helldivers," *Iowa Engineer* (September 1944): 35–36; Surveys of former Cadettes, 1991–1992, University of Minnesota archives.

67. Jean Kneeland Geelan, August 16, 1991, University of Minnesota archives; Surveys of former Cadettes, 1991–1992, University of Minnesota archives.

68. Jean Kneeland Geelan, August 16, 1991, University of Minnesota archives; Surveys of former Cadettes, 1991–1992, University of Minnesota archives.

69. Surveys of former Cadettes, 1991–1992, University of Minnesota archives.

70. Newsletter by and for RPI Cadettes, ca. late spring 1945 and May 1946, and RPI newsletter, "The Katy Kadetter," ca. late spring 1945, RPI archives.

71. Newsletter by and for RPI Cadettes, ca. late spring 1945 and May 1946, and RPI newsletter, "The Katy Kadetter," ca. late spring 1945, RPI archives; William Price, "What Happened to the Girls?," *RPI Alumni News* (September 1952): 16.

72. Iowa State Cadette biographies, ca. 1993, Iowa State archives; *Cadette Gazette* 1980, University of Minnesota archives; Surveys of former Cadettes, 1991–1992, University of Minnesota archives.

73. Surveys of former Cadettes, 1991–1992, University of Minnesota archives.

74. Iowa State Cadette biographies, ca. 1993, Iowa State University archives.

75. Margaret Rossiter, ed., *Women Scientists in America: Struggles and Strategies to 1940* (Baltimore: Johns Hopkins University Press, 1982).

76. *Cadette Gazette*, 1980, University of Minnesota archives; Surveys of former Cadettes, 1991–1992, University of Minnesota archives.

77. "Radio Corporation to Train Cadettes in Engineering," *Purdue Exponent*, January 22, 1943; "RCA Plans Training of Cadettes," *Purdue Exponent*, March 18, 1943; "RCA Victor Will Open Cadette Program to Teach Eighty Women Radio Work," *Cornell Daily Sun*, March 18, 1943, 3.

78. Gretchen Sherry and Ann Hathaway, "The RCA Cadettes," *Purdue Engineer* (September 1943): 244–245; "RCA Officials Visit Campus Cadettes during School Tour," *Purdue Exponent*, June 24, 1943, 1, 4.

79. Mary Jo Johnson, "Cadette Log," *The Cadetter*, May 1943, Iowa State University archives; "Girls Invade AIEE Meeting, Hear Hale Speak on Airfields," *Purdue Exponent*, May 29, 1943, 5; "RCA Cadettes Honor Trebby," *Purdue Exponent*, June 3, 1943; "Cadettes Receive Army-Navy 'E' Pins from RCA Manager," *Purdue Exponent*, June 12, 1943; "Cadette Engineering Society Hears Dean A. A. Potter Speak," *Purdue Exponent*, May 20, 1943.

80. Ruth Freehafer, *R. B. Stewart and Purdue University* (West Lafayette, IN: Purdue University, 1983), 91; "RCA, Curtiss-Wright Cadettes Graduate Tonight," *Purdue Exponent*, December 21, 1944, 1.

81. "Adventure in Success: Aircraft Radio War Training Program for Women," University of Texas archives; "Shorts" column, *Minnesota Technolog* (December 1943): 112; Jeanette Croonquist, "Oh, for the Life of the High Voltage Cadettes," *Minnesota Technolog* (March 1944): 185; Patricia Simmons, "The Wright Field Cadettes," *Purdue Engineer* (September 1943): 245–246.

82. *The Debris*, Purdue yearbook, August 1943, 169.

83. Simmons, "The Wright Field Cadettes"; *The Debris*, Purdue yearbook, August 1943, 13, 20, 169.

84. Annual report, Office of the Dean of Women, 1943–1944, University of Illinois, 41/3/2, box 1, University of Illinois archives; University War Committee, "The Student in Wartime," *University of Illinois Bulletin* 40 (40) (May 25, 1943); H. M. Horner, May 14, 1943, Willard papers, box 70, folder "engineering college," 2/9/1, University Illinois archives; *Annual Report of the College of Engineering*, 1943–1944, University of Illinois, 41/3/2, box 1, University Illinois archives; Minutes, University of Illinois board meeting, April 29, 1943, Willard papers, box 70, folder "engineering college," 2/9/1, University Illinois archives.

85. "A College Goes to War: Lady Engineers' Upset Tradition," *Pittsburgh Press*, May 17, 1944; H. P. Hammond to Marian Nagle, July 9, 1943, Pennsylvania State University archives, file "war/ WWII/misc"; Betty Lehr, "New Arrivals," *The Cadetter*, July 1943, Iowa State University archives; "College Trains Fifteen Aircraft Engineers," *Centre Daily Times* (PA), June 2, 1943; Michael Bezilla, *Engineering Education at Penn State* (University Park: Pennsylvania State University Press, 1981), 142–145.

86. "New York State Education Department's Program of Training for War Production Workers," November 1942, RPI archives, Houston collection, box 62, folder 645; "Typical Case Histories of Female Enrollees," State Education Department's National Defense Training Program at RPI, ca. 1942–1943, RPI archives, AC1, box 11, folder 128; Stanley B. Wiltse, "Report on War Training Courses," November 30, 1942, RPI archives, Houston collection, box 62, folder 645; Livingston Houston, "Activities at RPI and Their Relation to the War: RPI under War conditions," *RPI Alumni News* (November 1942): 2.

87. Max Gartenberg, "Forty-five Women to Study Aeronautics at Heights under Chance-Vought Scholarships," *Heights Daily News*, February 26, 1943, 1, 4; Thomas J. Frusciano and Marilyn H. Pettit, *New York University and the City* (New Brunswick, NJ: Rutgers University Press, 192); "Vought Girls to Graduate," *Heights Daily News*, October 8, 1943, 1; "Chance Vought Girls Graduate," *Heights Daily News*, October 19, 1943, 1; "Chance Vought Offers Scholarships to Coeds for Aeronautical Training," *Cornell Daily Sun*, February 6, 1943, 8.

88. E. V. Gustavson, "Engineers Made to Order," *Aviation*, November 1943, 167, 169, 253–257; "The Boeing Aircraft Company Needs Men and Women Draftsmen," pamphlet, National Archives; Lucy Greenbaum, "Learning New Skills," *New York Times,* January 31, 1943.

89. "Girls, Girls, Girls: G-E Campus News," *Cornell Engineer* (November 1942): back cover.

90. "Interview Women for Naval Jobs," *Iowa State Daily Student*, October 9, 1942, 3; "Require Equipment Majors to Add Five-Hour Course," *Iowa State Daily Student*, October 15, 1942, 4; Ben Willis, "The Wires Take Over," *Iowa Engineer*, October 1943, 41; "Dr. Peet Visits Iowa State Women Doing Engineering Work," *Iowa State Daily*, November 7, 1942, 4; "EE Course for Women," *Iowa Engineer* (January 1943): 107.

91. Clifford B. Holt Jr., Oral history interview with Michael Bezilla, September 24, 1979, file AX-26282, Pennsylvania State University archive.

CHAPTER 3

1. Elsie Eaves, "Wanted: Women Engineers," *Independent Woman* (May 1942): 132–133, 158–159.

2. "Here Are the Engineers of 1946; Five Coeds, Too," *Minnesota Technolog* (December 1942): 86.

3. "They'll Help Play 'Taps' for Japs," *Daily Texan,* January 6, 1942, 1; "Wimmin,'" *Froth* (August 1942): 6.

4. G. A. McConnell to Kenneth Meade, December 19, 1942, and March 29, 1943, Willard papers, box 70, folder "dean of women," 2/9/1, University of Illinois archives; Cornelia Tuttle to Alice Howe, July 28, 1942, MIT archives.

5. "Enrollment of Undergraduate Civilian Engineering Students, as of November 5, 1945," *Journal of Engineering Education* 36 (1946): 388–394.

6. "Guard Pledges Woman," *Iowa Engineer* (January 1944): 100; "Pi Tau Sigma Elects," *Iowa Engineer* (April 1944): 155; "Woman Leads Engineers," *Iowa State Daily,* May 22, 1943, 5; "Graduates in Chem. E.," *Iowa Engineer* (July 1945): 22.

7. "James Awarded Badge," *Iowa Engineer* (May 1945): 184.

8. Mary Krumholtz, "Women in Engineering," *Iowa Engineer* (May 1945): 176; "Engineer Staff Announced," *Iowa Engineer* (July 1945): 20.

9. "Navy Girl," *Michigan Technic* (December 1943): 18; Melvin Long, "Women Architects and Engineers Organize Association," *Illinois Technograph* (March 1945): 14.

10. Mary Klein, "I. T. Women: Looking Back," *Minnesota Technolog* (winter 1976): 8–11.

11. "There's Glam in Engineering!," *Michigan Technic* (January 1944): 14–15.

12. "Lend a Helping Hand," *Daily Collegian,* September 22, 1942, 2.

13. Michael Bezilla, *Engineering Education at Penn State: A Century in the Land-Grant Tradition* (University Park: Pennsylvania State University Press, 1981), 142–145.

14. Phyllis Mann, "Coeds Train for War Effort," *Purdue Exponent,* February 14, 1943.

15. "War Goes to College," *Time,* February 15, 1943, 89–90; Pearl Bernstein, Letter to the editor, *New York Times,* January 14, 1943, 20.

16. Margaret Barnard Pickel, "A Warning to the Career Woman," *New York Times,* July 16, 1944.

17. Peggy Stefen, "Coed Engineer Comes Through over Comments of ME Faculty," *Purdue Exponent,* July 22, 1943, 1; "University's Thirty Girl Engineers Have Gripe; Male Engineers Scorn

Their Big Ambitions," *Purdue Exponent*, August 9, 1944; *Purdue Alumnus* (January–February 1944): cover.

18. Genevieve Husted, "AWS Will Hold First Vocational Meeting for Coed Scientists," *Purdue Exponent*, October 15, 1942.

19. "Josephine Webb, E. E. '40," *Purdue Engineer* (May 1943): 136.

20. "Purdue Personalities: Betty," *Purdue Engineer* (December 1945): 22, 24, 28.

21. Ellen Zeigler, "Women in War Industry," *Purdue Engineer* (July 1942): 26–27.

22. "University's Thirty Girl Engineers Have Gripe"; Stefen, "Coed Engineer Comes Through over Comments of ME Faculty."

23. "What about the Women?," *Purdue Engineer*, December 1945.

24. A. A. Potter to Karl Compton, April 25, 1945, and Karl Compton to A. A. Potter, May 7, 1945, both at MIT Archives.

25. Martha Lee Riggs, "Coeds Prove Themselves Good Engineering Students," *Purdue Exponent*, December 11, 1943, 3.

26. Martha McCulloch, "Lascoe Organizes New GE 41 Course for Campus Coeds," *Purdue Exponent*, August 26, 1944, 1; Elinor Hilton, "Shop Course Alleviates Christmas Present Problems of Coed Students," *Purdue Exponent*, December 5, 1944, 1, 4.

27. "RPI Opens Doors to Women: Institute Breaks 116 Year Old Rule Due to War Need," *Times Record* (Troy, NY), September 12, 1942, 1; "Rensselaer Breaks Long Standing Tradition; Opens Registration to Women for First Time," *Rensselaer Polytechnic*, September 15, 1942, 1, 4; Report of the President to the Board of Trustees for the year ending November 1, 1942, RPI archives, subject file "women—sources"; Ray Palmer Baker, October 12, 1942, RPI archives, subject file "women—statistics 2/2."

28. "Activities at RPI and Their Relation to the War: RPI under War Conditions," *RPI Alumni News*, November 1942, 2; *The Transit*, RPI yearbook, 1943, 20; *The Transit*, RIY yearbook, 1944, 16–18.

29. Rosalind Rosenberg, "Virginia Gildersleeve: Opening the Gates," *Columbia Magazine* (summer 2001), http://www.columbia.edu/cu/alumni/Magazine/Summer2001/Gildersleeve.html; Rosalind Rosenberg, "The Legacy of Dean Gildersleeve," http://beatl.barnard.columbia.edu/learn/documents/gildersleeve.htm; Jia Ahmad, "Leading Ladies," *The Eye: Columbia Daily Spectator*, December 3, 2009; Columbia Web site, http://www.columbia.edu/~rr91/3082_lectures/new_woman.htm and history timeline, http://beatl.barnard.columbia.edu/stand_columbia/Timeline-Women.htm.

30. "War Industries View for Trained Women," *New York Times*, October 7, 1942.

31. Memo: School of Engineering, Office of the Dean, Columbia University, May 15, 1942, Columbia University archives.

32. "Columbia Gives New Recognition to Women Studying the Sciences," *New York Times,* September 8, 1942.

33. James Kip Finch to Provost F. D. Fackenthal, Columbia University, August 26, 1942; J. K. Finch to Columbia's faculty of engineering, October 6, 1942, Columbia University archives.

34. J. K. Finch to the joint committee representing faculty, alumni, and students on the admission of women to the school of engineering, September 9, 1942, Columbia University archives.

35. J. K. Finch to Columbia's faculty of engineering, October 6, 1942, and J. K. Finch to the joint committee representing faculty, alumni, and students on the admission of women to the school of engineering, September 9, 1942, both in Columbia University archives.

36. J. K. Finch to Columbia's faculty of engineering, October 6, 1942, and J. K. Finch to T. T. Read, November 20, 1942, both in Columbia University archives.

37. "Saving of Culture Put to Colleges," *New York Times,* October 8, 1943; "Barrier–Breaking Gloria, Engineering's First Woman," *Columbia Engineering* (spring 2000): 29; "Pioneering Women Engineers at SEAS," *Columbia Engineering* (spring 2004): 29; "Engineering Runs in Bergen Family," *New York Times*, October 13, 1974; "Pioneering Bioengineering," *Columbia Engineering* (spring 2010): 29.

38. "First Woman Chosen for Chemistry Honor," *New York Times*, June 27, 1945.

39. "NYU to Be Co-ed in Bronx in Fall," *New York Times*, December 5, 1958.

40. "WOES (Women of Engineering Schools) Are Here," *Cornell Engineer* (January 1944): 13.

41. George Sabine to S. C. Hollister, November 22, 1944, and July 4, 1945, and Hollister to Sabine, July 2, 1945, both in Cornell University archives 16/2/2077, box 44, 71.

42. *The Transit*, RPI yearbook, 1946, 27, 37, 76; "Opening Doors for Women in Engineering," *At Rensselaer* (summer 1979): 10.

43. Edward Dion to Richard Schmelzer, January 24, 1958, and Richard Schmelzer to Edward Dion, January 28, 1958, both in RPI archives, subject file "Women—institute policies."

44. "Pat," *Alumni News* (November 1960); Roger Lourie, "RPI Plans Joint Program with Sage; Policy Endorses Female Admissions," *Rensselaer Polytechnic* (October 19, 1960): 1, 5.

45. Lourie, "RPI Plans Joint Program with Sage; Policy Endorses Female Admissions"; Sherry Pelton, "The Rensselaer Coeds," *Rensselaer Polytechnic*, March 20, 1963, 3.

46. Myles Brand, "A Coeducational Rensselaer?," *Rensselaer Engineer* (March 1963): 3.

47. Pelton, "The Rensselaer Coeds," 3.

48. Samuel Reznbeck, *Education for a Technological Society: A Sesquicentennial History of Rensselaer Polytechnic Institute* (Troy, NY: RPI, 1968), 180, 320–321, 347, 392, 415–416; Melanie McCulley, "The History of Women at Rensselaer," 1991, RPI archives, subject file "women—institute policies"; "School letters," ca. 1992, RPI archives, subject file "women—articles 1/2"; Pelton, "The Rensselaer Coeds," 3.

49. "First Woman Chosen for Chemistry Honor," *New York Times*, June 27, 1945; "Pioneering Women Engineers at SEAS," *Columbia Engineering* (spring 2004): 29.

50. Michael Bennett, *When Dreams Came True: The G.I. Bill and the Making of Modern America* (Washington, DC: Brassey's, 1996).

51. Linda Eisenmann, *Higher Education for Women in Postwar America* (Baltimore: Johns Hopkins University Press, 2006).

52. Lee Fuhrman, "Tech Girls Carve Academic Niche: Doing Fine Job, Officials Find," *Atlanta Constitution*, November 26, 1953.

53. Ibid.

54. University of Illinois, *Careers in Engineering* 47 (7) (August 1949) and 49 (71) (June 1952), 51 (57) (April 1954), both in University of Illinois archives, 11/1/870; Engineering Alumni Committee manual, *Liaison Men with High School Teachers and Administrators*, 1954, University of Illinois archives, 11/1/829.

55. Wayne Urban, *More Than Science and Sputnik* (Tuscaloosa: University of Alabama Press, 2010).

56. Fred Morris, "A Plan for Training Women in Engineering," *Journal of Engineering Education* (November 1952): 174–176.

57. William Everitt, "Dean Everitt Finds Russian Engineering Education Sound, But Too Specialized," *Illinois Alumni News* (February 1959): 3.

58. President's Committee on Scientists and Engineers, *Scientific and Technological Manpower News Round-up*, November 15, 1957; President's Committee on Scientists and Engineers, *Final Report to the President*, December 1958.

59. Fred Morris, "A Plan for Training Women in Engineering," *Journal of Engineering Education* (November 1952): 174–176.

60. Ibid.

61. Eric Walter, "Women Are NOT for Engineering," *Penn State Engineer* (May 1955): 9, 20.

62. Wilma Smith, "Women Are for Engineering," and Penelope Hester, "What Really Counts?," *Penn State Engineer* (April 1956): 18–19.

63. Burke Arehart, "A Place for Women Engineers: Industry Reluctant to Hire Women Engineers," *Purdue Engineer* (May 1957): 72; Betty Lou Bailey, "Women Can Be Engineers," speech, October 27, 1962, University of Illinois archives, 11/1/810.

64. Bailey, "Women Can Be Engineers"; "Careers in Engineering," *University of Illinois Bulletin* 62 (59) (February 1965): 12; Ann McGreaham, "The Opposite Sex in Engineering," *Purdue Engineer* (May 1963): 20–24.

65. "Purdue Personalities: Anna," *Purdue Engineer* (December 1951): 46, 48.

66. Louis Monson, "Should Women Be Elected to Membership?," *The Bent of Tau Beta Pi* (December 1945).

67. "New Society Organizes," *Iowa Engineer* (May 1946): 222; "Pi Omicron," *Purdue Engineer* (January 1948): 16.

68. "Purdue Personalities: Eleanor," *Purdue Engineer* (May 1948): 44, 46; "Purdue Personalities: Anna," *Purdue Engineer* (December 1951): 46, 48.

69. Iris Ashwell to Malcolm Willey, November 11, 1950, University of Minnesota archives, AB1.1, Pres papers, box 225, file "women 1947–55"; "By-laws of the Society of Women Engineers," May 1958, University of Illinois archives, 41/2/41.

70. Irene Carswell Peden, *Women in Engineering Careers*, 1965 booklet, SWE collection, Wayne State University archives; Joy Miller, "Women Engineers: They're Feminine and So Bright," *Perth Amboy NJ News*, July 30, 1964.

71. Peden, "Women in Engineering Careers"; Miller, "Women Engineers: They're Feminine and So Bright."

72. Alta Rutherford, "Women Engineers in Redlands Spotlight," *Detroit News*, April 19, 1954.

73. University of Illinois, *Careers in Engineering* 56 (53) (March 1959), University of Illinois archives, 11/1/870.

74. Helen O'Bannon, "The Social Scene: Isolation and Frustration," box 128, file "Women in Engineering—Beyond Recruitment Conference Proceedings, June 22–25, 1975," SWE collection, Wayne State University archives; Mildred Dresselhaus, "A Constructive Approach to the Education of Women Engineers," box 128, file "Women in Engineering—Beyond Recruitment Conference Proceedings, June 22–25, 1975," SWE collection, Wayne State University archives; Mildred Dresselhaus, "Some Personal Views on Engineering Education for Women," *IEEE Transactions on Education* 18 (1) (February 1975): 30–34.

75. *Engineering Outlook at the University of Illinois* 1 (4) (July 1960): 4; "Newsletter—Dept. of General Engineering, University of Illinois" 8 (1) (October 1965), University of Illinois archives, 11/7/809; *Engineering Outlook at the University of Illinois* 7 (3) (March 1966): 4.

76. "General Engineering at the University of Illinois," *University of Illinois Bulletin* 59 (47) (January 1962): 18.

77. Jan Schwarz, Karen Schulte, and Joanne Warner, "Student Opinion Page," *Minnesota Technolog* (November 1966): 8; "A Student's Observations," n.d., University of Illinois archives, 11/1/810, possibly by Illinois student Katherine Miller at a conference for high school principals and counselors at Illinois in October 1962.

78. Undated program for the University of Illinois open house, University of Illinois archives, 11/1/805, box. 1.

79. Jan Schwarz, Karen Schulte, and Joanne Warner, "Student Opinion Page," *Minnesota Technolog* (November 1966): 8; Betty Lou Bailey, "Women Can Be Engineers," speech, October 27, 1962, University of Illinois archives, 11/1/810.

80. "Purdue Personalities: Jay," *Purdue Engineer* (November 1961): 62; "Boilermakers: Pedji," *Purdue Engineer* (May 1963): 55.

81. Bailey, "Women Can Be Engineers."

82. "Women Engineers," *Purdue Engineering* (February 1967): 13.

83. "What Students Want to Know about Careers in Engineering," *University of Illinois Bulletin* 62 (73) (April 1965); *Engineering Outlook at the University of Illinois* 5 (1) (November 1964): 2; "Society of Women Engineers," *The Illio*, University of Illinois yearbook, 1966, 1967, and 1968.

CHAPTER 4

1. B. Eugene Griessman, Sarah E. Jackson, and Annibel Jenkins, *Images and Memories: Georgia Tech 1885–1985* (Atlanta: Georgia Tech Foundation, 1985), 66, 100, 144.

2. John Dunn, "Women: Thirty Years at Tech," *Georgia Tech Alumni Magazine* (fall 1982): 9–13.

3. Marion Smith to M. L. Brittain, July 26, 1937, box 9, folder "coed," 85-11-01, Georgia Tech archives (hereafter cited as GTA).

4. "Want Coeds? Debaters Say No!," *Technique*, November 9, 1939.

5. "Poll Shows That a Wife's Place Is Home, Not Office," *Technique*, January 23, 1942, 6.

6. Lamar Q. Ball, "It's Really War, Boys! Girls Enroll at Tech," *Atlanta Constitution*, April 17, 1942; Louise Mackay Carlton, "Thirty Girl Graduates Learning Intricacies of Defense Tools: Atlantans Training for Positions as Factory Supervisor Inspectors," *Atlanta Journal,* April 19, 1942.

7. Ball, "It's Really War, Boys!"

8. Bob Elder, "Thirty Lovely Defense Girls Attend Gauge Classes at Tech: Defense Trainees Are College Grads," *Technique,* April 24, 1942, 1; Ball, "It's Really War, Boys! Girls Enroll at Tech."

9. "Girl Meets Boy! Defense Damsel Meets M.E. Student and Flees," *Technique*, April 17, 1942.

10. Elder, "Thirty Lovely Defense Girls Attend Gauge Classes at Tech."

11. "Tech War Courses Hum: Girls Take Engineering Courses to Prepare for Jobs as Engineers' Aides," *Technique*, March 12, 1943, 5; "WAVES 'Like Tech': Fifty Have 'Chow' at Brittain Hall," *Technique,* March 12, 1943, 1, 3; Griessman, Jackson, and Jenkins, *Images and Memories: Georgia Tech 1885–1985*, 160, 162.

12. "Women Invading Engineering Field," *Technique*, March 24, 1945; "Tech Coed? Never! Registrar's Office Foils Coed Attempts to Enter," *Technique*, September 8, 1945.

13. "How's This for a New Start in Life at Tech: Coeds, Love-life, Fun, Main Worry of Officials," *Technique*, April 1, 1944; Bob November, "Planners Should Consult Students on Tech Future," *Technique*, June 3, 1944, 3, 4.

14. *Yellow Jacket* 50 (March 1947).

15. Ibid.

16. Martin L. Gursky, "Coeds at GAT? Never! Clearly Invitation to Disaster," *Technique*, March 8, 1947.

17. Henry Caulkins, "The Surveyor: Coeds at Tech," *Technique*, March 8, 1947.

18. John Couric, "'Women at Tech' Plan Pondered by Regents," *Atlanta Constitution*, May 8, 1947, 1, 6.

19. Buddy Fiske, "Coeds at Tech Would Cause Many Changes to Be Made," *Technique*, May 10, 1947.

20. Anne Wood, "Co-ed Plan Takes Tech Aback: What! What! Rah! Rah! Woe! Woe!," *Atlanta Constitution,* May 8, 1947, 6.

21. Jim Williford, "Coeducation—Never," *Technique*, May 17, 1947.

22. Griessman, Jackson, and Jenkins, *Images and Memories: Georgia Tech 1885–1985*, 168, 207; Blake R. Van Leer to M. A. Dickert [chair], L. W. Chapin, G. W. Rainey, Mrs. J. H. Crosland, W. A. Alexander, L. Mitchell, J. K. Moore, May 23, 1947. box 9, folder "coed," 85-11-01, GTA; Fiske, "Coeds at Tech Would Cause Many Changes to Be Made"; "Tech Has Room Space for Co-eds in Six Courses," *Atlanta Constitution*, May 9, 1947.

23. Hamilton Lokey to Herman A. Dickert, June 13, 1947, box 9, folder "coed," 85-11-01, GTA.

24. L. R. Siebert to Blake R. Van Leer, Guy H. Wells, and Frank R. Reade, July 7, 1947, box 9, folder "coed," 85-11-01, GTA; Rebecca Franklin, "Tech Coed-Student Question Revived: Letter from late Marion Smith Reveals He Favored Plan Presented to Regents," *Atlanta Journal,* November 22, 1947.

25. Walter Deyerle, "Spring Fever: Revealing Facts about Co-eds," *Yellow Jacket* 51 (April 1948): 8, 21.

26. "Tech Coeds: No Foolin'," *Atlanta Constitution,* April 1, 1948; Olive Ann Burns, "Coed Ramblin' Wreck," *Atlanta Journal,* November 7, 1948.

27. "The Gal Goes West: Belle of an Engineer Turned Down by Tech," *Atlanta Journal,* September 18, 1947, 38; L. R. Siebert to members of the Committee on Organization and Law, October 28, 1948, box 9, folder "coed," 85-11-01, GTA.

28. Sandy Beaver to L. R. Siebert, October 29, 1948, box 2, folder "9," 86-01-08, GTA; J. R. Anthony to Blake R. Van Leer, November 2, 1948, box 2, folder "9," 86-01-08, GTA; L. R. Siebert to Anne Bonds, November 10, 1948, box 9, folder "coed," 85-11-01, GTA; see also Rebecca Franklin, "Atlanta Girl Denied Engineer Course Fee: Regents Refuse to Pay Auburn Tuition for Anne Bonds, Barred as Tech Student," *Atlanta Journal,* November 22, 1948.

29. Blake R. Van Leer to Sandy Beaver, November 8, 1948, box 2, folder "9," 86-01-08, GTA.

30. Lee Fuhrman, "Tech Girls Carve Academic Niche: Doing Fine Job, Officials Find," *Atlanta Constitution,* November 26, 1953; Dunn, "Women: Thirty Years at Tech." In 1951, Maryly Van Leer became the first woman to earn a chemical engineering degree from Vanderbilt, which she completed magna cum laude, at the top of her class in chemical engineering, and the first female Vanderbilt student to be given the Tau Beta Pi award. She was subsequently the first woman to earn a master's of science and a doctorate in chemical engineering at the University of Florida. She became national vice president of the Society of Women Engineers, had four children, and worked as a research engineer at Rocketdyne, helping develop engine and solid-fuel technologies for the space program. Maryly Van Leer Peck later entered university administration and became president of Florida's Polk Community College in 1982, that state's first female higher-education president.

31. Blake R. Van Leer to Sandy Beaver, November 8, 1948, box 2, folder "9," 86-01-08, GTA.

32. Blake R. Van Leer to Harmon Caldwell, December 28, 1948; Russell A. Smith to C. L. Emerson, December 13, 1948; C. L. Emerson to Blake R. Van Leer, December 27, 1948; all in box 2, folder "9," 86-01-08, GTA.

33. Lloyd W. Chapin to President Van Leer, November 1, 1948, box 2, folder "9," 86-01-08, GTA; Ed Danforth, "Tech Cheer Leaders in Skirts? Decadent Idea Meets Opposition," *Atlanta Journal,* August 11, 1949.

34. Flyer, ca. fall 1950, box 5, folder 9, 87-01-01, GTA; Annie Lou Hardy, "Engineer Society Takes First Femme Member," *Atlanta Constitution,* October 29, 1950; Betsy Hopkins, "First Tech Girl Student Studies in Man's World," *Atlanta Constitution,* February 11, 1950. Barbara Hudson married a fellow student from the Georgia Technical Institute and postponed her plans for studying architecture in favor of motherhood. "Shop Talk," *Atlanta Constitution,* September 14, 1952.

35. Ruth McMillan to Board of Regents, February 7, 1952. box 9, folder "coed," 85-11-01, GTA; Fred Bennett, "Co-ed Enrollment at Tech? Atlanta Women Petition for Female Enrollment at Engineering Institution," *Technique,* February 26, 1952, 1, 2.

36. Minutes, meeting of the Board of Regents of the University System of GA, February 13, 1952, box 9, folder "coed," 85-11-01, GTA; Bennett, "Co-ed Enrollment at Tech?," 1, 2.

37. "Student Council Minutes," *Technique*, February 29, 1952; Arthur Bruckner II, letter to the editor, *Technique*, March 7, 1952.

38. "*The Technique* . . . Favors Coeds Only on Same Basis as Male Students," *Technique*, February 29, 1952, 4.

39. Roane Beard, "From the Secretary's Desk," *Georgia Tech Alumnus* 30 (4) (March–April 1952): 5.

40. Blake R. Van Leer to Harmon Caldwell, March 6, 1952; box 9, folder "coed," 85-11-01, GTA.

41. Ibid.

42. Amber W. Anderson to Blake R. Van Leer, April 6, 1952, box 9, folder "coed," 85-11-01, GTA.

43. Dorothy Crosland, no date; Dorothy M. Crosland to Rutherford L. Ellis, March 20, 1952; Dorothy M. Crosland to Rutherford L. Ellis, March 27, 1952; all in box 3, folder 36A "Coeducation," 84-06-01, GTA.

44. E. F. Bradford to Dorothy M. Crosland, March 29, 1952; E. C. Seyler to Dorothy M. Crosland, March 31, 1952; Arthur M. Gowan to Dorothy M. Crosland, April 3, 1952; A. W. Stewart to Dorothy M. Crosland, April 3, 1952; Lilyan B. Bradshaw to Dorothy M. Crosland, March 31, 1952; W. R. Woolrich to Dorothy M. Crosland, March 31, 1952; Clercie Small to Dorothy M. Crosland, April 1, 1952; Juanita Stott to Dorothy M. Crosland, April 4, 1952; Clarice Slusher to Dorothy M. Crosland, April 1, 1952; E. K. Collins to Dorothy M. Crosland, April 4, 1952; Ray Sommer to Dorothy M. Crosland, March 31, 1952; Mrs. F. W. Toppino to Dorothy M. Crosland, March 31, 1952; J. Harvey Croy to Dorothy M. Crosland, April 1, 1952; Dorothy M. Crosland to Rutherford L. Ellis, April 4, 1952; all in box 3, folder 36A "Coeducation," 84-06-01, GTA.

45. Minutes, meeting of the Board of Regents of the University System of Georgia, April 9, 1952, box 9, folder "coed," 85-11-01, GTA.

46. "Regents Vote Tech as Coed: Seven-Five Okay Given after Hot Debate; Only Engineering Degrees Offered," *Atlanta Journal*, April 9, 1952.

47. Albert Riley, "Tech Grants Admission to Women," *Technique*, April 10, 1952; Blake R. Van Leer to R. L. Ellis, April 11, 1952, box 9, folder "coed," 85-11-01, GTA.

48. Albert Riley, "Tech Grants Admission to Women," *Technique*, April 10, 1952; John Pennington, "State Law Cited in Tech Coed Plan: Would It Be Legal?," *Atlanta Journal*, April 10, 1952.

49. Blake R. Van Leer to W. L. Carmichael, April 11, 1952, box 9, folder "coed," 85-11-01, GTA; Blake R. Van Leer to Colonel A. D. Amoroso, April 16, 1952, box 9, folder "coed," 85-11-01, GTA.

50. John Pennington, "State Law Cited in Tech Coed Plan," *Atlanta Journal,* April 10, 1952; Dunn, "Women: Thirty Years at Tech."

51. Leo Aikman, "Will She Survey the Tech Field?," *Atlanta Constitution,* Apr. 16, 1952.

52. Ibid.

53. Bob McNatt, "Petite Blonde Is First Tech Coed Candidate," *Atlanta Journal and Constitution,* April 13, 1952; Jean Rooney, "Coed Prepares to Enter Tech," *Atlanta Constitution,* May 6, 1952.

54. Roane Beard, "From the Secretary's Desk," *Georgia Tech Alumnus* 30 (5) (May–June, 1952): 5.

55. Roane Beard, "From the Secretary's Desk," 5; Louise Harkrader, "The Women," *Georgia Tech Alumnus,* 30 (5) (May–June 1952): 17.

56. Lois Norvel,, "Engineering Calling Frantically, Hoping a Woman Answers: Tech Signs Four," *Atlanta Journal,* July 14, 1952.

57. Lois Norvell, "Atlanta Mother Becomes Georgia Tech's First Coed," *Atlanta Journal,* July 28, 1952; "Tech Accepts Four Coeds for Registration This Fall," *Technique,* August 8, 1952, 4.

58. Dotty Lundy, "Horrors! Girlie Touch Looms for Rat Caps," *Atlanta Journal and Constitution,* August 3, 1952; Jean Rooney, "Two Lone Co-eds Don Tech Rat Caps Monday," *Atlanta Constitution,* September 23, 1952; Nora Key, "Engineers in Petticoats—But Coeds Won't Wreck Tech," *Atlanta Constitution-Journal,* September 7, 1952.

59. "Tech Accepts Four Coeds for Registration This Fall," *Technique,* August 8, 1952, 4; Key, "Engineers in Petticoats"; Bill Paradice Jr., "Registration Planned to Be Held Monday: Large Freshman Class Includes Three Coeds," *Technique,* September 23, 1952, 1; Rooney, "Two Lone Co-eds Don Tech Rat Caps Monday"; "The First of Many," *Georgia Tech Alumnus* 34 (8) (July 1956): 7.

60. Rooney, "Two Lone Co-eds Don Tech Rat Caps Monday."

61. "Georgia Tech's First Coed: Traditional 'For-Men-Only' School Opens Its Doors to Women for First Time This Year," December 1952; *Georgia Tech Alumnus* 31 (2) (November–December, 1952): 5; Marjory Smith, "Tech's First Betty Coed Plans Electronics Career: Feels at Home," *Atlanta Constitution,* December 26, 1952; "From All of Us: We Do Mean 'Welcome'" *Technique,* October 24, 1952.

62. "From All of Us: We Do Mean 'Welcome'"; "Co-eds Too? Composition of Females," *Technique,* October 10, 1952.

63. Shelton Till, "Honorary Degrees Given to Wives of Graduates," *Technique,* January 30, 1953.

64. Rooney, "Two Lone Co-eds Don Tech Rat Caps Monday"; Yolande Gwin, "Still Definitely a Minority: Tech Coeds to Number Fourteen during Fall Term," *Atlanta Journal,* ca. summer 1954; William G. Humphrey, letter to the editor, *Technique,* November 25, 1952; Elizabeth Cofer Herndon, letter to the editor, *Technique,* November 25, 1952.

65. Smith, "Tech's First Betty Coed Plans Electronics Career: Feels at Home"; "Georgia Tech's First Coed: Traditional 'For-Men-Only' School Opens Its Doors to Women for First Time This Year."

66. Georgia Tech yearbook, *Blueprint,* 1953; Lee Fuhrman, "Tech Girls Carve Academic Niche: Doing Fine Job, Officials Find," *Atlanta Constitution,* November 26, 1953; Jean Butts, "A Very Civil Engineer: Tech's First Coed to Wed Fellow Student," *Atlanta Journal,* March 27, 1953.

67. "On the Hill: Co-eds Contingent Doubles," *Georgia Tech Alumnus* 31 (3) (January–February 1953): 12; "100% Increase: Two New Co-eds Register Here," *Technique,* January 9, 1953, 1.

68. Fran Lillard, "Attention Esquire: A Co-ed Looks at Fashions Here," *Technique,* February 26, 1953, 1.

69. Name withheld, letter to the editor, *Technique,* February 3, 1953; George Greenacre, letter to the editor, *Technique,* February 3, 1953; Roy R. Turner, letter to the editor, *Technique,* February 3, 1953; Sheldon Till, "How Techmen Justify Their Attire: Comfort, Not Fashions," *Technique,* March 6, 1953, 3, 8; John B. Wise, letter to the editor, *Technique,* February 3, 1953 (emphasis in the original).

70. Blake R. Van Leer to Chancellor Harmon Caldwell, September 27, 1952, box 9, folder "coed," 85-11-01, GTA; Minutes, meeting of the Board of Regents of the University System of Georgia, October 8, 1952, box 9, folder "coed," 85-11-01, GTA.

71. Elizabeth J. Newbury to Blake Van Leer, February 22, 1953; Blake R. Van Leer to Chancellor Harmon Caldwell, February 23, 1953; Minutes, meeting of the Board of Regents of the University System of Georgia, March 11, 1953; Harmon Caldwell to Blake Van Leer, March 13, 1953; Blake R. Van Leer to Elizabeth Newbury, March 13, 1953; Amber W. Anderson to Rutherford L. Ellis, Chairman Committee on Education, Board of Regents, University System of Georgia, April 23, 1953, all in box 9, folder "coed," 85-11-01, GTA.

72. Harmon Caldwell to Blake Van Leer, March 13, 1953, box 9, folder "coed," 85-11-01, GTA; "Atlanta Women's Organization Sponsors Co-ed Scholarship," *Technique,* May 22, 1953, 1, 2.

73. "To all the Recks: Let's Sell the Product—Then the Brand," *Georgia Tech Alumnus* (summer 1953): 2A; "Engineering: One of the Last Frontiers," *Georgia Tech Alumnus* (summer 1953): 3.

74. "1,200 Freshmen Enroll," *Georgia Tech Alumnus* 32 (1) (September–October 1953): 13; "1,125 Freshman Apply for Admission in Fall: Total Enrollment Rises 200 over Last Year," *Technique,* June 9, 1953, 1; Gordon Oliver, "Seven Coeds Register here, Total Enrollment Now Up 15%," *Technique,* October 2, 1953, 1; Bob Jessup, "Twelve Hundred Enroll in Large Frosh Class," *Technique,* September 9, 1953, 1; Despo Vacalis, "A Belle of an Engineer: T-Cut for Coed? Architect Ann's Not Scared," *Atlanta Constitution,* September 27, 1953; "Co-ed Caroline Seale Is First to Receive Woman's Scholarship," *Technique,* October 6, 1953; John Shelnutt, "One of Family: Tech Co-ed Aims to Be Engineer," *Atlanta Constitution-Journal,* September 7, 1953.

75. Marvin Gechman, "Lone Co-ed Enrolls in Summer Session," *Technique,* July 2, 1953, 1, 4; Vacalis, "A Belle of an Engineer."

76. Gechman, "Lone Co-ed Enrolls in Summer Session"; John Shelnutt, "One of Family," *Atlanta Constitution-Journal*, September 7, 1953; Vacalis, "A Belle of an Engineer"; "Trailblazer Shirley Clements Mewborn," *Georgia Tech Alumni Magazine* (fall 1982): 16–17.

77. Vacalis, "A Belle of an Engineer."

78. "Techsters Stare: A Sorority?," *Atlanta Constitution*, April 21, 1954; "Tech Coeds," *Atlanta Constitution*, April 27, 1954; Margaret Turner, "Coeds All Steamed Up for Tech Wreck Race," *Atlanta Journal*, ca. fall 1954; John Dunn, "Women: Thirty Years at Tech," *Georgia Tech Alumni Magazine* (fall 1982): 9–13. Even after her graduation, Shirley Clements Mewborn remained involved with Tech's chapter of Alpha Xi Delta, heading its building corporation and otherwise supporting its growth. "Trailblazer Shirley Clements Mewborn," *Georgia Tech Alumni Magazine*.

79. Lee Fuhrman, "Tech Girls Carve Academic Niche: Doing Fine Job, Officials Find," *Atlanta Constitution*, November 26, 1953; "Number of Day Students Tops Last Year by Three Hundred: Freshman Class and Coeds Reach Record Enrollments," *Technique*, October 5, 1954, 1; "Seven New Co-eds Plan to Attend Tech Classes," *Technique*, August 26, 1954; Harry Phipps, "Facts and Figures on Coeds," *Technique*, October 5, 1954, 4. Tech's 1954 freshman class hit a new high of 1,693 male students and total enrollment of 4,675 male students, so women's already low percentage representation at Tech actually dropped a bit in 1954.

80. Yolande Gwin, "Still Definitely a Minority," *Atlanta Journal*, ca. summer 1954.

81. Gwin, "Still Definitely a Minority"; "Precedent Broken at Tech," *Atlanta Constitution*, October 4, 1954; Frances Cawthon, "Pioneering Girls Crash Tech's Air Force ROTC," *Atlanta Journal*, January 25, 1955, 24.

82. "The First of Many," *Georgia Tech Alumnus*.

83. *Georgia Tech Blueprint*, 1956.

84. "The First of Many," *Georgia Tech Alumnus*; "Trailblazer Shirley Clements Mewborn," *Georgia Tech Alumni Magazine*.

85. "The First of Many," *Georgia Tech Alumnus*.

86. "Trailblazer Shirley Clements Mewborn," *Georgia Tech Alumni Magazine*; Bob Wallace Jr., "Ramblin': The Editor's Note," *Georgia Tech Alumnus* 35 (6) (March 1957): 2; *Georgia Tech Alumnus* 35 (6) (March 1957): 26. Michel also later married.

87. Paula Stevenson. "A Woman in a Man's World," *Georgia Tech Alumnus* 36 (4) (December 1957): 7–8.

88. Ibid, 8.

89. Ibid.; Robert McKee, "Tech Women Are Making History," *Atlanta Constitution*, October 26, 1962.

90. Stevenson, "A Woman in a Man's World," 7–8.

91. Ibid.

92. "Paula Stevenson Humphreys: It Was the Best Finishing School," *Georgia Tech Alumni Magazine* (fall 1982): 18–20.

93. Stevenson, "A Woman in a Man's World"; "Paula Stevenson Humphreys: It Was the Best Finishing School."

94. Stevenson, "A Woman in a Man's World."

95. "Need Coed Facilities," *Technique*, May 30, 1958, 4.

96. "Lack of Housing Facilities Limits Increase in Coeds," *Technique*, May 30, 1958, 1; "Need Coed Facilities," *Technique*, May 30, 1958, 4; Rodney House, "Poll Reveals Students Support Better Co-ed Housing Facilities," *Technique*, July 4, 1958; "Dorms Pose Question," *Technique*, August 1, 1958.

97. *The Freshman Girls' Handbook*, Georgia Tech, 1966–67.

98. Mary O'Kon, "*Technique's* Mary O'Kon Welcomes Incoming Coeds," *Technique*, September 5, 1958; Barbara Carpenter, "Coed Chatter: Frosh Coeds Steal Pigskins, Join AFROTC, Run Races," *Technique*, October 10, 1958; "New Coed Majorettes, Herald Trumpets to Represent Jacket Band at Halftime," *Technique*, October 17, 1958; Barbara Carpenter, "Coed Chatter: WSA Holds First Meeting; Barbara Cass to Attend SC," *Technique*, October 17, 1958.

99. "Coeds Excel Men in Fall Averages," *Technique*, January 9, 1959, 1; "Coeds Surpass Males," *Technique*, January 9, 1959; Bill Cox, "Undergraduate Scholarship Reaches Lowest Average since Winter of '51," *Technique*, January 15, 1960, 1; Howard Arnold, "Rumblings: Coed's Average Hits New Low," *Technique*, January 22, 1960, 8.

100. "Subject: Engineering Education and the Future," *Georgia Tech Alumnus* 36 (6) (March 1958): 6–9.

101. Arnold Kleine to Edwin Harrison, October 21, 1959; Edwin Harrison to Arnold Kleine, October 23, 1959; Mary Morris to E. D. Harrison, January 6, 1958; Edwin D. Harrison to Mary Morris, January 16, 1959; all in box 9, folder "coed," 85-11-01, GTA.

102. Walter J. Skelly to the Regents, University System of Georgia, November 25, 1957; Walter J. Skelly to Edwin D. Harrison, November 25, 1957; Harrison to Harmon Caldwell, December 16, 1957; all in box 9, folder "coed," 85-11-01, GTA.

103. Edwin D. Harrison to Fuller E. Callaway, June 19, 1961, and Fuller E. Callaway Jr. to R. O. Arnold, June 26, 1961, both in box 9, folder "coed," 85-11-01, GTA.

104. Robert D. Arnold to Fuller E. Callaway Jr., June 27, 1961, box 9, folder "coed," 85-11-01, GTA.

105. Fuller E. Callaway Jr. to Robert O. Arnold, June 28, 1961, box 9, folder "coed," 85-11-01, GTA.

106. "Women: An Historical View of the Females at Tech," *Georgia Tech Alumnus* 48 (2) (January–February 1970): 6–13; M. J. Goglia to E. D. Harrison, February 12, 1962, and Edwin D. Harrison to E. J. Goglia, February 14, 1962, both in box 9, folder "coed," 85-11-01, GTA.

107. Edwin D. Harrison to Robert O. Arnold, September 4, 1962; Dorothy Jean McDowell to Robert O. Arnold, April 23, 1962; Robert O. Arnold to Dorothy Jean McDowell, May 1, 1962; Edwin D. Harrison to Dorothy Jean McDowell, May 17, 1962; all in box 9, folder "coed," 85-11-01, GTA. The Board of Regents agreed to let McDowell major in chemistry, which was an exception to the usual ban on female enrollment.

108. Julia Dallas Bouchelle to Robert O. Arnold, November 7, 1962; Robert O. Arnold to Edwin Harrison, November 8, 1962; Edwin D. Harrison to Julia D. Bouchelle, May 7, 1963; Carolyn E. Williams to Edwin Harrison, August 24, 1964; Varner Jo Blackshaw to Ralph A. Hefner, May 13, 1963; Ralph A. Hefner to Edwin D. Harrison, May 21, 1963; Edwin D. Harrison to Harmon Caldwell, June 12, 1963; all in box 9, folder "coed," 85-11-01, GTA.

109. Sylvianne Bassett to Edwin D. Harrison, June 30, 1963; Edwin D. Harrison to Harmon W. Caldwell, July 2, 1963; Edwin D. Harrison to Harmon W. Caldwell, July 2, 1963; Edwin D. Harrison to L. R. Siebert, February 23, 1963; Barbara J. Daniels to M. J. Goglia, October 31, 1962; M. J. Goglia to Paul Weber, November 20, 1962; Edwin D. Harrison to Harmon Caldwell, April 1, 1963; Edwin D. Harrison to Harmon Caldwell, December 3, 1962; Edwin D. Harrison to L. R. Siebert, February 23, 1963; Harmon Caldwell to E. D. Harrison, December 13, 1962; all in box 9, folder "coed," 85-11-01, GTA.

110. Vernon Crawford to S. Walter Martin, November 17, 1964, box 3, folder 36A, "Coeducation," 84-06-01, GTA.

111. Vernon Crawford to S. Walter Martin, November 17, 1964; "Institutional Self-Study program," Georgia Tech, 1963, 1–3.

112. Edwin D. Harrison to S. Walter Martin, February 15, 1965, box 9, folder "coed," 85-11-01, GTA.

113. James A. Dunlap to Edwin D. Harrison, January 4, 1965; Edwin D. Harrison to S. Walter Martin, January 27, 1965; Edwin D. Harrison to S. Walter Martin, February 15, 1965; S. Walter Martin to Edwin D. Harrison, March 10, 1965; all in box 9, folder "coed," 85-11-01, GTA; Mike Peach, "IM Department Opens Up for Women, Four Co-eds Switch Fields of Study," *Technique*, February 9, 1968.

114. J. R. Anthony to James E. Dull and R. B. Logan, March 27, 1964, box 9, folder "coed," 85-11-01, GTA; Jim Long, "Coeds Make Their Mark at Tech," *Atlanta Journal and Constitution*, April 29, 1962, 22, 24, 26; "Girls in the Techman's Life: Coeds Exposed as Real Females," *Technique,* January 29, 1965; *The Freshman Girls' Handbook*, Georgia Tech, 1966–67.

115. Ron Vinson, "Tech Coeds Discuss Struggle, Analyze Progress on Campus: Opening Up Engineering," *Technique*, March 1, 1968, 8.

116. "Girls in the Techman's Life," *Technique,* January 29, 1965; Marion Woll, "Education at Georgia Tech Worth Effort, Coeds Say: One Hundred Studying There," *Atlanta Constitution,* ca. 1965–1966; Long, "Coeds Make Their Mark at Tech";Vinson, "Tech Coeds Discuss Struggle," 8; Howard Collins, "She's No Ordinary Engineer: Georgia Tech–Bound Maconite Is a Blonde," *Macon Georgia Telegraph,* April 15, 1968.

117. "Institutional Self-Study Program: Fifth Year Interim Report, to the Southern Association of Colleges and Schools," Georgia Tech, November 1968, 51–52; Lorraine M. Bennett, "Will Girls End Tech Wreck?," *Atlanta Constitution,* ca. 1968; "Kathy: A Freshman Coed with Beauty and Style," *Georgia Tech Alumnus* 48 (2) (January–February 1970): 5;Vinson, "Tech Coeds Discuss Struggle"; Carolyn Marvin, "Tech Graduating How Many Girls?," *Atlanta Constitution,* June 12, 1970, 3B.

118. Long, "Coeds Make Their Mark at Tech"; Dunn, "Women: Thirty Years at Tech"; "Kathy: A Freshman Coed with Beauty and Style," *Georgia Tech Alumnus;*Vinson, "Tech Coeds Discuss Struggle."

119. "Women: An Historical View of the Females at Tech," *Georgia Tech Alumnus.*

120. James T. Patterson, *Brown v. Board of Education: A Civil Rights Milestone and Its Troubled Legacy* (New York: Oxford University Press, 2001). For background on the history of black college education, see John R. Thelin, *A History of American Higher Education,* 2nd ed. (Baltimore: Johns Hopkins University Press, 2011).

121. "Education: Shame in Georgia," *Time,* January 20, 1961; see also Calvin Trillin, *An Education in Georgia; The Integration of Charlayne Hunter and Hamilton Holmes* (New York:Viking Press, 1964); Donald L. Grant and Jonathan Grant, *The Way It Was in the South: The Black Experience in Georgia* (Athens: University of Georgia Press, 2001); Thomas G. Dyer, *The University of Georgia: A Bicentennial History, 1785–1985* (Athens: University of Georgia Press, 1985); Robert A. Pratt, *We Shall Not Be Moved: The Desegregation of the University of Georgia* (Athens: University of Georgia Press, 2002).

122. "Georgia Tech Is Desegregated with No Fuss," *Modesto Bee,* September 18, 1961, A4; "Segregation Ends Peacefully at Georgia Tech," *Rome News-Tribune,* September 18, 1961, 1; "Desegregation, Activism, and a New Direction," *Georgia Tech Alumni Magazine* (spring 1998); Robert C. McMath, *Engineering the New South: Georgia Tech, 1885–1985* (Athens: University of Georgia Press, 1985).

123. Kevin M. Kruse, *White Flight: Atlanta and the Making of Modern Conservatism* (Princeton, NJ: Princeton University Press, 2005): 40;Virginia H. Hein, "The Image of 'A City Too Busy to Hate': Atlanta in the 1960s," *Phylon* 33 (3) (1972): 205–221; Stephen G. N. Tuck, *Beyond Atlanta: The Struggle for Racial Equality In Georgia, 1940–1980* (Athens: University of Georgia Press, 2003); James C. Cobb, *Selling of the South: The Southern Crusade for Industrial Development, 1936–90* (Urbana: University of Illinois Press, 1993); George McMillan, "With the Police on an Integration Job," *Life,* September 15, 1961, 35. For more on the complex interactions of race, politics,

business, civic life, and southern culture in Atlanta, see Ronald H. Bayor, *Race and the Shaping of Twentieth-Century Atlanta* (Chapel Hill: University of North Carolina Press, 2000).

124. "Living History," http://gtalumni.org/Publications/techtopics/spr06/livinghistory.html, accessed September 6, 2012; Liz Burnett, "Students Reflect on Black History at Tech," *Technique*, February 2, 2007, 9; "Georgia Tech Graduates Its First Negro," *Morning Record* (Meriden, CT), June 15, 1965, 20.

125. Burnett, "Students Reflect on Black History at Tech"; "Living History," http://gtalumni .org/Publications/techtopics/sum03/livinghistory.html, accessed September 9, 2012; Edward Davis, "Achieving Siblings Gathering to Celebrate Family Matriarch," *CrossRoads News*, http:// crossroadsnews.com/view/full_story/1445385/article-Achieving-siblings-gathering-to-cele brate-family-matriarch, accessed September 9, 2012.

126. Kimberly Link-Wills, "Women of Distinction Heralded for Contributions," December 20, 2010, http://gtalumnimag.com/2010/12/women-of-distinction-heralded-for-contributions, accessed September 5, 2012; Charles Seabrook, "Ms. Engineer? Tech's Trying; Minorities Sought," *Atlanta Constitution*, October 21, 1973, 1-C; Lavice Deal, "'More Books Than Boys': Social Life Lags, Tech Coeds Declare," *Atlanta Journal*, May 2, 1973.

127. *The Freshman Girls' Handbook*, Georgia Tech, 1966–67; "Women Engineers," *Southern Engineering*, April 1968, in "GTA Subject File—Women Graduates," GTA; "Institutional Self-Study Program—Fifth Year Interim Report, to the Southern Association of Colleges and Schools," Georgia Tech, November 1968, 51–52; Seabrook, "Ms. Engineer?"

128. "Women: An Historical View of the Females at Tech," *Georgia Tech Alumnus*; Linda Street, "Coeds Compete with Men," *Atlanta Journal*, November 7, 1961, 30; "Kathy: A Freshman Coed with Beauty and Style," *Georgia Tech Alumnus*.

129. Vinson, "Tech Coeds Discuss Struggle."

130. Street, "Coeds Compete with Men."

131. *The Freshman Girls' Handbook*, Georgia Tech, 1966–67.

132. Ibid.

133. Ibid.

134. "Women: An Historical View of the Females at Tech," *Georgia Tech Alumnus*.

135. Cathy Yarbrough, "Competing in Man's World Is Frustrating," *Free-Lance Star*, April 27, 1973, 17.

136. Woll, "Education at Georgia Tech Worth Effort, Coeds Say"; Marvin, "Tech Graduating How Many Girls?"; Deal, "More Books Than Boys"; "Kathy: A Freshman Coed with Beauty and Style," *Georgia Tech Alumnus*.

137. Long, "Coeds Make Their Mark at Tech"; McKee, "Tech Women Are Making History."

138. Long, "Coeds Make Their Mark at Tech"; "Girls in the Techman's Life: Coeds Exposed as Real Females," *Technique,* January 29, 1965.

139. Vinson, "Tech Coeds Discuss Struggle."

140. Michelle Murphy, "Life Looks Up for Tech Women," *Technique,* September 23, 1970; Deal, "'More Books Than Boys'"; Marvin, "Tech Graduating How Many Girls?"

141. Vinson, "Tech Coeds Discuss Struggle."

142. Vinson, "Tech Coeds Discuss Struggle"; Linda Craine, letter to the editor, "Coeds Pursue Learning—Not Men," *Technique,* July 10, 1970; Murphy, "Life Looks Up for Tech Women."

143. Murphy, "Life Looks Up for Tech Women"; Lee Denny, "Tech Coeds Stay Home; Most Men Afraid to Ask," *Technique,* October 11, 1974.

144. "Girls in the Techman's Life," *Technique,* January 29, 1965; Deal, "'More Books Than Boys.'"

145. McKee, "Tech Women Are Making History"; Woll, "Education at Georgia Tech Worth Effort, Coeds Say"; "Girls in the Techman's Life: Coeds Exposed as Real Females," *Technique,* January 29, 1965; "Kathy: A Freshman Coed with Beauty and Style," *Georgia Tech Alumnus.*

146. Denny, "Tech Coeds Stay Home"; Deal, "'More Books Than Boys'"; Cathy Yarbrough, "Competing in Man's World Is Frustrating," *Free-Lance Star*, April 27, 1973, 17.

147. Long, "Coeds Make Their Mark at Tech."

148. Craine, letter to the editor, "Coeds Pursue Learning—Not Men."

149. "Kathy: A Freshman Coed with Beauty and Style," *Georgia Tech Alumnus*; 5; Deal, "'More Books Than Boys.'"

150. Deal, "'More Books Than Boys'"; "Kathy: A Freshman Coed with Beauty and Style."

151. "Kathy: A Freshman Coed with Beauty and Style"; Howard Collins, "She's No Ordinary Engineer: GAT-Bound Maconite Is a Blonde," *Macon Georgia Telegraph*, April 15, 1968; Sam Hopkins, "Anita, Twenty-one, a Girl in a Thousand," *Atlanta Journal and Constitution,* November 22, 1962, 56-A.

152. Woll, "Education at Georgia Tech Worth Effort, Coeds Say"; "Kathy: A Freshman Coed with Beauty and Style."

153. Craine, letter to the editor, "Coeds Pursue Learning—Not Men"; Murphy, "Life Looks Up for Tech Women."

154. "Mary Anne: A Ph.D. Aspirant in a Complicated Field," *Georgia Tech Alumnus* 48 (2) (January–February 1970): 17.

155. Street, "Coeds Compete with Men"; Woll, "Education at Georgia Tech Worth Effort, Coeds Say"; "Mary Anne: A Ph.D. Aspirant in a Complicated Field."

156. Street, "Coeds Compete with Men."

157. Hopkins, "Anita, Twenty-one, a Girl in a Thousand."

158. Vinson, "Tech Coeds Discuss Struggle"; "The Technological Labor Market," *Georgia Tech Alumnus* 48 (6) (July–August 1970): 4–9; Marvin, "Tech Graduating How Many Girls?"

159. Jane Leonard, "Women Grads at Tech Come on Strong in Job Picture," *Atlanta Journal and Constitution,* April 2, 1972; Marvin, "Tech Graduating How Many Girls?"; Seabrook, "Ms. Engineer?"

160. Leonard, "Women Grads at Tech Come on Strong in Job Picture"; Marvin, "Tech Graduating How Many Girls?"

161. Georgia Tech "I Can" leaflet, 1973, box 9, folder "coed," 85-11-01, GTA; Seabrook, "Ms. Engineer?"

162. Esther Lee Burks, special assistant to the dean of engineering for women, to Administrative Council, subject "I can," August 23, 1973, box 9, folder "coed," 85-11-01, GTA.

163. Georgia Tech "I Can" leaflet, 1973, box 9, folder "coed," 85-11-01, GTA.

164. "Women in Engineering at Georgia Tech" brochure, no date; Georgia Tech Subject file, "Women and engineering," GTA.

165. "Women in Engineering at Georgia Tech" brochure.

166. Ibid.

167. Dunn, "Women: Thirty Years at Tech."

CHAPTER 5

1. For more on the history of Caltech coeducation within the context of the development of technology in the state of California, see Amy Sue Bix, "Coeducation at 'Millikan's Monastery': Caltech's Masculine Identity and the Tensions of Women's Technical Minds," in *Where Minds and Matters Meet: Technology in California and the West*, ed. Volker Janssen, 197–230 (Berkeley: University of California Press, 2012).

2. Sally Hacker, "Mathematization of Engineering: Limits on Women and the Field," in *Machina ex Dea*, ed. Joan Rothschild, 38–58 (New York: Teachers College Press, 1983); Amy Slaton, *Reinforced Concrete and the Modernization of American Building, 1900–1930* (Baltimore: Johns Hopkins University Press, 2001.

3. Bruce Sinclair, "Engineering the Golden State: Technics, Politics, and Culture in Progressive Era California," *Where Minds and Matters Meet: Technology in California and the West*, ed. Volker Janssen, 43–70 (Berkeley: University of California Press, 2012).

4. Donald Pisani, *From Family Farm to Agribusiness: The Irrigation Crusade in California and the West* (Berkeley: University of California Press, 1984).

5. Gerald Nash, *The American West Transformed: The Impact of the Second World War* (Lincoln: University of Nebraska Press, 1990).

6. Helen Rice to Shelley Erwin, May 15, 1989, file X4.1, CIT historical files, Caltech archives.

7. Kevin Starr, *California: A History* (New York: Modern Library, 2005).

8. Kevin Starr, *Golden Dreams: California in an Age of Abundance, 1950–1963* (New York: Oxford University Press, 2009).

9. Mark Carnes, *Secret Ritual and Manhood in Victorian America* (New Haven, CT: Yale University Press, 1989).

10. G. W. Beadle to Dean E. C. Watson, May 31, 1947, file X4.1, Caltech archives.

11. Lee DuBridge to James Killian Jr., November 18. 1947, file X4.1, Caltech archives.

12. James Killian Jr. to Lee DuBridge, November 25, 1947, file X4.1, Caltech archives.

13. Margaret W. Rossiter, *Women Scientists in America: Before Affirmative Action, 1940–1972* (Baltimore: Johns Hopkins University Press, 1998), 85.

14. "Women—Now Five Percent of the Graduate Student Body—Find a Happy Home at Caltech," *Caltech News* (June 1967): 4–5.

15. Charles Newton, June 17, 1976, file X4.1, Caltech archives.

16. A. B. Ruddock to Charles Newton, ca. 1951, file X4.1, Caltech archives.

17. No author, ca. 1951, file X4.1, Caltech archives.

18. "Coed Graduates Set Tech Record," *California Tech*, October 17, 1963, 2.

19. "History of the Proposal for the Admission of Women Undergraduates," ca. fall 1968, file L4.6, Caltech archives.

20. "Women—Now Five Percent of the Graduate Student Body."

21. "Alumni," *Bulletin of the California Institute of Technology* 74 (1) (February 1965): 23–24.

22. "Sex on Campus," *California Tech*, February 18, 1965, 5.

23. "Coed: New Student House Opens for Business," *California Tech*, June 3, 1965, 1.

24. Ira Herskowitz, "Houses Breed Boors," *California Tech*, October 6, 1966, 2.

25. John Middleditch, "Editorial: 'Eunuchs of Science,'" *California Tech*, October 6, 1966, 2.

26. Lee DuBridge to Janet Louise Mangold, May 1, 1967, file 106.4, DuBridge papers, Caltech archives

27. "Appendix V: Reflections on Several Worlds," September 17, 1968, file L3.9, Caltech archives.

28. Ibid.

29. *California Tech*, January 30, 1969, 6.

30. "Appendix V," Caltech archives.

31. Ibid.

32. James Gilbert, *Men in the Middle: Searching for Masculinity in the 1950s* (Chicago, IL: University of Chicago Press, 2005); Michael Kimmel, *Manhood in America: A Cultural History* (New York: Oxford University Press, 2006).

33. Kevin White, *The First Sexual Revolution: Male Heterosexuality in Modern America* (New York: New York University Press, 1992).

34. "Appendix V," Caltech archives.

35. Memo, November 20, 1969, file L4.1, Caltech archives.

36. Interview with Abraham Kaplan, file L3.9, Caltech archives.

37. "The Carl Rogers Student-Faculty group meeting of 1 March 1967," "Appendix V," Caltech archives.

38. "Faculty Committee Proposes Female Undergrad Admission," *California Tech*, October 5, 1967, 1.

39. Faculty discussion group meeting, November 10, 1967 agenda, file L4.6, Caltech archives.

40. Lee Austin, "Can Women Break Caltech Barrier?," *Los Angeles Times*, November 5, 1967, 1, 6.

41. "Some Notes Regarding the Admission of Women to Caltech from [the] Ad Hoc Committee on the Admission of Women," ca. fall 1967; file L4.6, Caltech archives.

42. Ibid.

43. Richard Flammang, "A Not-Unbiased Report from Caltech's Undergraduate Ad Hoc Committee on the Admission of Women Which Says, in Effect: Bring On the Girls!," *Engineering and Science* (December 1967): 38–39.

44. "EPC Results: Tech Wants Coed Campus," *California Tech*, October 19, 1967, 1.

45. Jim Cooper, "Faculty Board Votes for Undergrad Girls," *California Tech*, November 16, 1967, 1.

46. Norman Davidson to Arnold Beckman, December 1, 1967, file L4.6, Caltech archives.

47. Jim Cooper, "Faculty Votes for Women," *California Tech*, November 30, 1967, 1.

48. Lee DuBridge to Arnold Beckman, January 2, 1968, DuBridge papers.

49. Jim Cooper, "Female Admissions Late But Inevitable," *California Tech*, December 7, 1967, 1.

50. "A Study of High-Level Girl Science Students," January 12, 1968, file L4.7, Caltech archives.

51. Ibid.

52. "A Study of Caltech Women Graduate Students," January 30, 1968, file L4.6; Caltech archives.

53. Ibid.

54. Ibid.

55. Minutes committee on admitting women undergraduates meeting, January 23, 1968, file L4.2, Caltech archives.

56. Robert Alexander to L. Winchester Jones, January 26, 1968, file L4.2, Caltech archives.

57. "Appendix IV—freshman admissions committee minutes, February 14, 1968 meeting," file L3.9; Charles Newton to Lee DuBridge, September 17, 1968, file L3.9, Caltech archives.

58. Minutes, May 2, 1968, file L4.72, Caltech archives.

59. "Appendix IV," Caltech archives.

60. Berto Kaufman, "Move to Push for Minority Enrollees," *California Tech*, November 7, 1968, 1, 5.

61. Joseph Rhodes, "The Caltech Myth," *California Tech*, May 16, 1968, 3, 4.

62. James Rawls and Walton Bean, *California: An Interpretive History* (Boston, MA: McGraw Hill, 2003); Robert Cherny, Richard Griswold delCastillo, and Gretchen Lemke-Santangelo, *Competing Visions: A History of California* (Boston, MA: Houghton Mifflin, 2005).

63. John Healy, "ARPettes Tell All: Hi-Life in ASCIT," *California Tech*, October 31, 1968, 1, 4.

64. *Engineering and Science* (October 1968): 34–35.

65. Healy, "ARPettes Tell All," 1, 4.

66. Ned Munger, "Prof Suggests Way to Get Girls," *California Tech*, May 16, 1968, 2.

67. "Girls Coming to Tech! Board of Trustees Vote Coeds for 1970," *California Tech*, November 7, 1968, 1.

68. "History of the Proposal for the Admission of Women Undergraduates," ca. fall 1968, file L4.6, Caltech archives.

69. Lee DuBridge to Dean L. W. Jones, January 9, 1968, box 107.1, DuBridge papers.

70. Lee DuBridge to Arnold Beckman, September 10, 1968, box 107.9, DuBridge papers.

71. Charles Newton to Lee DuBridge, September 17, 1968, file L3.9, Caltech archives.

72. Ibid.

73. Quotations from student questionnaire, October 10, 1968, file L4.1, Caltech archives.

74. Judith Goodstein, *Millikan's School: A History of the California Institute of Technology* (New York: Norton, 1991).

75. Quotations from student questionnaire, October 10, 1968.

76. Lee Dubridge to the Board of Trustees, October 25, 1968, box L4.2, DuBridge papers.

77. Norman Davidson to Arnold Beckman, October 28, 1968, file L4.6, Caltech archives.

78. Lee DuBridge to Peter Miller, November 11, 1968, file 108.1, DuBridge papers.

79. Lee Dubridge to Ray Owen, November 7, 1968, file 108.1, DuBridge papers.

80. "Women Trigger Reactions," *California Tech*, November 7, 1968, 1, 5.

81. "Girls Coming to Tech!," *California Tech*, November 7, 1968, 1.

82. Lee DuBridge to Peter Miller, November 11, 1968, DuBridge papers.

83. John Healy, "Caltech Welcomes First Lady RA," *California Tech*, November 7, 1968, 1, 3.

84. Nick Smith, "'Hey There, Dabney Girl . . .' Great Experiment Hasn't Happened Before," *California Tech*, May 1, 1969, 1, 4.

85. Ira Moskatel and Craig Sarazin, "Editorial: A Little Variety, a Lot of Normalcy," *California Tech*, May 1, 1969, 2; Marc Aaronson, "Aaronson Outlines Need for Coeducational Housing," *California Tech*, May 1, 1969, 4.

86. Charles Newton, June 17, 1976, file X4.1, Caltech archives.

87. "Coeds Coming Next Year—Trustees Give OK," *California Tech*, October 16, 1969, 1.

88. *California Tech*, February 5, 1970, 10.

89. "218 Accept Admission to Tech," *California Tech*, May 7, 1970, 1.

90. Debbie Dison and Marion Movius, "Confessions of Two Co-Techs," ca. 1974, box X4, Caltech archives.

91. "Report on the ACE Freshman Questionnaire," ca. 1976, file L4.2, Caltech archives.

92. Barbara Brown to Ray Owen, November 5, 1975, file L4.7, Caltech archives.

93. Memo, May 20, 1976, file L4.1, Caltech archives.

94. "Notes on women admitted to Caltech in 1977 who chose to go elsewhere," file L4.1, Caltech archives.

95. "Letters women undergraduates wrote to women applicants who had been offered admission to the freshman class in Sept. 1978," no date, file L4.7, Caltech archives.

96. Brochure, "What's a nice girl like you doing in a place like Caltech?," ca. 1978, file L4.7, Caltech archives.

97. Ibid.

98. Ibid.

99. Sharon Long, "'Being a Woman at Caltech Has Its Own Special Set of Challenges,'" *Engineering and Science* (June 1973): 22–23.

100. "Comments pages and questionnaire," spring 1979, and Karen Hellgren, "Undergraduate Women at Caltech: A Survey of Attitudes," June 30, 1980, both in file X4.2.2, Caltech archives.

101. Ibid.

102. Ibid.

103. *California Tech*, October 15, 1970, 2.

104. Ibid.

105. Ibid.

106. Ibid.

107. Ibid.

108. Judith Cohen, "Women's issues: A presentation to the visiting committee of the division of physics, mathematics, and astronomy of the California Institute of Technology," May 1995, file M5.3, Caltech archives.

109. Ibid.

110. Brochure, "Women at Caltech," ca. 1980s, Caltech archives.

## CHAPTER 6

1. A. J. Angulo, *William Barton Rogers and the Idea of MIT* (Baltimore: Johns Hopkins University Press, 2008); David Kaiser, ed., *Becoming MIT* (Cambridge, MA: MIT Press, 2010); Philip Alexander, *A Widening Sphere* (Cambridge, MA: MIT Press, 2011).

2. *The Handbook*, 1943, and Association of Women Students, *Cheney and All That: Freshman Coed Handbook*, ca. 1956, both in AC 4, box 1 f7. All archival material in this chapter, unless specified otherwise, is from the MIT archives.

3. *Report of the President* (Cambridge, MA: MIT Press, 1944) (various years).

4. Florence Stiles to Karl Compton, February 3, 1945, and memo from Florence Stiles, February 22, 1945, both in AC4, box 210, folder 18.

5. L. F. Hamilton to Julius Stratton, October 24, 1956; memo from J. A. Stratton, "A Statement of Policy on Women Students," January 24, 1957; Roland B. Greeley to Devrie Shapiro, October 4, 1961; all in AC134, box 116, folder "women students."

6. "New Tech Coeds: Freshman Coeds Reveal Their Interests Are in Men as Well as in Chemistry," *The Tech*, August 3, 1945, 1; Memo from Everett Baker, January 26, 1947, AC 4, box 26, folder 12; Emily Wick, "Proposal for a New Policy for Admission of Women Undergraduate Students at MIT," March 9, 1970.

7. Florence Stiles to Carroll Webber Jr., March 28, 1946, AC220, box 2, folder 2.

8. E. Francis Bowditch to J. R. Killian Jr. and J. A. Stratton, April 24, 1952, AC220, box 2, folder 2.

9. L. F. Hamilton to Julius A. Stratton, November 14, 1956, AC220, box 2, folder 2.

10. Oral history interview with Sheila Widnall by Shirlee Sherkow, November, 1976, MC86, box 8.

11. Margaret Alvort to L. F. Hamilton, June 21, 1956.

12. James Killian to Lee duBridge, November 25, 1947.

13. L. F. Hamilton to Julius A. Stratton, November 14, 1956; *Report of the President*, 1957, 235–236; Herbert I. Harris to L. F. Hamilton, July 31, 1956; all quoted in Evelyn Fox Keller, "New Faces in Science and Technology: A Study of Women Students at MIT," August 1981.

14. Survey of former Women Students at the Massachusetts Institute of Technology, conducted by the Registration committee of the MIT women's Association, May 1953, MC65, box 1, f. 13; *Report of the President*, 1954.

15. Interview of Christina Jansen by Shirlee Shirkow, 1977, MC86, box 9.

16. Memo from J. A. Stratton, January 24, 1957, "A Statement of Policy on Women Students," AC 4, box 1 f7; *Report of the President*, 1951 and 1962.

17. Ruth Bean to John T. Rule, September 14, 1959, AC 134, box 116, folder "women students"; memo from Ruth Bean, November 23, 1959, MC65, box 1, folder 16.

18. J. R. Killian Jr. to J. A. Stratton, October 22, 1956, and M. D. Rivkin to J. A. Stratton, October 15, 1959, both in AC134, box 116, folder "women students"; *Report of the President*, 1957, 1959.

19. Ibid. See also *Report of the President*, 1959, 219–220; M. D. Rivkin to J. A. Stratton, etc., October 15, 1959, AC134, box 116, folder "women students."

20. "Where the Brains Are," *Time*, October 18, 1963, 51; *Seventeen* (October 1964): 44, 46.

21. *This Is MIT*, 1963–1964; K. R. Wadleigh to the Academic Council, March 6, 1964; Jacquelyn A. Mattfeld to Malcolm G. Kispert etc., January 21, 1964; all in AC220, box 4, folder 2.

22. Jacquelyn A. Mattfield to Lois Pratt, February 3, 1964, AC134, box 48, folder "Dean of student affairs," and Office of the Dean of Student Affairs, "A Program for the Women of MIT," March 16, 1964, both in AC134, box 116, folder "women students."

23. Office of the Dean of Student Affairs, "A Program for the Women of MIT," March 16, 1964, AC134, box 116, folder "women students."

24. Ibid.

25. Mona Dickson, essay on MIT life for *Seventeen*, ca. spring 1964; AC220, box 4, folder 1.

26. Steve Carhart, "Myth Conference Discusses Modern Institute Images," *The Tech*, October 24, 1967, 1, 3; Questionnaires, May 1964, AC220, box 2, folder 1; Dickson, essay on MIT life for *Seventeen*, ca. spring 1964; Jacquelyn Mattfeld, "Summary of Info on the Women Students," ca. 1963–1964, AC220, box 4, folder 1.

27. Interview of Christina Jansen by Shirlee Shirkow, 1977, MC86, box 9.

28. Memo from Judy Risinger, December 13, 1963, MC65, box 5, file "Women's Symposium 1964 news releases."

29. "Female Scientist Image Blasted," *Michigan State News*, November 4, 1964, and MIT press releases, October 8 and October 15, 1964, both in MC65, box 5, file "Women's Symposium 1964 news releases"; see also Margaret W. Rossiter, *Women Scientists in America: Before Affirmative Action, 1940–1972* (Baltimore: Johns Hopkins University Press, 1998), 366–368.

30. Timothy Leland, "Over the Din of Babes at MIT," *Boston Globe*, October 24, 1964; Cynthia Parsons, "Women in Science Spar at Conclave," *Christian Science Monitor*, November 3, 1964; see also Jacquelyn Mattfeld and Carol Van Aken, eds., *Women and the Scientific Professions: The MIT Symposium on American Women in Science and Engineering* (Cambridge, MA: MIT Press, 1965).

31. "Sugar and Spice Analysis: Good Girls Won't Become Scientists," *Washington DC Post-Times Herald*, October 28, 1964; Arline Grimes, "Science Role for Women Is Discussed," *Boston Herald*, October 24, 1964; Jacquelyn Mattfeld to Carroll Bowen, March 2, 1965, AC134, box 116, folder "women students."

32. Memo from Jacquelyn Mattfeld, "Information on Women's Program, MIT, 1964–65," July 1, 1965, MC65, box 2, folder 25.

33. Academic Council notes on the housing and admission of undergraduate women, March 2, 1965, AC134, box 116, folder "women students"; *AMITA Newsletter*, March 1965.

34. Memo from Jacquelyn Mattfeld, "Information on Women's Program, MIT, 1964–65," July 1, 1965, MC65, box 2, folder 25; J. A. Mattfeld to K. R. Wadleigh, April 27, 1965, AC134, box 116, folder "women students."

35. J. A. Mattfeld to Killian and Stratton, November 24, 1964, AC220, box 4, folder 1; Memo from Jacquelyn Mattfeld, "Information on Women's Program, MIT, 1964–65," July 1, 1965, MC65, box 2, folder 25.

36. Academic Council, minutes of March 2, 1965, and notes of Jacquelyn Mattfeld, both in AC134, box 1, folder "Academic council 6/64–6/65."

37. Jacquelyn Mattfeld to Philip Stoddard, May 26, 1965; Benson Snyder to Mattfeld, May 26, 1965; Roland B. Greeley to Jacquelyn Mattfeld, May 26, 1965; all in AC220, box 4, folder 1.

38. Minutes of the Wellesley-MIT planning committee, September 22, 1967, AC220, box 1, folder 34; [No author], Notes, "Reactions on Wellesley Affiliation," May 18, 1967, AC220, box 1, folder 34; Cartoon, *The Tech*, March 15, 1968; Paul Johnson, "Students' Ideas on New Program Hit All Extremes," *The Tech*, May 19, 1967, 1, 6; "Letters to the Editor," *The Tech*, March 19, 1968, 4.

39. Vera Kistiakowsky to R. A. Alberty, September 21, 1972, MC 485, box 9, folder "Corresp 1972–73"; Minutes of the Wellesley-MIT Joint Committee, May 20, 1968, AC118, box 85, folder 16; Joel Orlen to Malcolm Kispert, September 30, 1968; R. A. Alberty to MIT Academic Council, December 18, 1969, April 21, 1970, and April 6, 1971; all in AC118, box 85, folder 16; Emily Wick, "Proposal for a New Policy for Admission of Women Undergraduate Students at MIT," March 9, 1970; *MIT Provost's Report* (Cambridge, MA: MIT Press, 1974–75), 136.

40. Monroe Benaim, "An Analysis of the MIT-Wellesley Cross-Registration Program," January 14, 1969, and Richard Douglas to Robert Alberty, March 29, 1968, both in AC118, box 85, folder 16.

41. *President's Report*, 1972–73, 39; "Reactions on Wellesley Affiliation," May 18, 1967, AC220, box 1, folder 34; *The Social Beaver*, MIT, 1969.

42. *The Social Beaver*, 1969.

43. *The Social Beaver*, 1969; Office of the Dean of Student Affairs, "A Program for the Women of MIT," March 16, 1964, AC134, box 116, folder "women students."

44. Gail Halpern, "The Coed Mystique," *MIT Technique*, 1967, 97.

45. Emily Wick, "Proposal for a New Policy for Admission of Women Undergraduate Students at MIT," March 9, 1970; "Karen Wattel, "Coeds Seek Privacy, Convenience," *The Tech*, October 6, 1967, 1; *AMITA Newsletter*, fall 1980; Interview of Mildred Dresselhaus by Shirlee Sherkow, 1976; all in MC86, box 8.

46. Emily Wick to Paul Gray, November 16, 1971, and the Ad Hoc Committee on the Role of Women at MIT to J. Daniel Nyhart, February 28, 1972, both in MC485, box 13, file "MIT."

47. Interview of Mildred Dresselhaus by Shirlee Sherkow, 1976, MC86, box 8; Joanne Miller, "IAP Forum Explores Areas of Concern to Women at MIT," *Tech Talk*, January 12, 1972, 3.

48. *The Ad Hoc Committee on the Role of Women at MIT*, report, 1972, MC485, box 13, file "MIT."

49. Interview of Mildred Dresselhaus by Shirlee Sherkow, 1976, MC86, box 8; Mildred Dresselhaus, "Some Personal Views on Engineering Education for Women," *IEEE Transactions on Education* (February 1975): 30–34; Interview of Sheila Widnall by Shirlee Shirkow, 1976, MC86, box 8.

50. Association of Women Students, *This Is MIT for Women*, no date, AC220, box 4, folder 2.

51. Interview of Mildred Dresselhaus by Shirlee Sherkow, 1976, MC86, box 8; Notes taken at Women's Forum meetings, IAP, 1972, AC 220 box 1 folder 27; Carolyn Scheer to Gloria Steinem, June 28, 1974, MC485, box 17, "folder "Gloria Steinem"; *How to Get Around MIT*, September 1973; Minutes, Women's Advisory Group, May 8, 1975; all in AC220, box 17, folder "MIT—women's advisory group, 1972–75."

52. Mildred Dresselhaus, "Some Personal Views on Engineering Education for Women," *IEEE Transactions on Education* (February 1975): 30–34.

53. Mark Crane, "Undergraduate Cancellation Study," November 1977, and H. Dany Siler and Debye Meadows to Peter Richardson, August 9, 1978, both in MC485, box 13, file "MIT."

54. Walter Mckay to Roland Greeley, May 4, 1972; "Women in Engineering," a film draft proposal, June 27, 1974; Lisa C. Klein to Myron Tribus, May 30, 1975; Sheila Widnall to Alfred Keil, September 30, 1975; all in AC12, box 57, file "films—Women in engineering."

55. Marilyn S. Swartz to Joann Miller etc., August 8, 1972.

56. *Women in Science and Technology: A Report on the Workshop on Women in Science and Technology*, 2–4; Anne Hirsch to ILOs, March 13, 1973; Mary S. McNulty to Ms. Shur, May 3, 1973; all in AC12, box 57, file "films—Women in engineering."

57. Diane White, "A Centennial for the Women of MIT," *Boston Evening Globe*, June 4, 1973, 16; "MIT Vice-President's Report, 1974–75," 382–383, MC485 box 29.

58. Mildred Dresselhaus, ca. spring 1974, AC220, box 2, folder 19.

59. Carolyn Scheer to Gloria Steinem, June 28, 1974, MC485, box 17, "folder "Gloria Steinem"; Margaret Brandeau, "Steinem: Castes Trap Women," *The Tech*, January 15, 1975, 1.

60. Minutes, Women's Advisory Group, May 8, 1975, AC220, box 17, folder "MIT—women's advisory group, 1972–75."

61. Vera Kistiakowsky to Holly Heine, August 5, 1977, AC12, box 56, file "Engineering Education."

62. H. Dany Siler to Peter Richardson, November 14, 1978, MC485, box 13, file "MIT."

63. Memo from unknown students to Donna Baranski, April, 1978, and The Michigan Project Committee to the MIT Steward committee, April 2, 1979, both in MC485, box 13, file "MIT";

Holliday Heine to James Mar, November 14, 1979, MC485, box 13, folder "Ad Hoc committee on women's admission."

64. *AMITA Newsletter*, fall 1980.

65. Oral history interview with Ellen Henderson, by Shirlee Shirkow, March 28, 1977; MC 86, box 8.

66. Oral history interview with Sheila Widnall by Shirlee Sherkow, March, 1977, MC86 box 8; Vera Kistiakowsky, "The Next President of MIT," February 23, 1979, MC485, box 29, no folder; Interview of Vera Kistiakowsky by Shirlee Shirkow, 1976, MC86, box 8.

67. Minutes, Women's Advisory Group, January 20, 1976, AC220, box 2, folder 8, Vera Kistiakowsky, "The Next President of MIT," February 23, 1979, MC485, box 29, no folder; Holliday Heine to James Mar, November 14, 1979, MC485, box 13, folder "Ad Hoc committee on women's admission."

68. Letter from female grad students and technical staff, area II, to colleagues, March 25, 1982, AC 12, box 86, file "Committee on women."

69. Letter from female grad students and technical staff, area II, to colleagues, March 25, 1982, and Peter Elias to Area II Faculty, March 25, 1982, both in AC 12, box 86, file "Committee on women"; *AMITA Newsletter*, fall 1980.

CHAPTER 7

1. "There's Glam in Engineering," *Michigan Technic* (January 1944): 14–15.

2. "Girl of the Month," *Penn State Engineer* (March 1956): 21; "Miss Spectrum," *Penn State Spectrum* (November 1967).

3. "Miss Queenie Sliderule," *Purdue Engineer* (March 1968).

4. "Co-ed of the Month," *Purdue Engineer* (April 1956); "Playboy Engineer," *Purdue Engineer* (March 1970); "Miss Queenie Sliderule," *Purdue Engineer* (March 1968).

5. "Brainzella," *Purdue Engineer* (January 1971).

6. *Tech Engineering News* (February 1966): 20; *Tech Engineering News* (November 1966); *Tech Engineering News* (November 1968): 37.

7. *SWE Newsletter* (May 1978): 7; *SWE Newsletter* (January–February 1979): 11; *Tech Engineering News* (January 1975): back cover.

8. Elsie Eaves to Rose Mankofsky, April 22, 1954, file "Misc. Corr. Eaves, 1942, 1946, 1951–57," box 187, SWE, Wayne St.

9. Karen Laferty Instedt, "How Should SWE Serve Undergraduates?," *SWE Newsletter* (June–July 1978); "Dear John" pamphlet, University of Missouri at Rolla, spring 1974, Missouri University

of Science and Technology archives, box R: 9/53/5, "Student Affairs: Student Organizations: Society of Women Engineers."

10. Beginning in the 1960s, SWE also presented certificates of merit and financial prizes to high school women who demonstrated excellence in math or science or who presented outstanding technical exhibits at local and national science fairs.

11. Purdue flyer, ca. 1970s, box #70, file "student activities 1974–75" and "Progress Report: Women in Engineering at Purdue Univ.," ca. 1978, both in box "SWE bio/subj.," file "women engineering students," SWE, Wayne St. Many other universities established programs with similar elements. For example, Ohio State, North Dakota, and Lehigh's SWE chapters ran "big and little sisters" programs during the 1970s.

12. Press release, May 29, 1974, box 129, file "Henniker IV," and booklet "Women in Engineering: Role Models from Henniker 3," both in box 119, file "Role Models," SWE, Wayne St.

13. "Women in Engineering: Why Not You?," ca. 1974, box #139 file "Iowa, Univ. of," SWE, Wayne St.

14. NJIT press release, ca. April 1976, box 140, file "Newark College of Engineering," and "A Symposium on the Opportunities for Today's Woman," ca. 1974, both in box #138, file "Florida," SWE, Wayne St.

15. SWE Student Section—University of California, Berkeley, *Junior High School Outreach: A Practical Guide*, 1980, box 118, file "Junior High School Outreach 1980," Brochure "Tinker . . . Toys . . . Technology," ca. fall 1982, file "A-V material," box 133, SWE, Wayne St.; Deborah S. Franzblau, "Have You Considered Outreach?," *U.S. Woman Engineer* (December 1980): 15. To note just two similar examples among many, the student SWE section of Lawrence Institute of Technology gave presentations to high school women in Detroit, while University of Michigan's SWE worked with the Ann Arbor school system's Career Planning Office to offer lectures on engineering at elementary and secondary high schools.

16. *Terry's Trip*, ca. 1979, box #131, file "Terry's trip"; *Betsy and Robbie*, ca. 1983, box #119, file "Betsy and Robbie," SWE, Wayne St.; see also Sarah Sloan, "Terry's Trip," *SWE Newsletter* (November–December 1979).

17. Elizabeth O'Callaghan and Natalie Enright Jerger, "Women and Girls in Science and Engineering: Understanding the Barriers to Recruitment, Retention and Persistence across the Educational Trajectory," *Journal of Women and Minorities in Science and Engineering* 12 (2006): 209–232; see also Gerhard Sonnert, "Women in Science and Engineering: Advances, Challenges, and Solutions," *Annals of the New York Academy of Sciences* 869 (April 1999): 34–57.

18. Karen Tonso, "The Impact of Cultural Norms on Women," *Journal of Engineering Education* (July 1996): 217–225; Karen Tonso, "Engineering Gender—Gendering Engineering: A Cultural Model for Belonging," *Journal of Women and Minorities in Science and Engineering* 5 (4) (1999): 365–404.

19. Sue Rosser, "Attracting and Retaining Women in Science and Engineering," *Academe* 89 (4) (2003): 25–29; Judith McIlwee and Gregg Robinson, *Women in Engineering: Gender Power and Workplace Culture* (Albany: State University of New York Press, 1992); C. Burack and S. Franks, "Telling Stories about Engineering: Group Dynamics and Resistance to Siversity," *National Women's Studies Association Journal* 16 (1) (2004): 79–95; see also Barbara Whitten et al., "'Like a Family': What Works to Create Friendly and Respectful Student-Faculty Interactions," *Journal of Women and Minorities in Science and Engineering* 10 (2004): 229–242.

20. Brenda Capobianco, "Undergraduate Women Engineering Their Professional Identities," *Journal of Women and Minorities in Science and Engineering* 12 (2006): 95–117.

21. For instance, see Richard Felder et al., "A Longitudinal Study of Engineering Student Performance and Retention. III. Gender Differences in Student Performance and Attitudes," *Journal of Engineering Education* (April 1995): 151–163; see also Suzanne Brainard and Linda Carlin, "A Six-Year Longitudinal Study of Undergraduate Women in Engineering and Science," *Journal of Engineering Education* (October 1998): 369–375.

22. Complicating discussions of "stereotype threat," some researchers found that certain circumstances might create instead a "stereotype boost," where women in nontraditional fields such as engineering took pride in defying assumptions. But achieving such confidence took time and effort. See R. Crisp, L. Bache, and A. T. Maitner, "Dynamics of Social Comparison in Counterstereotypic Domains: Stereotype Boost, Not Stereotype Threat, for Women Engineering Majors," *Social Influence* 4 (2009): 171–184; see also Annique Smeding, "Women in Science, Technology, Engineering, and Mathematics (STEM): An Investigation of Their Implicit Gender Stereotypes and Stereotypes' Connectedness to Math Performance," *Sex Roles* (December 2012): 617–629.

23. Jane Margolis, Allan Fisher, and Faye Miller, "Living among the 'Programming Gods': The Nexus of Confidence and Interest for Undergraduate Women in Computer Science," working paper of the Carnegie Mellon Project on Gender and Computer Science, http://www.cs.cmu.edu/afs/cs/project/gendergap/www/confidence.html, accessed January 14, 2012.

24. A. C. Strenta, R. Elliott, R. Adair, M. Matier, M., and J. Scott, "Choosing and Leaving Science in Highly Selective Institutions," *Research in Higher Education* 35 (5) (1994): 513–547; "Women Aren't Becoming Engineers Because of Confidence Issues, Study Suggests," *Science News*, October 25, 2011; Erin Cech, Brian Rubineau, Susan Silbey, and Caroll Seron, "Professional Role Confidence and Gendered Persistence in Engineering," *American Sociological Review* 76 (5) (2011): 641–666; Amy Bell, Steven Spencer, Emma Iserman, and Christine Logel, "Stereotype Threat and Women's Performance in Engineering," *Journal of Engineering Education* (October 2003): 307–312.

25. W. S. Smith, and T. O. Erb, "Effect of Women Science Career Role Models on Early Adolescents' Attitudes toward Scientists and Women in Science," *Journal of Research in Science Teaching* 23 (8) (1986): 667–676; Cynthia C. Fry, Jessica Davis, and Yasaman Shirazi-Fard, "Recruitment and Retention of Females in the STEM Disciplines: The Annual Girl Scout Day Camp at Baylor University," *ASEE/IEEE Frontiers in Education Conference* (2006): S3D-1–S3D-5.

26. WEPAN, "The First Ten Years, 1990–2000," http://www.wepan.affiniscape.com/associa tions/5413/files/The%201st%2010%20Years.pdf, accessed October 5, 2012.

{COMP: FOLLOWING NOTE HAS A WORD WITH A STRIKEOUT (MID-NOTE, "OUTSIDE")}
27. E. Seymour, "The Loss of Women from Science, Mathematics, and Engineering Undergraduate Majors: An Explanatory Account," *Science Education* 79 (4) (1995): 437–473; S. L. Murray, C. Meinholdt, and L. S. Bergmann, "Addressing Gender Issues in the Engineering Classroom," *Feminist Teacher* 12 (3) (1999): 169–183; Yevgeniya V. Zastavker, Maria Ong, and Lindsay Page, "Women in Engineering: Exploring the Effects of Project-Based Learning in a First-Year Undergraduate Engineering Program," *ASEE/IEEE Frontiers in Education Conference* (2006): S3G-1–S3G-6; Peter Shull and Michael Weiner, "'Thinking Outside Inside of the Box': Retention of Women in Engineering," *ASEE/IEEE Frontiers in Education Conference* (2000): F2F-13–F2F-16; Jerald Henderson et al., "Building the Confidence of Women Engineering Students with a New Course to Increase Understanding of Physical Devices," *Journal of Engineering Education* (October 1994): 1–6.

28. For example, see Elaine Seymour, "The Role of Socialization in Shaping the Career-Related Choices of Undergraduate Women in Science, Mathematics, and Engineering Majors," *Annals of the New York Academy of Sciences* 869 (1) (1999): 118–126.

29. Abigail Powell, Barbara Bagilhole and Andrew Dainty, "How Women Engineers Do and Undo Gender: Consequences for Gender Equality," *Gender, Work and Organization* 16 (4) (July 2009): 411–428.

30. Lisa McLoughlin, "Spotlighting: Emergent Gender Bias in Undergraduate Engineering Education," *Journal of Engineering Education* (October 2005): 373–381.

31. National Science Foundation, "Advance at a Glance," http://www.nsf.gov/crssprgm/advance/index.jsp, accessed September 29, 2012; "Georgia Tech ADVANCE," http://www.advance.gatech.edu/index.html, accessed September 29, 2012.

32. William Wulf, "The Image of Engineering," *Issues in Science and Technology* (winter 1998): 23–24; National Academy of Engineering, *Diversity in Engineering: Managing the Workforce of the Future* (Washington, DC: NAE, 2002); Committee on Maximizing the Potential of Women in Academic Science and Engineering, National Academy of Sciences, National Academy of Engineering, and Institute of Medicine, *Beyond Bias and Barriers: Fulfilling the Potential of Women in Academic Science and Engineering* (Washington, DC: National Academies Press, 2007).

33. *EngineerGirl* Web site, http://www.engineergirl.com, and *Engineer Your Life* Web site, http://www.engineeryourlife.org, accessed September 30, 2012.

34. http://www.nerdgirls.com; Harriet King, "Nerd Girls: The Beauty of Brains," http://www.diversitycareers.com/articles/pro/09-decjan/mentors_nerdgirls.htm; John R. Platt, "How Do You Get Women to Stay in Engineering? Nerd Girls Has the Answer," *IEEE USA Today's Engineer* (July 2010), http://www.todaysengineer.org/2010/Jul/Nerd-Girls.asp; "Karen Panetta Makes

Engineering Cool," *Live Science*, May 10, 2012, http://www.livescience.com/20221-computer-engineering-panetta-nsf-sl.html; Helene Ragovin, "Presidential Applause," *Tufts Now*, November 18, 2011, http://now.tufts.edu/articles/presidential-applause; all accessed October 12, 2012.

35. http://www.nerdgirls.com; King, "Nerd Girls: The Beauty of Brains"; Platt, "How Do You Get Women to Stay in Engineering?"; "'Nerd Girls' Out to Prove That Beauties Can Be Brainy," *NBCToday.com*, July 18, 2008, http://today.msnbc.msn.com/id/25736678/ns/today-money/t/nerd-girls-out-prove-beauties-can-be-brainy/#.UHgazI5uGlI; Jessica Bennett, "Revenge of the Nerdette," *Newsweek*, June 7, 2008, http://www.thedailybeast.com/newsweek/2008/06/07/revenge-of-the-nerdette.html; "Karen Panetta Makes Engineering Cool," *Live Science*; "Meet the Nerd Girls,"YouTube, http://www.youtube.com/watch?v=JRG18wYmKJ4; all accessed October 12, 2012.

36. Larry Gordon, "Caltech Chemistry Improves," *Los Angeles Times*, August 6, 2007, http://articles.latimes.com/2007/aug/06/local/me-caltech6, accessed September 28, 2012.

37. "WITI Hall of Fame," http://www.witi.com/center/witimuseum/halloffame/247621/Dr.-Eleanor-Baum-Dean-of-Engineering-Cooper-Union-Engineering-school; Cooper Union for the Advancement of Science and Art, *The Cooper Union 1989 National Survey of Women Engineers* (New York: Cooper Union, 1989); Eleanor Baum, "Women in Engineering: Creating a Professional Workforce for the Twenty-first Century," 1991, http://gos.sbc.edu/b/baum.html, accessed September 29, 2012; John Lienhard, "Women Engineers: A Survey," 1991, http://www.uh.edu/engines/epi568.htm, accessed September 29, 2012.

38. Cecilia M. Vega and Jaxon VanDerbeken, "UC Santa Cruz Chancellor Jumps to Her Death in S.F.," *San Francisco Chronicle*, June 24, 2006; Jonathan Glater, "Stunned Campus Mourns Its Chief, an Apparent Suicide," *New York Times,* June 30, 2006; M.R.C. Greenwood and Anne Peterson, "Let Denton's Legacy Be an End to Wounds from Glass Ceilings," *Mercury News*, July 2, 2006; Adam Dylewski, "Friends, Colleagues Remember the Late Denice Denton," *Unversity of Wisconsin–Madison News*, July 25, 2007; "Denice Denton '82," *MIT Technology Review*, November 14, 2006; Scott Jaschik, "Suicide of a Chancellor," *Inside Higher Ed*, June 26, 2006, http://www.insidehighered.com/news/2006/06/26/denton, accessed September 29, 2012.

39. "Women Engineering Deans," *SWE Magazine* 56 (3) (summer 2010): 49; Peggy Layne, "Women Engineering Leaders in Academe," *SWE Magazine* 56 (4) (fall 2010): 36.

40. "Enrollment of Minorities and Women," http://www.hmc.edu/about1/administrative offices/registrar1/registrarstats1/minoritieswomen.html, accessed September 28, 2012.

41. Pauline Rose Clance, *The Impostor Phenomenon: When Success Makes You Feel Like a Fake* (New York: Bantam, 1986); see also Hannah Frankel, "I'm Not Good Enough," *Times Educational Supplement*, January 15, 2010, 31–33.

42. Katie Hafner, "Giving Women the Access Code," *New York Times*, April 2, 2012; Christine Alvarado and Zachary Dodds, "Women in CS: An Evaluation of Three Promising Practices," *Proceedings of SIGCSE 2010*, http://www.cs.hmc.edu/~alvarado/papers/fp068-alvarado.pdf,

accessed September 28, 2012; "Lack of Women Does Not Compute," http://cascade.uoregon
.edu/spring2011/natural-sciences/lack-of-women-does-not-compute, accessed September 28,
2012; Angela Haines, "How One College President Is Breaking Down Barriers for Women in
Tech," *ForbesWoman*, December 12, 2011, http://www.forbes.com/sites/85broads/2011/12/12/
how-one-college-president-is-breaking-down-barriers-for-women-in-tech, accessed September
28, 2012.

43. Sue Rosser, "Will EC 2000 Make Engineering More Female Friendly?," *Women's Studies
Quarterly* 29 (3–4) (fall–winter, 2001): 164–186.

44. Margaret Rossiter, *Women Scientists in America: Struggles and Strategies to 1940* (Baltimore: Johns
Hopkins University Press, 1982).

45. Ibid.

46. Ethan Bronner, "Women's College to Diversify via Engineering," *New York Times*, February
20, 1999; Domenico Grasso and Joseph Helble, "Holistic Engineering and Educational Reform,"
in *Holistic Engineering Education: Beyond Technology*, ed. Domenico Grasso and Melody Burkins
(New York: Springer, 2009); John Sippel, "Taking on Engineering's Gender Gap," *NewsSmith*,
1999, http://www.smith.edu/newssmith/NSSpring99/cover.html, accessed October 15, 2012.

47. "Blazing a Trail for Women, Smith College Prepares to Graduate the Country's First All-
Female Class of Engineers," Smith College news release, April 20, 2004, http://www.smith.edu/
newsoffice/releases/03-070.html, accessed October 15, 2012; "Engineering Receives $1 Million
for Technology," Smith College news release, January 5, 2009, http://www.smith.edu/newsoffice/
NewsOffice09-001.html, accessed October 15, 2012; Elizabeth Farrell, "Smith College's First
Engineers Feel Like 'Rock Stars," *Chronicle of Higher Education*, May 28, 2004; "The Bridge: An
Annual Newsletter for Alumnae," issue 1, August 2010, http://www.science.smith.edu/depart
ments/Engin/pdf/200910Newsletter.pdf, accessed October 15, 2012.

48. Robert VerBruggen, "Breaking: Equations Not All That Important in Engineering," *National
Review Online,* February 14, 2008, http://www.nationalreview.com/phi-beta-cons/42826/break
ing-equations-not-all-important-engineering, accessed October 15, 2012; Robert VerBruggen,
"Defending Smith College on Engineering," *National Review Online,* February 19, 2008, http://
www.nationalreview.com/phi-beta-cons/42808/defending-smith-college-engineering, accessed
October 15, 2012.

49. Stacy Teicher Khadaroo, "How to Reengineer an Engineering Major at a Women's College,"
*Christian Science Monitor*, February 14, 2008; Ursula Gross, "Engineering a Liberal Education,"
*AAC&U News,* August 2007, http://www.aacu.org/aacu_news/AACUNews07/august07/
feature.cfm, accessed October 15, 2012; Smith College, "Picker Engineering Program," http://
www.science.smith.edu/departments/Engin, accessed October 15, 2012.

50. Smith College, "Picker Engineering Program," http://www.science.smith.edu/departments/
Engin, accessed October 15, 2012.

51. *Talk to Me* Web site, http://talk2mebook.com/index.html?5241ee98, accessed October 15, 2012.

52. Larry Gordon, "Caltech Chemistry Improves," *Los Angeles Times*, August 6, 2007, http://articles.latimes.com/2007/aug/06/local/me-caltech6, accessed September 28, 2012.

53. Brian Yoder, "Engineering by the Numbers," American Society for Engineering Education, 2012.

54. U.S. Department of Education, *The Condition of Education 2010* (Washington, DC: U.S. Government Printing Office, May 2010), 36, 76, 297; see also Peter Schmidt, "Men's Share of College Enrollments Will Continue to Dwindle, Federal Report Says," *Chronicle of Higher Education*, May 27, 2010, http://chronicle.com/article/Mens-Share-of-College/65693, accessed September 28, 2012.

55. Lawrence Summers, "Remarks at NBER Conference on Diversifying the Science and Engineering Workforce," January 14, 2005, http://www.harvard.edu/president/speeches/summers_2005/nber.php, accessed May 28, 2013.

56. Carol Muller, Sally Rice, et al., "Gender Differences and Performance in Science," *Science* 307 (5712), February 18, 2005, 1043.

57. "A Study on the Status of Women at MIT," 1999, http://web.mit.edu/fnl/women/women.html, accessed October 13, 2012; Brendan Koerner, "The Boys' Club Persists," *U.S. News and World Report*, March 28, 1999.

58. In 2008, a singing group with Georgia Tech connections produced a song and music video titled "The Ratio." http://www.youtube.com/watch?v=YLAvhE4a_hI, accessed September 23, 2012.

59. "Georgia Institute of Technology," *Students' Guide to Colleges* (New York: Penguin, 2005), 227.

60. James Martin, "Men Are Not Always Hitting on Women," *Technique*, February 9, 2001, 10; Katie O'Connor, "Focus on Dating Ignores Bigger Picture," *Technique*, February 2, 2001, 10; David Flowers, "Time for People to Find Middle Ground in Dating War," *Technique*, February 16, 2001, 10; Alex Salazar, "Women Are People, Not Simply a Thing to 'Date,'" *Technique*, February 2, 2001, 10.

61. Kathleen Richter, "*Ms.* Goes to College," *Ms.* (August–September 2001), http://www.msmagazine.com/aug01/college3.html; "WITI Hall of Fame Ceremony," http://www.theory.caltech.edu/people/patricia/change/halloffame.html, accessed September 29, 2012.

62. http://factbook.gatech.edu/content/class-enrollment-genderethnicity, accessed August 6, 2012; http://www.coe.gatech.edu/content/stats-rankings-women, accessed September 24, 2012.

63. "InGear Report on the Status of Women, Georgia Institute of Technology, 1993–1998," Georgia Tech, 1998, http://www.academic.gatech.edu/study/report.htm, accessed October 9,

2012; Brian Yoder, "Engineering by the Numbers," American Society for Engineering Education, 2012.

64. InGear Report on the Status of Women," Georgia Tech, 1998.

65. Andy Guess, "Enrollment Surge for Women," *Inside Higher Ed*, August 7, 2007, http://www
.insidehighered.com/news/2007/08/07/enrollment, accessed September 29, 2012; "Can the
New Yahoo CEO Attract More Women to Engineering?," Georgia Tech Web site, July 19, 2012,
http://www.amplifier.gatech.edu/articles/2012/07/can-new-yahoo-ceo-attract-more-women-
engineering, accessed September 29, 2012; Ian Bailie, "Lack of Women at Tech More Than Joke,"
*Technique*, December 2, 2010.

66. April Brown, Donna Llewellyn, and Marion Usselman, "Institutional Self-Assessments as
Change Agents: Georgia Tech's Two-Year Experience," in *Proceedings of the 2001 American Society
for Engineering Education Annual Conference and Exposition* (Washington, DC: American Society for
Engineering Education, 2001).

67. "Student Ambassador Program," https://login.gatech.edu:443/cas/login?service=http%3A%
2F%2Fcoe.gatech.edu%2FStudent-Ambassador-Program-Article&gateway=true, accessed September 29, 2012; "Women's Resource Center, Academic Programs," http://www.womens
center.gatech.edu/plugins/content/index.php?id=11, accessed September 29, 2012.

68. Goodman Research Group, "Final Report of the Women's Experiences in College Engineering Project," April 2002, http://www.grginc.com/WECE_FINAL_REPORT.pdf, accessed
January 14, 2012, vii, x, 179.

69. "Women in Technology Sharing Online," https://piazza.com/witson, accessed October 12,
2012.

70. "Overview: Psychological Sense of Community for Women in Engineering," Assessing
Women in Engineering (AWE) Project 2005, *AWE Research Overviews*, http://www.engr.psu
.edu/awe/misc/ARPs/SenseOfCommunityWeb_03_11_05.pdf, accessed May 28, 2013; Nilan-
jana Dasgupta, "Ingroup Experts and Peers as Social Vaccines Who Inoculate the Self-Concept:
The Stereotype Inoculation Model," *Psychological Inquiry* 22 (4) (2011): 231–246; see also Stephen
J. Ceci, Wendy M. Williams, Rachel A. Sumner, and William C. DeFraine, "Do Subtle Cues about
Belongingness Constrain Women's Career Choices?," *Psychological Inquiry* 22 (4) (2011):
255–258.

71. Phi Sigma Rho, http://www.phisigmarho.org, accessed September 29, 2012; Bevlee A.
Watford and Sharnnia Artis, "Hypatia: A Residential Program for Freshman Women in Engineering," *Proceedings of the 2004 American Society for Engineering Education Annual Conference*, http://
search.asee.org/search/fetch;jsessionid=4aumhi7qb38lt?url=file%3A%2F%2Flocalhost%2FE%3A
%2Fsearch%2Fconference%2F28%2FAC%25202004Paper384.pdf&index=conference_papers&
space=1297467972036057917166676178&type=application%2Fpdf&charset=, accessed September 29, 2012; "Virginia Tech's Successful Residential Learning Community for Women Engineering Students," *Women in Academia Report*, September 19, 2012, http://www.wiareport

.com/2012/09/virginia-techs-successful-residential-learning-community-for-women-engineer ing-students/?utm_source=Women+In+Academia+Report&utm_campaign=f7037d4b3a -Women_in_Academia_Report_6_14_116_13_2011&utm_medium=email, accessed September 29, 2012; Sarah Bruyn Jones, "Virginia Tech Residential Program Helps Retain Female Students," *Roanoke Times*, August 12, 2012, http://www.roanoke.com/business/wb/312737, accessed September 29, 2012.

72. http://admissions.caltech.edu/about/stats, accessed July 30, 2012.

73. Larry Gordon, "Caltech Chemistry Improves," *Los Angeles Times*, August 6, 2007, http:// articles.latimes.com/2007/aug/06/local/me-caltech6, accessed September 28, 2012.

74. Brian Yoder, "Engineering by the Numbers," American Society for Engineering Education, 2012.

75. MIT, "Report of the School of Engineering," *Reports of the Committees on the Status of Women Faculty*, March 2002, http://web.mit.edu/faculty/reports/soe.html, accessed October 13, 2012.

76. MIT, "A Report on the Status of Women Faculty in the Schools of Science and Engineering at MIT, 2011," http://web.mit.edu/faculty/reports/pdf/women_faculty.pdf, accessed October 13, 2012.

77. "A Culture of Collaboration," July 16, 2012, http://www.achievement.org/autodoc/page/ hoc0bio-1, accessed September 29, 2012.

Printed in the United States
by Baker & Taylor Publisher Services